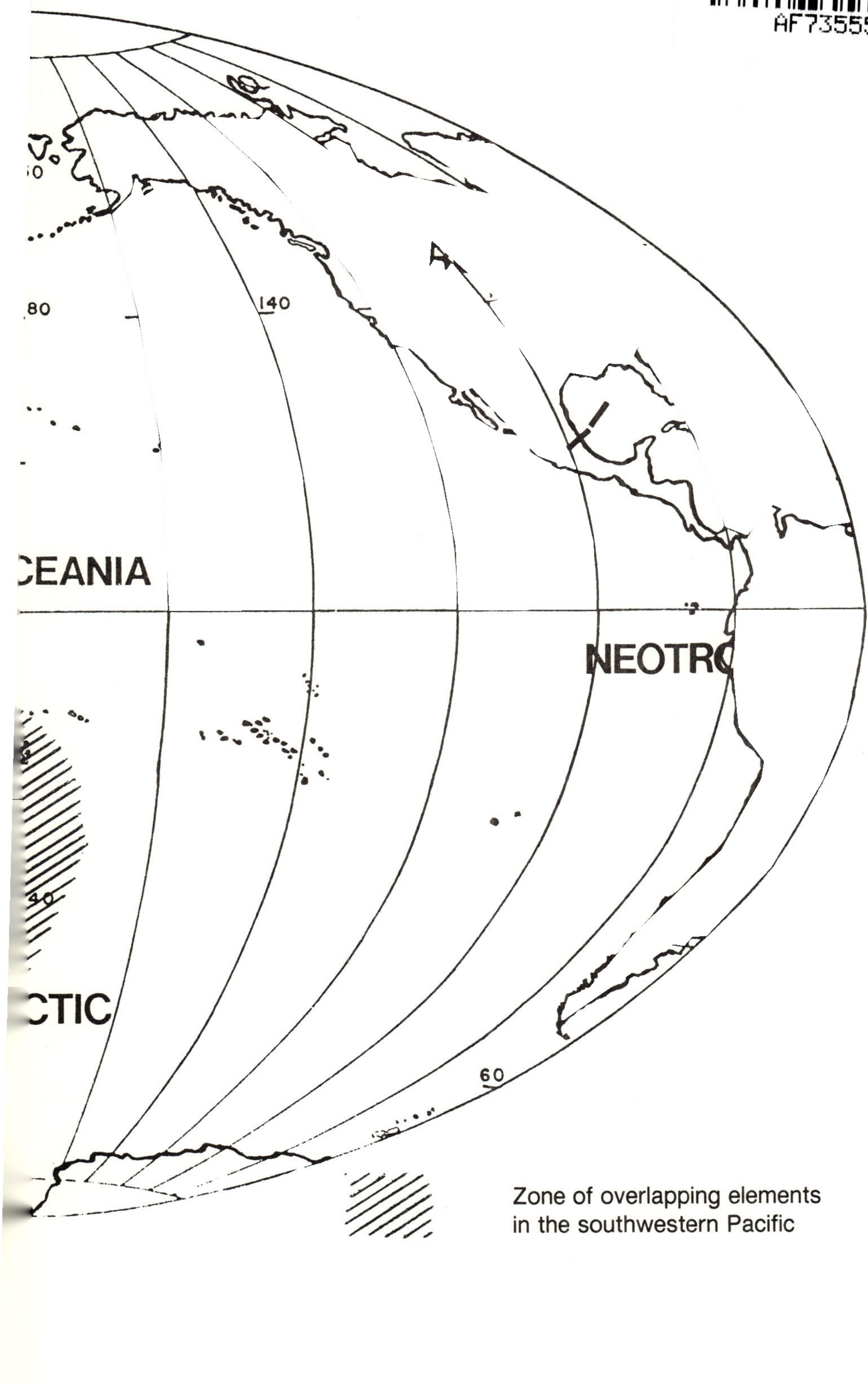

Zone of overlapping elements
in the southwestern Pacific

BIOGEOGRAPHY OF THE TROPICAL PACIFIC

BIOGEOGRAPHY OF THE TROPICAL PACIFIC

Proceedings of a Symposium

Edited by

Frank J. Radovsky,

Peter H. Raven, and S. H. Sohmer

Dedicated
to
J. Linsley Gressitt

Published jointly by the
Association of Systematics Collections and the
Bernice P. Bishop Museum as
Bishop Museum Special Publication No. 72

BIOGEOGRAPHY OF THE TROPICAL PACIFIC

Published December 28, 1984

ISBN Hardback: 0-942924-08-8
ISBN Softcover: 0-942924-09-6

Copies of BIOGEOGRAPHY OF THE TROPICAL PACIFIC
can be ordered from:

ASC
Museum of Natural History
University of Kansas
Lawrence, Kansas 66045
U.S.A.
(913) 864-4867

This volume is dedicated to
Judson Linsley Gressitt 1914–1982.

Contents

Foreword

(These remarks were included in the application to the National Science Foundation for support of the 1982 symposium reported in this volume. They were written some 6 months prior to Dr. Gressitt's death on April 26, 1982.)

Twenty years ago at the Tenth Pacific Science Congress in Honolulu, August 1961, a major symposium, "Pacific Basin Biogeography," was held. The proceedings of the symposium were published as a congress volume, of the same title, by Bishop Museum (Gressitt, ed., 1963). That symposium (and volume) had a very broad scope, covering biology and some geology/geography for the entire Pacific basin, from Asia to the Americas and the Arctic to the Antarctic. It also had a great many contributors (45). It was predominantly on terrestrial biota.

The timing of that symposium was rather notable. It was in the early phase of the geophysical research (partly involving the IGY) leading to the modern documentation of plate tectonics. The proceedings of that 1961 symposium embraced highly divergent views, mostly of the old school, and the term "tectonics" was probably not even mentioned. In fact, one of the best known Pacific biogeographers, giving the final summary presentation, quoted a poem to emphasize that the "continental drifters" had been wrong "all their lives"! The "land-bridge" advocates, some of whom participated, were also maligned; not to mention advocates of long distance air dispersal, including the organizer both of the above and of the here-proposed symposium, who had a mixed reception.

In the 20 years that have elapsed since that Congress, revolutionary discoveries and extensive documentation were brought forth relating to the age of the earth, the dynamism of the ocean bottoms, the nature and role of Antarctica and its fossils, the general history of Pacific islands, and the dynamics of insular populations. New information has revolutionized our thinking in these and related fields.

Also during the elapsed two decades, a number of textbooks of biogeography and related subjects have disseminated the new perspectives on various aspects of the dynamics of these subjects. Likewise many comprehensive insular studies of various sorts have appeared. As examples that may be cited are new volumes of the "Insects of Hawaii" series by Hardy and Zimmerman (Zimmerman, 1978), many issues of the "Insects of Micronesia" series (Bishop Museum), various works on Micronesian, Marquesan, Fijian, New Caledonian, and other floras, (Fosberg, Sachet, Smith, etc.), Insects of Campbell Island (Gressitt, ed., 1964), Biogeography of New Caledonia (Holloway, 1979), Biogeography and Ecology of New Zealand (Kuschel, ed., 1975), Ecological Biogeography of Australia (Keast, ed., 1981), Biogeography and Ecology of New Guinea (Gressitt, ed., 1981), (Hawaii) Island Ecosystems (Mueller-Dombois, Bridges, & Carson, eds., 1981), works on Hawaii marine and shore fauna, and many others.

In view of the changes in thought and documentation of former "wild theories" and the greatly increased knowledge of the history of the Pacific and its insular biota, the elucidation of changes in continental orientation, the improved documentation of dispersal and the effects of glacial cycles and many other factors, the improved inventorying of insular flora and fauna, the development of insular biogeographic theory (MacArthur & Wilson 1967), including population dynamics, species replacement, evolution and extinction on islands (Diamond, 1972, etc.), it seems logical to hold another Pacific biogeography symposium. . .

Hawaii is a most logical spot for such a symposium. Not only is Hawaii relatively central to the area of concern, but it is also a center for various phases of such research, including astronomical, deep sea, littoral, volcanological, climatological, ecological, systematic, and especially on the nature of terrestrial evolution on isolated islands. . .

It is proposed to publish the proceedings of the symposium. . .

J. Linsley Gressitt
Honolulu, Hawaii
1981

References Cited

Diamond, J. 1972. Geographic kinetics; estimation of relaxation times for avifaunas of southwest Pacific islands. *Proc. Natl. Acad. Sci. USA* 69(11): 3199–203.

Gressitt, J. L., ed. 1954–present. *Insects of Micronesia.* Bishop Museum. (70 nos. published to date.)

———. 1963. *Pacific Basin Biogeography.* Bishop Museum Press, 563 p.

———. 1964. Insects of Campbell Island. *Pac. Insects Monogr.* 7:1–663.

———. 1981. Biogeography and ecology of New Guinea. *Monogr. Biol.* 42: 1–983, W. Junk.

Holloway, J. 1979. Lepidoptera and biogeography and ecology of New Caledonia. *Series Entomologica,* 15: 1–588, W. Junk.

Keast, A., ed. 1981. Ecological biogeography of Australia. *Monogr. Biol.* Vol. 41., W. Junk.

Kuschel, G., ed. 1975. Biogeography and ecology of New Zealand. *Monogr. Biol.,* W. Junk.

MacArthur, R. and **E. O. Wilson.** 1967. *The theory of island biogeography.* Princeton Univ. Press.

Mueller-Dombois, D., K. W. Bridges & H. L. Carson, eds. 1981. *Island ecosystems: biological organization in selected Hawaiian communities.* Hutchinson Ross Publ. Co., 583 p.

Zimmerman, E. C. 1978. *Insects of Hawaii: Microlepidoptera.* Vol. 9. Univ. Press of Hawaii, 1903 p.

Preface

The symposium of which the proceedings are presented in this volume was conceived and organized by the late J. Linsley Gressitt, to whom both the symposium and the proceedings have been dedicated.

When initial preparations were being made, in the spring of 1981, for the meetings of the Association of Systematics Collections (ASC) at the Bishop Museum in 1982, Dr. Gressitt proposed the symposium to be part of the program. The symposium along with a Forum on Pacific Systematics Collections, proposed by S. H. Sohmer, represented a departure from the form of prior ASC meetings in that the duration and the scientific content of the meetings increased. The ASC gave approval, and an application for support of the symposium was submitted to the National Science Foundation in the fall of 1981. Prior to this grant application, Dr. Gressitt contacted the biologists whom he proposed to include in the symposium and obtained their agreement to participate and to speak on specified topics. He also asked Peter Raven to co-chair the symposium with him.

In April 1982, after award of the symposium grant had been received and arrangements for the ASC meetings were completed, Dr. Gressitt and his wife left for China, with the first stop to be as an invited lecturer and as guests of Zhongshan (Sun Yat-Sen) University in Guangzhou (Canton). The Gressitts had spent many years there before and after World War II, when the same campus was Lingnan University and Dr. Gressitt was both a faculty member and responsible for the biological collections. Colleagues and friends around the world were shocked and stunned when news came that the Gressitts were among the passengers on a flight from Guangzhou to Guilin that crashed without survivors. This happened on April 26, 1982, just one month prior to the date of the biogeography symposium.

The symposium was held as scheduled, and Dr. Raven chaired the sessions.

From the first proposal that the symposium be held, publication of the Proceedings was anticipated. The specific mode of publication and the question of whether ASC or Bishop Museum would publish it were left open. However, funds were requested and awarded, as part of the NSF grant, to aid in publication costs. At the 1982 meetings (May 23–26), the governing board of the ASC and representatives of the museum agreed that this should be a joint publication of the two organizations, and plans were begun to accomplish this.

In Dr. Gressitt's unusually productive career, one of many areas of outstanding contributions was in publishing. We refer not to the long and impressive bibliography of papers and monographs that he wrote, but rather to the journals and other serials that he initiated and nurtured and to the many thematic volumes that he organized and edited. This volume is one of several for which he had laid the groundwork and that are being published as memorials to him.

A number of obituary notices for Dr. Gressitt have appeared. Perhaps the most complete to date in tracing his career is in the *International Journal of Entomology* (formerly *Pacific Insects*) 25(1):1–10 (1983). A list of his publications is in the following issue of that journal [*Int. J. Entomol.* 25(2–3):87–102].

Frank J. Radovsky
Honolulu, Hawaii
1984

Biogeographic Mosaics in the Pacific

Jared Diamond
Department of Physiology
University of California Medical Center
Los Angeles, California 90024 USA

INTRODUCTION

Biogeographers, like most humans, love to generalize. Despite experience to the contrary, we still hope that distributional patterns detected in one species group will also apply to another.

In fact, many or most biotas are mosaics of astonishingly dissimilar elements. Consider the following examples. Hawaii has birds, land snails, and lobelioids that offer textbook examples of adaptive radiation, living beside butterflies, ferns, and a bat that radiated scarcely or not at all. New Caledonia has ancient, distinctive plants and insects along with recent immigrant, undistinctive birds and bats. New Guinea has numerous Australian birds and flightless mammals, feeding on predominantly Asian plants and insects. New Guinea *Nothofagus* is an old Gondwanan relict, but the insects associated with New Guinea *Nothofagus* are predominantly recent and Asian. What happened to the insects that must have been associated with *Nothofagus* in Gondwanan times?

These examples illustrate that different species groups living in the same region may have very different histories. The biogeographer's task is to interpret these different histories in terms of the different biological attributes of the species. I shall show that species differences in dispersal ability, in proneness to extinction, and in the evolutionary capacity to alter dispersal ability and habitat preference have contributed importantly to mosaic patterns in biogeography. I first discuss New Guinea as an example of a mosaic, then other tropical Pacific islands more briefly. The phenomena on which I shall focus are distributional range, species ages, endemism, and species radiations. This discussion is based on the following source references of Pacific biogeography, and on other papers cited in these references: Balgooy, 1966, 1971; Carlquist, 1965, 1970, 1974; Cracraft, 1973; Darlington, 1957, 1965; Diamond, 1977; Diamond and Marshall, 1976; Gressitt, 1963, 1981; Kay, 1980; Keast, 1981; Mayr, 1941a, 1941b; Raven, 1979; Raven and Axelrod, 1972, 1974; and Rich, 1975.

NEW GUINEA, A SPECIES MELTING POT

I begin with New Guinea because it has the richest biota of the tropical Pacific and has served as the major source for other Pacific biotas. The elements of New

Guinea's biota differ markedly with respect to species diversity, age, and Oriental vs. Australian affinities (see Gressitt, 1981 for details).

Suppose one compares species diversity in New Guinea with that in climatically similar parts of the Oriental region, such as Borneo or the Malay Peninsula. One will find that New Guinea is rich in birds, beetles, and reef fish; very poor in the fish, molluscs, and insects of fresh water; and devoid of flightless native placental mammals except rodents. These differences correspond to differences in species' ability to cross the water gaps separating New Guinea from Asia.

New Guinea is a melting pot of young and old species. Its alpine flora is a young mixture of species from Malaysian, Australian, and New Zealand sources, with little generic endemism. Its rich assemblage of montane Ericaceae consists almost entirely of endemic species but not a single endemic genus, and has probably arrived from Malaysia in the past 10 million years. At the opposite extreme, New Guinea also has an old relict biota that has persisted as a legacy from Gondwanaland, including *Nothofagus, Araucaria, Agathis,* podocarps, cycads, and *Weinmannia* among the plants, and marsupials, hylid frogs, chelid tortoises, galaxiid fishes, migadopine carabid beetles, and peloridiid bugs among the animals. While attempts have been made to reinterpret its avifauna and that of Australia in terms of continental drift (Cracraft, 1973; Rich, 1975), and it seems probable that some of its birds (e.g. ratites) will prove to have been derived from Gondwanan ancestors, the fact remains that there is not a single bird group with the classic Gondwanan relict distribution (i.e., South America, New Zealand, New Caledonia, Australia, New Guinea). I shall postpone interpreting these differences in species ages until page 6.

New Guinea is also a melting pot for Asian and Australian species, with the result that it is considered part of the Australian region by vertebrate zoogeographers but part of the Oriental region by botanists and entomologists. Conspicuous Australian elements include marsupials, monotremes, birds of paradise, bowerbirds, honey eaters, and many songbird genera recently recognized to be part of a marsupial-like radiation on Australia and New Guinea (Sibley, 1976). Those who place New Guinea in the Australian region also can point to the absence of species-rich Asian groups such as flightless placentals (except rodents), woodpeckers, bulbuls, babblers, and cyprinid fish, and the relative paucity of dipterocarps. Yet New Guinea is equally poor in some old and species-rich Australian groups such as leptodactylid frogs and mygalomorph spiders. And its richness in Indomalaysian coral-reef fish, Pandanaceae, Ericaceae, phytophagous beetles, and bloodsucking dipteran flies (Pupipara) would, if taken alone, place it in the Oriental region.

These anomalies have puzzled me for many years, and other biogeographers even longer. The best sense I can make of them at present is as follows. New Guinea is close to tropical southeast Asia in its wet tropical climate, but separated from Asia by many water barriers. Conversely, New Guinea is close to Australia geographically (no water barrier at all during the Pleistocene), but very different in its wet tropical climate from predominantly arid Australia. Of all living things, birds and mammals as homeotherms are least tied to specific habitats or climate zones, as reflected in the numerous homeotherm families with vicariant species or genera in rain forest and in arid habitats (e.g. naming in each case the rain forest taxon before the arid-zone taxon: the bowerbirds *Sericulus* vs. *Chlamydera,* the butcher birds *Cracticus cassicus* vs. *C. nigrogularis,* the quail-thrush *Cinclosoma ajax* vs. *C. cinnamomeum,* the marsupial mice *Antechinus* vs. *Antechinomys,* the bandicoots *Echymipera* vs. *Perameles,* and the kangaroos *Dendrolagus* vs. *Macropus).* Thus, Australian birds and mammals could take advantage of New Guinea's proximity as New Guinea developed in the last 30 million years, and could adapt (at least in evolutionary time) to New Guinea's different habitats and climate. In contrast, plants and insects of arid Australia have been held out of New Guinea by climate. New Guinea plants and insects came to consist of those Oriental species capable of leaping water barriers, plus Gondwanan relicts of temperate Australia that found similar habitat in the mountains of New Guinea. The arid savanna of south New Guinea has a largely Australian biota, while the tropical rainforest of north Cape York has a largely New Guinea biota. This account is a gross oversimplification of the biogeographic details summarized in Walker (1972), Keast (1981), and Gressitt (1981), but I think that it helps make the main pattern less mysterious.

PACIFIC ISLAND BIRDS

For tropical Pacific islands east of New Guinea I shall briefly summarize the land and fresh-water avifaunas, against which other species groups can then be compared.

Fig. 1 summarizes the colonization routes (Mayr, 1941a, 1941b; Diamond, 1972; Diamond and Marshall, 1976). All Galapagos birds, many or most Hawaiian birds, and possibly two sandpipers of the Tuamotus are derived from the New World. Otherwise, all tropical Pacific land birds originated from the west, mainly from New Guinea. The main pathway leads from New Guinea to the Bismarcks, Solomons, New Hebrides and New Caledonia, Fiji, and Samoa to the Societies, thence to the Tuamotus and Marquesas. Even the most remote islands of the tropical Pacific, Pitcairn and Hawaii, have land birds derived from New Guinea. (We do not know whether remote Easter Island also formerly had land birds, as the Polynesian settlers removed Easter's forest long before the first European ornithologist arrived.) However, for islands of comparable area the number of land-bird species decreases exponentially with distance from New Guinea, by a factor of 2 for each 1600 mi (2575 km) (Fig. 2). Australian birds constitute the probable ancestors of all New Zealand birds, many birds of New Caledonia and the New Hebrides, and a few birds of the Solomons and Fiji. Micronesia received birds both from New Guinea and from Asia.

Endemism among Pacific birds east of New Guinea is low. Only New Zealand, Hawaii, and the Galapagos have endemic families of two or more species. In addition, a family Rhynchocetidae is considered endemic to New Caledonia in the form of a single species, the Kagu, and is believed related to the recent fossil *Aptornis* of New Zealand, formerly considered a rail but now placed in a monotypic family Aptornithidae (Olson, 1977). Most land birds of other tropical Pacific islands belong to New Guinea genera, and almost all of the remainder are in endemic genera closely related to New Guinea genera. Even the bird species of New Caledonia, an ancient island with old and distinct plants and insects, belong to New Guinea and/or Australian genera except for three monotypic endemic

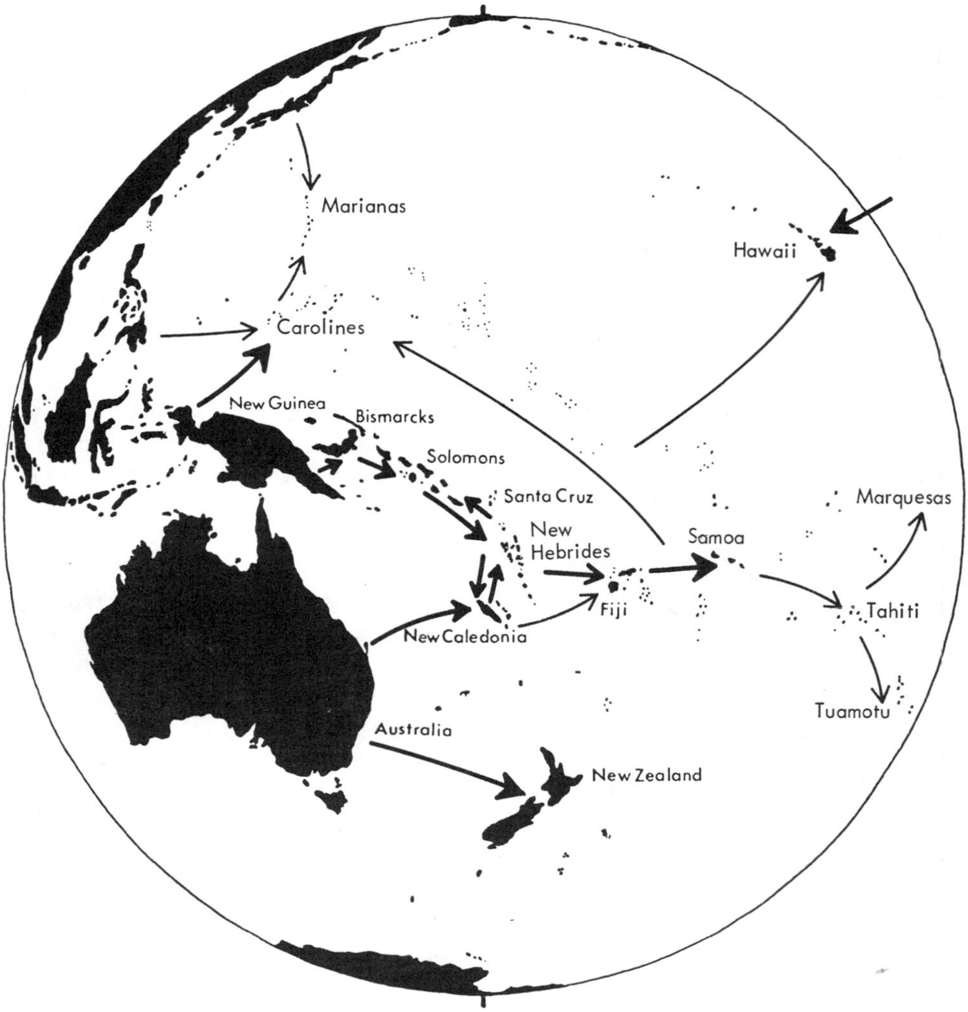

Fig. 1. Main routes by which land and fresh-water birds colonized Pacific islands. See text for details.

genera (*Rhynchocetos* and two others). These statements need to be qualified with the comment that bird fossils have yet to be sought on most Pacific islands. A rich subfossil avifauna that succumbed to Polynesian settlement has long been known for New Zealand and is presently being studied for Hawaii (Olson and James, 1982). Whether there were similar extinctions on New Caledonia and other Pacific islands is still unknown and is a question of much interest. At present there are not proven modern or fossil Gondwanan bird relicts known from any Pacific island, nor even any plausible candidates except *Rynchocetos* and the moas, kiwis, and penguins of New Zealand. (By "Gondwanan relicts" in this sentence, I mean bird populations on present-day islands derived from ancestral populations that lived on the same land mass when it was part of Gondwanaland, as opposed to populations derived by overwater colonization from New Guinea and/or Australia after the breakup of Gondwanaland.)

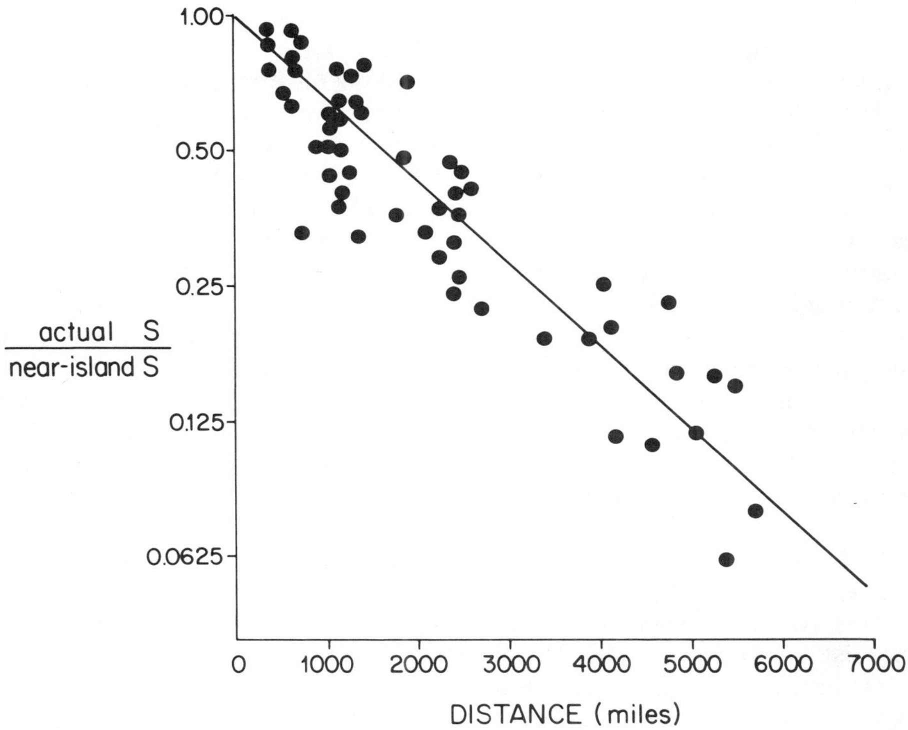

Fig. 2. Impoverishment of Pacific island avifaunas with distance. Ordinate (logarithmic scale): number of resident land and fresh-water birds on tropical Southwest Pacific islands more than 300 mi (480 km) from New Guinea, divided by number of species on islands of equivalent area less than 300 mi from New Guinea. Abscissa: island distance from New Guinea. Note that species number decreases by a factor of 2 per 1600 mi (2575 km). (From Diamond, 1972.)

Hawaii, the Galapagos, and New Zealand are famous for the adaptive radiations of the Hawaiian honey creepers (Drepanididae), Darwin's finches (Geospizinae), and moas (Dinornithidae), respectively. In contrast, other Pacific islands east of New Guinea lack anything comparable; there are only one or two endemic species pairs each on the more remote archipelagoes [Fiji, Samoa, the Societies, and the Marquesas (Diamond, 1977)].

WHO GOT HOW FAR?

Let us now compare these patterns with those for other plant and animal groups, beginning with the question how far each group has penetrated the Pacific from New Guinea. For other groups as for birds, the biotas are derived entirely or overwhelmingly from the west on Pacific islands other than Hawaii and the Galapagos. Even the Tuamotus and Pitcairn, which are closer to the New World than to New Guinea, lack New World species. The probable reason is that the numerous island chains eastward from New Guinea served as stepping-stones and also selected species for overwater dispersal. Beyond the Galapagos, Juan

Fernandez, and the few inshore islands along the Pacific coast of the New World, there is no land until one reaches Easter and Hawaii.

As we have seen, New Guinea birds reached the most remote scraps of land in the Pacific. For other species groups the pattern of distribution is very different (e.g. Darlington, 1957; Balgooy, 1966, 1971; Kay, 1980). The following details apply only to native species. Fresh-water fish and turtles are confined to New Guinea and did not even reach the Bismarcks, whose largest island (New Britain) lies only 55 miles east of New Guinea. Frogs and a few flightless mammals extend to the Solomons. (Two dubiously native frogs are on Fiji.) Snakes, *Agathis,* and the molluscan genera *Melo, Voluta,* and *Nautilus* reached Fiji. Lizards reached Fiji and Tonga. Bats, anemonefish (*Amphiprion*), caddisflies, mayflies, termites, *Pometia,* and *Intsia* reached Samoa. (Hawaii and the Galapagos were each reached by a bat from the New World but not from the west.) Giant clams and abalones reached the Societies. Figs (*Ficus*) reached the Marquesas and Pitcairn. Ferns, damselfish (*Pomacentridae*), and cone shells (*Conus*) reached Easter and Hawaii, the tropical Pacific islands most remote from New Guinea except for the Galapagos.

The long list of other plant and animal groups that failed to reach Hawaii includes groupers (Serranidae), sea-perches (*Lutjanus*), mangroves, gymnosperms, Cunoniaceae, fleas, cockroaches, earwigs, and mantids (Zimmermann, 1948; Carlquist, 1970).

These differences can largely be understood in terms of the species' varying adaptations for transoceanic dispersal, such as flight, seed characteristics favoring wind or bird transport, tolerance of salt water or salt spray, and length of planktonic larval stage (see e.g. Carlquist, 1965, 1974). Thus, differences in overwater dispersal ability provide part of the explanation why biotas are mosaics.

WHO LASTS HOW LONG?

A second part of the explanation has to do with species differences in proneness to extinction. Perhaps the most striking mosaic on earth with respect to species' ages is New Caledonia.

New Caledonia is an ancient land mass that has not been submerged or subject to volcanism for at least 80 million years. This is reflected in its numerous groups of old, very distinctive plants and insects that are relicts from Gondwanaland. Yet New Caledonia has no primarily flightless terrestrial vertebrates (i.e. ignoring two secondarily flightless birds) except lizards, half of them in endemic genera but none in endemic families and none especially ancient. Except for the Kagu, New Caledonia's birds are no more distinctive than those of the New Hebrides, the next archipelago to the north, most of whose land area is only a few million years old. From its vertebrate fauna, one could not have guessed that New Caledonia even existed before the Pleistocene. What happened to the frogs, lizards, birds, and mammals that fed on New Caledonia's Gondwanan carabids and lived in its Gondwanan *Nothofagus, Araucaria,* and *Dacrydium* when New Caledonia was still part of Gondwanaland? Why did some of the carabids and plants survive, but none of the vertebrates?

An answer is suggested by Mayr's (1965) recognition that the percentage of endemic forms in an island fauna increases with island land area, suggesting that

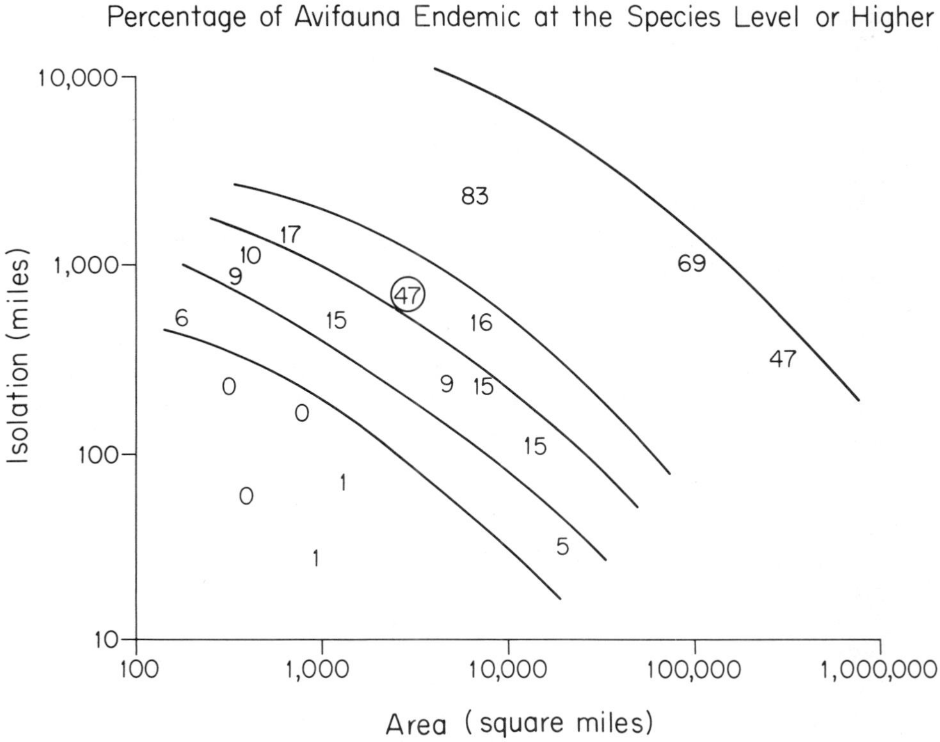

Fig. 3. Numbers on the graph give, for some Pacific island or archipelago, the percentage of its breeding land bird species that are endemic at the level of full species or higher. Location of number gives the island's area (square miles, abscissa) and distance from nearest major colonization source (miles, ordinate). For example, 16% of the species of Fiji, with an area of 7055 mi² (18,275 km²) and distance of 520 mi (835 km) from the New Hebrides, are endemic full species or belong to endemic genera. Curves are drawn by hand to group islands with similar endemism. Circled "47" refers to the Galapagos Islands, which have higher avifaunal endemism than do Southwest Pacific islands (because Neotropical birds, the source for the Galapagos, are poorer colonizers than New Guinea birds, the source for the Southwest Pacific). (From Diamond, 1980.)

extinction rates decrease with island area in geological time as they do in ecological time (e.g. MacArthur and Wilson, 1967; Jones and Diamond, 1976; Terborgh and Winter, 1980). Fig. 3 expands on Mayr's insight by depicting the percentage of endemic species in Pacific island avifaunas, as a function of island age and isolation. Endemism is seen to increase with island area and isolation. The only archipelagoes with more than 50% of their bird species endemic are two medium-sized very remote ones (Hawaii and Galapagos), big and somewhat isolated New Zealand, and accessible but huge New Guinea.

The message of Fig. 3 is that bird species have sufficiently short average lifetimes that most bird populations are much younger than their islands and give no indication of island age. The average age of a species is evidently greatest on the largest and/or most remote islands, as one would expect from large population sizes on large islands and paucity of competitors on remote islands. As species go extinct, they are continuously replaced by new arrivals. When a new island arises from the sea, its avifauna quickly (on a geological time scale) attains a standing age distribution appropriate to that island's age and isolation. By the same token, the percentage of voters over the age of 40 in the American electorate tells a lot about birth, death, and immigration rates in America's human populations, and nothing about the age of the American continent. Any differences in the standing-age distribution of the electorate on Big Island (700,000 years old) and Kauai (5 million years old) in the Hawaiian Archipelago reflect the modern demography of these electorates, not the differing number of millenia elapsed since these two islands arose from the sea.

One expects different species groups to differ greatly in the average lifetimes of their species on a land mass of a given area. One major variable is population size: lines consisting of many individuals should survive longer than those consisting of few individuals. This consideration would tend to make insect lines more long-lived than vertebrate lines. Another variable is the ability to survive for many centuries through difficult times in the form of dormant seeds or long-lived adults, and this consideration favors the survival of plants in general, and trees in particular, over vertebrates. This reasoning may explain why New Caledonia retained ancient insects and plants but not vertebrates.

More generally, for each species group one expects a characteristic dependence of a species line's lifetime on area. Of the modern Pacific land masses derived from Gondwanaland, only the largest (Australia) was large enough to retain marsupials for the tens of millions of years since separation from other land masses; only the two largest (Australia and New Zealand) were large enough to retain frogs; but both of these land masses plus the smaller New Caledonia could retain Gondwanan insects and plants.[1]

Thus, a reflection of a fauna's age is its degree of endemism, since differentiation to the level of endemic species, genus, or family requires that a line survive long enough to reach that degree of differentiation. As Table 1 illustrates, different species groups differ greatly in their proportion of endemic forms in Hawaii, and similar tables could be assembled for other Pacific archipelagoes. The high proportion of endemic genera among Hawaiian insects and land snails may reflect their high population densities, hence low extinction rates, hence likelihood of many lines surviving to reach the level of endemic genera. However, at least two other factors may contribute to these differences in proportion of endemic forms. First, high immigration rates of conspecifics from other archipelagoes will inhibit differentiation of a newly founded population, so that proportion of endemics should decrease with immigration rates. This may explain why ferns and inshore fishes, with their wind-borne spores and water-borne larvae respectively, have so

[1]Two alternative explanations advanced by Raven and Axelrod (1974) for the puzzling near absence of old vertebrates on New Zealand and New Caledonia are that they never arrived or else were kept out by climate. For example, marsupials are not known even as fossils on New Zealand. Neither explanation could apply to birds; the climate explanation is unlikely to apply to New Caledonia. It is unclear how much weight to place on the lack of marsupial fossils for New Zealand in view of their lack also for Africa despite their presumed West Gondwanaland origin.

Table 1. Percentage of endemic taxa in the Hawaiian biota.*

Group	Species	Genera
Insects	>99	54
Land molluscs	>99	51
Land birds	99	38
Angiosperms	94	13
Ferns	71	8
Mosses	51	2
Inshore fish	29	<1
Marine molluscs	20	<1

* For each taxon group the table gives the percentage of Hawaiian species and genera that are endemic to Hawaii. From various sources (mainly Zimmerman, 1948; Randall, 1976; and Kay, 1979).

few Hawaiian endemic genera. Second, lines differ in their rates of morphological evolution: *Limulus* is more conservative than *Homo*.

Thus, biotas contain elements of very different age and distinctness.

WHO RADIATES WHERE?

Adaptive radiation, the multiplication of species from the same ancestral stock, has proceeded to different degrees on different islands and in different taxa. Taxa evidently differ in the area and isolation of island required for radiation.

As background for understanding these difference, Fig. 4 illustrates three modes by which species can arise through geographic isolation (Diamond, 1977). "Continental speciation" is the mode prevailing on continents and large islands, where the geographic barrier separating ancestral sister populations is within a single land mass. This mode is exemplified by the radiation in Pleistocene rain forest refugia that produced the bewildering species diversity of Amazonia (Haffer, 1969). In "inter-archipelagal radiation," as exemplified by the triple invasion of Norfolk Island from Australia by white-eyes of the *Zosterops lateralis* group, the isolating geographic barrier is the water gap between two archipelagoes. Finally, "intra-archipelagal speciation," the mode made famous by the Galapagos finches and Hawaiian honey creepers, depends on water gaps between islands of the same archipelago.

The largest tropical Pacific island, New Guinea, has behaved as a continent to most taxa and permitted adaptive radiations of bats, birds, rats, plants, and insects within a single land mass. The even larger Australia, conventionally termed a continent, has also harbored continental speciation in all these groups.

For birds, only the second largest Pacific island, New Zealand, has also been large enough to permit speciation within a land mass. [It is not known whether speciation of moas proceeded intra-archipelagally across the intermittent water gap between New Zealand's North and South Islands, or continentally within one of these islands or an intermittently fused Greater New Zealand. However, present distributions of kiwis (*Apteryx*) and wekas (*Gallirallus*) suggest speciation within North or South Island].

For bats, no Pacific island besides New Guinea was big enough to permit continental speciation. In fact, bats have failed to radiate in other Pacific archi-

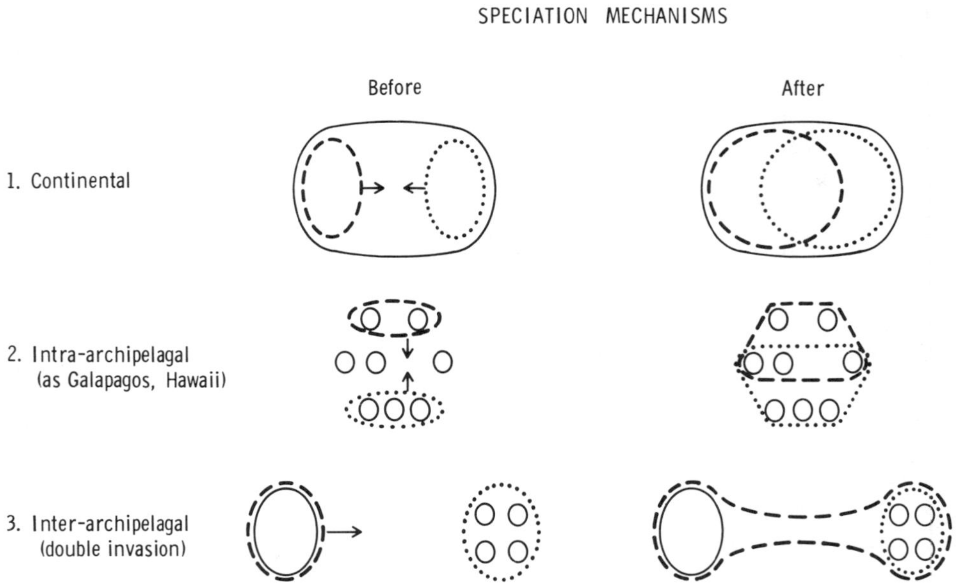

Fig. 4. Three alternative speciation mechanisms, which differ as to whether the barrier permitting divergence is within a single land mass (above: "continental speciation"), between islands of the same archipelago (middle: "intra-archipelagal speciation"), or between two archipelagoes (below: "inter-archipelagal speciation"). In each case the sketches on the left and right illustrate distributions of two related taxa respectively before and after an expansion of one taxon resulted in sympatry. Island borders are shown by solid lines, the range of one taxon by dashes, the range of the other taxon by dots, and the arrows indicate directions of expansion. (From Diamond, 1977.)

pelagoes by any mode: there are endemic bat species, but apparently no pairs or clusters of species that arose in a Pacific archipelago from a common ancestor, unless in the Bismarck or Solomon bat faunas. Similarly, outlying Pacific archipelagoes have endemic species of marine fishes and marine molluscs, but no radiations of these taxa.

The Solomons supported a radiation of *Sphenomorphus* skinks (10 endemic species) and the beginnings of a radiation of giant rats (three sympatric endemic species of *Uromys*).

New Caledonia was large enough for continental speciation of skinks (11 endemic species of *Lygosoma),* geckoes (10 endemic species in three endemic genera), *Araucaria* (eight species), *Dacrydium, Pancheria* (30 species), and carabids. However, New Caledonia was too small for any adaptive radiation in birds. There is no known case of a bird species pair that originated on New Caledonia, nor of a bird species that breaks up into multiple subspecies on the main island of New Caledonia itself.

Hawaii has supported famous radiations of land snails, tarweeds, *Drosophila, Cyrtandra, Bidens,* and a single bird stock (Drepanididae); very little radiation of ferns; and no radiation at all of its butterflies or its bat.

How can one account for such marked differences among archipelagoes and among taxa? To begin with, radiation involves the formation of endemic species and thus requires time. We therefore expect radiation, by whatever mechanism, only in those archipelagoes large enough to guarantee a long lifetime to that group of taxa. For the same reasons as those discussed in the section "Who Lasts How Long?," this consideration alone suffices to indicate why small archipelagoes may support radiations of plants and insects but not of vertebrates. However, it does not explain why lizards but not birds could radiate in the Solomons and New Caledonia, and why birds could radiate intra-archipelagally but not continentally in Hawaii and the Galapagos. For further understanding, we need to consider each mode of speciation separately.

Since continental speciation involves formation of geographic isolates within a single land mass, it requires that the land mass be large compared to the mean dispersal distances of the group in question. Mammals are relatively mobile overland; birds, even more so. As a result, mammals and birds have radiated continentally only on the world's largest land masses (Soulé, 1980). Specifically, the smallest islands (areas as mi^2/km^2) on which there is evidence for continental speciation in birds and large mammals are New Guinea (306,000/793,000) and Madagascar (227,000/588,000); for small mammals, Cuba (44,000/114,000) and Luzon (40,000/104,000). Among groups that disperse less well overland, lizards have radiated on New Caledonia (7360/19,000), the Compositae on Masafuera (18/47) and Masatierra (35/90) of the Juan Fernandez group, and notoriously slow-moving land snails on many small islands of the Pacific (e.g. Cooke and Kondo, 1960).

Inter-archipelagal speciation requires that an archipelago be moderately isolated by the standards of the dispersal ability of the taxon in question: sufficiently close that the ancestral stock can invade more than once, but sufficiently remote that the second invasion is unlikely to have arrived before descendants of the first invasion have had time to achieve reproductive isolation. The cases of speciation by double invasion in Pacific birds involve interarchipelagal distances of several hundred to 1000 mi: e.g. the invasions from Australia that produced the white-eyes *Zosterops lateralis, Z. tenuirostris* and *Z. albogularis* on Norfolk Island 900 mi (1450 km) distant; the invasions from Australia that produced the gallinules *Porphyrio ("Notornis") mantelli* and *P. porphyrio,* the ducks *Anas gibberifrons* and *A. aucklandica,* and the ducks *Aythya novaeseelandiae* and *A. australis* on New Zealand 1000 mi (1600 km) distant; and the invasions from Fiji that produced the trillers *Lalage maculosa* and *L. sharpei* on Samoa 500 mi (800 km) distant. I have not been able to find data to illustrate the distance required for inter-archipelagal speciation in other taxa.

Intra-archipelagal speciation to produce a whole cluster of species requires an archipelago sufficiently remote that niches are unlikely to be filled by other stocks and that speciation is unlikely to be by inter-archipelagal invasions. In practice, only the most remote Pacific archipelagoes, Hawaii and the Galapagos, were sufficently remote for inter-archipelagal speciation in birds. In contrast, *Sphenomorphus* lizards, being poorer water-crossers than birds, were able to speciate intra-archipelagally in the Solomons, only 100 mi (160 km) from the Bismarcks and 400 mi (640 km) from New Guinea.

Intra-archipelagal speciation on remote archipelagoes presents a paradox. Given that a species could jump a broad water gap to reach a remote archipelago, how could the much narrower interisland gaps within the archipelago permit the development of reproductive isolation necessary for species? For example, the ancestor of Darwin's finches crossed the 650 mi (1045 km) of ocean between the Galapagos and South America. How could the water gaps of only 10–100 mi (16–160 km) within the Galapagos ever have retarded gene flow enough to permit speciation?

The explanation is that dispersal ability is subject to natural selection and may undergo large increases or decreases in evolutionary time. Clear examples are provided by species that are confined to remote oceanic islands, have presently no means of overwater dispersal, but are derived from ancestors with efficient means of overwater dispersal. Examples include flightless island birds derived from flying ancestors; "psychologically flightless" island birds, that are mechanically capable of crossing water but refuse to do so; and island plants derived from ancestors with fleshy or barbed seeds adapted for bird dispersal or else winged seeds adapted for wind dispersal, but themselves lacking such adaptations (Carlquist, 1965, 1974; Diamond, 1981). I speculate that part of the reason why bats have not radiated on Pacific islands is that evolution has produced dozens of flightless birds but not a single flightless bat. How could a flightless bat forage?

SUMMARY

This paper examines why different elements of Pacific biotas differ in age, species richness, geographic source, and degree of radiation.

For example, New Guinea is rich in some groups but poor in others, and is a melting pot of old Gondwanan relicts and of recent immigrants. The relative importance of Asia and Australia as source areas for colonists of New Guinea differs among taxa, because of taxon differences in the relative importance of the water barriers between Asia and New Guinea and of the climate and habitat barriers between Australia and New Guinea.

Land birds of islands east of New Guinea are derived mainly from New Guinea. Marked endemism and adaptive radiation in these birds are found only on New Zealand, Hawaii, and the Galapagos. As reflected in degree of endemism, Pacific island avifaunas have achieved standing age distributions much younger than the age of the islands themselves.

Other taxa on Pacific islands differ in how far they spread from New Guinea, depending on their overwater dispersal ability; in how long they last, depending on island area, population density, and the ability to survive as dormant seeds or as long-lived adults; in degree of endemism, depending on how long they last, dilution by immigrants, and rate of evolution; and in degree of adaptive radiation, depending on how long they last and on overland and overwater dispersal ability. Three modes of speciation ("continental," "inter-archipelagal," and "intra-archipelagal") are distinguished, related to whether the geographic isolation required for speciation is within a single land mass or across water gaps. Dispersal ability of a species line often changes over evolutionary time.

LITERATURE CITED

Balgooy, M. M. J. van. 1966. Pacific plant areas. *Blumea* 5: supplement.
———. 1971. Plant geography of the Pacific. *Blumea* 6: supplement.
Carlquist, S. 1965. *Island life.* Natural History Press, New York.
———. 1970. *Hawaii, a natural history.* Natural History Press, New York. 463 p.
———. 1974. *Island biology.* Columbia University Press, New York.
Cooke, C. M. & Y. Kondo. 1960. Revision of Tornatellinidae and Achatinellidae (Gastropoda, Pulmonata). *Bernice P. Bishop Mus. Bull.* 221.
Cracraft, J. 1973. Continental drift, paleoclimatology, and the evolution and biogeography of birds. *J. Zool. Lond.* 169: 455–545.
Darlington, P. J. 1957. *Zoogeography: the geographical distribution of animals.* Wiley, New York.
———. 1965. *Biogeography of the southern end of the world.* Harvard University Press, Cambridge, Mass.
Diamond, J. M. 1972. Biogeographic kinetics: estimation of relaxation times for avifaunas of Southwest Pacific Islands. *Proc. Natl. Acad. Sci. USA* 69: 3199–203.
———. 1977. Continental and insular speciation in Pacific land birds. *Syst. Zool.* 26: 263–68.
———. 1980. Species turnover in island bird communities. *Acta 17th Int. Ornithol. Congr.* 2: 777–82.
———. 1981. Flightlessness and fear of flying in island species. *Nature* 293: 507–08.
Diamond, J. M. & A. G. Marshall. 1976. Origin of the New Hebridean avifauna. *Emu* 76: 187–200.
Gressitt, J. L., ed. 1963. *Pacific Basin biogeography.* Bishop Museum Press, Honolulu.
———. 1981. *Biogeography and ecology of New Guinea.* Junk, the Hague.
Haffer, J. 1969. Speciation in Amazonian forest birds. *Science* 165: 131-37.
Jones, H. L. & J. M. Diamond. 1976. Short-time-base studies of turnover in breeding bird populations on the California Channel Islands. *Condor* 78: 526–49.
Kay, E. A. 1979. *Marine shells of Hawaii.* Bishop Museum Press, Honolulu.
———. 1980. *Little world of the Pacific: an essay on Pacific Basin biogeography.* University of Hawaii, Lyon Arboretum, Lecture No. 9.
Keast, A., ed. 1981. *Ecological biogeography of Australia.* Junk, the Hague.
MacArthur, R. H. & E. O. Wilson. 1967. *The theory of island biogeography.* Princeton University Press, Princeton.
Mayr, E. 1941a. Borders and subdivision of the Polynesian Region as based on our knowledge of the distribution of birds. *Proc. 6th Pac. Sci. Congr.* 4: 191–95.
———. 1941b. The origin and history of the bird fauna of Polynesia. *Proc. 6th Pac. Sci. Congr.* 4: 197–216.
———. 1965. Avifauna: turnover on islands. *Science* 150: 1587–88.
Olson, S. L. 1977. A synopsis of the fossil Rallidae. p. 339–73 *In:* Ripley, S. D., ed., *Rails of the world.* Godine, Boston.
Olson, S. L. & H. F. James. 1982. Fossil birds from the Hawaiian Islands: evidence for wholesale extinction by man before western contact. *Science* 217: 633–35.
Randall, J. E. 1976. The endemic shore fishes of the Hawaiian Islands, Lord Howe Island. *Colloque Commerson 1973.* O.R.S.T.O.M. Travaux et Documents no. 47.
Raven, P. R. 1979. Plate tectonics and southern hemisphere biogeography. p. 3-24 *In:* Larsen, K. & L. B. Holm-Nielsen, eds., *Tropical botany.* Academic Press, London.
Raven, P. R. & D. I. Axelrod. 1972. Plate tectonics and Australasian paleobiogeography. *Science* 176: 1379–86.
———. 1974. Angiosperm biogeography and past continental movements. *Ann. Mo. Bot. Gard.* 61: 539–673.
Rich, P. V. 1975. Changing continental arrangements and the origin of Australia's non-passeriform continental avifauna. *Emu* 75: 97–112.
Sibley, C. G. 1976. Protein evidence of the relationships of some Australian passerine birds. p. 557–70 *In: Proc. 16th Int. Ornithol. Congr.*

Soulé, M. E. 1980. Thresholds for survival: maintaining fitness and evolutionary potential. p. 151–69 *In:* Soulé, M. E. & B. A. Wilcox, eds. *Conservation biology.* Sinauer Assoc., Sunderland, Massachusetts.

Terborgh, J. & B. Winter. 1980. Some causes of extinction. p. 119–33 *In:* Soulé, M. E. & B. A. Wilcox, eds., *Conservation Biology.* Sinauer Assoc., Sunderland, Massachusetts.

Walker, D., ed. 1972. *Bridge and barrier: the natural and cultural history of Torres Strait.* Australian National University, Canberra.

Zimmerman, E. C. 1948. *Insects of Hawaii,* Vol. 1. University of Hawaii Press, Honolulu.

Patterns of Speciation in the Indo-West Pacific

E. Alison Kay
Department of Zoology
University of Hawaii
Honolulu, Hawaii 96822 USA

INTRODUCTION

There can be no discussion of the distribution of tropical marine organisms or of their derivation without mention of the Indo-Pacific, the faunal province stretching between latitudes of about 30°N and 30°S from the east coast of Africa through the Indian Ocean and into the Pacific to the Hawaiian Islands, the Society Islands, the Marquesas, and Easter Island, a province covering more than one-fourth of the globe. The area was recognized in the mid-19th century when John Richardson (1845) described the region from the Red Sea to Polynesia as "one ichthyological province." James Dwight Dana (1853) called it "the great ocean." Charles Darwin (1859) described it as where ". . .many shells are common to the eastern islands of the Pacific and eastern shores of Africa, in almost exactly opposite meridians of latitude." Edward Forbes gave it its name, Indo-Pacific, about 1850 (Kay, 1980).

It was Sven Ekman, however, who did more than anyone else to impress the significance of the Indo-Pacific in the thinking of biologists. Ekman (1935, 1953) called it the Indo-west Pacific (as distinct from the other warm water shelf fauna, the Atlanto-east Pacific), setting its boundaries as East Africa and Hawaii, citing a host of references to the wealth of animal life within its compass, and pointing to a "centre and focus," the Indo-Malayan region, ". . .from which the others recruited the main contingent of its fauna. The further one moves from this centre in any direction, the more the fauna appears as a progressively impoverished Indo-Malayan (Indo-Australian) fauna. . .. The other faunas occupy marginal regions which can therefore not be ranked with the Indo-Malayan central region in zoogeographical importance."

Ekman's "other faunas" were the subregions recognized within the Indo-west Pacific, where, as he put it, "homogeneity can only be relative" (Ekman, 1953). In the Pacific, for example, Dana (1853) distinguished the Polynesian Province, the Hawaiian Province, and the Rarotongan Province based on the distribution of Crustacea; Knox (1958) used polychaetes to describe Micronesia, Melanesia, and Polynesia; Powell (1958) identified 11 subregions on the basis of molluscan distribution; Schilder (1961) recognized seven subregions utilizing cowrie distribution; and Stehli and Wells (1971) found three subprovinces of corals: a main Indo-Pacific subprovince, a distinct fauna in the Eastern Pacific from Midway to Easter Island, and a third subprovince in the southwestern Pacific and southern Indian Ocean. Ekman himself (1953) recognized Hawaii and the islands of the central Pacific as distinctive. Briggs (1974) talked of three "centres of endemism": Hawaii, Easter Island, and the Marquesas.

Despite the number of subregions that have been distinguished, however, there is no common denominator among them and, lacking a common theme, they remain merely lines on a map. Neither subdivisions nor numbers provide answers to questions that should be asked of the fauna. (1) Are there patterns of speciation both for the Indo-west Pacific as a whole and for the "faunas" or subdivisions within it that are different from those of a region with a highly diverse center and attenuate fringes? (2) If there are such patterns, what are the characteristics of the faunas that distinguish them, one from the other?

In this review, I first discuss patterns of speciation in the Indo-west Pacific as a whole, then I focus on patterns of speciation within the Pacific, and finally I attempt to interpret these patterns using one group of marine animals as an example.

MATERIALS AND METHODS

The data that comprise the base of this study are those for which I believe there exist reasonably reliable and comprehensive distributional lists. What I did was simply to compile lists of mollusks, echinoderms, crustaceans, and fishes—some 3000 species and 500 genera—from these lists. The data sources are listed in the Appendix.

Each taxon is assigned to one of four distribution categories. (1) If it is widely distributed in the Indo-west Pacific or found in either the Indian Ocean or the Pacific Ocean and the western Pacific, it is assigned the category of "Indo-west Pacific." (2) If it is restricted in its distribution to the Indo-Malayan archipelago, the Philippines, southern Japan or the east coast of Australia, it is termed "western Pacific" [Abbott (1960) has called this area the "western Pacific arc"]. (3) If the taxon is restricted to the Indian Ocean it is so designated. (4) If the taxon is found only in the Pacific Basin (that is, on the Pacific tectonic plate), it is so designated.

It must be recognized that the numbers represent but a fraction of the species and genera that are now believed to be found in the Indo-west Pacific, and the data are thus manifestly incomplete. And, because they come from different sources, they suffer from different interpretations and different approaches.

SPECIATION PATTERNS IN THE INDO-WEST PACIFIC

As a first approximation to describing speciation patterns within the Indo-west Pacific, families and genera of the major groups of marine animals were analyzed for their occurrence within the province. All families were determined to be widespread; none is endemic to any one of the three areas. The genera (Table 1), however, fall into four categories: 76.5% are widely spread through the Indo-west Pacific, 17% are restricted to the western Pacific, and 7% are found only in the Indian or Pacific oceans, 5% in the Indian Ocean, 2% in the Pacific Basin. In

Table 1. Generic and subgeneric endemism in the Indo-west Pacific.*

	No. genera	% Indo-west Pacific	% Western Pacific	% Indian Ocean	% Pacific Ocean
MOLLUSCA					
Gastropoda	62	91	5	2	2
Bivalvia	26	100	0	0	0
ECHINODERMATA					
Crinoidea	35	74	26	0	0
Asteroidea	59	61	25	12	2
Ophiuroidea	66	68	23	7	1.5
Echinoidea	59	81	8	5	5
Holothuroidea	70	64	28	7	0
FISHES	77	75	17	5	4
Averages		76.5	17	5	2

* Data from sources cited in the Appendix; figures rounded.

those few instances where there are genera endemic to the Indian or Pacific oceans, they are for the most part monotypic. Springer (1982) noted this pattern was characteristic of the fishes he studied in the Pacific Basin: no endemism at the family level but ten genera are endemic to the Pacific Plate and all are monotypic, with the possible exception of *Lentipes* (Gobiidae).

The picture for species is quite different. *A priori* it was anticipated that the greater proportion of species would be restricted to the western Pacific, consonant with the thesis of the high diversity in that region. Not so. An average of about 28% of the species is restricted to the western Pacific (Fig. 1) and, indeed, because the figures include both southern Japanese and some Australian faunas, as well as those of islands such as New Caledonia and Fiji, the proportion of species endemic to Ekman's Indo-Malayan center is probably less than the figure cited here. As for the others, about 37% are widespread Indo-west Pacific forms and 34% are endemic to the Indian or Pacific oceans, 18% in the Indian Ocean and 16% in the Pacific Ocean (Fig. 1).

Species Distributions

There are noticeable differences among the groups examined (Fig. 1). In the Arthropoda, the Crustacea give the impression of having a high proportion of widespread groups; the echinoderms have fewer widespread forms. The Bivalvia, some crustaceans, the Echinoidea among the echinoderms and a couple of families of fishes have high proportions of widespread species; the archaeogastropods, hapalocarcinids, crinoids, and some of the fish families have high proportions of species in the western Pacific. In the mollusks (Fig. 2), the neogastropod Mitridae and Conidae show relatively high proportions of Indo-west Pacific species as do the Cerithiidae in the mesogastropods. There would also appear to be substantially less endemism in the mollusks in the Pacific Ocean than in the Indian Ocean (Fig. 2).

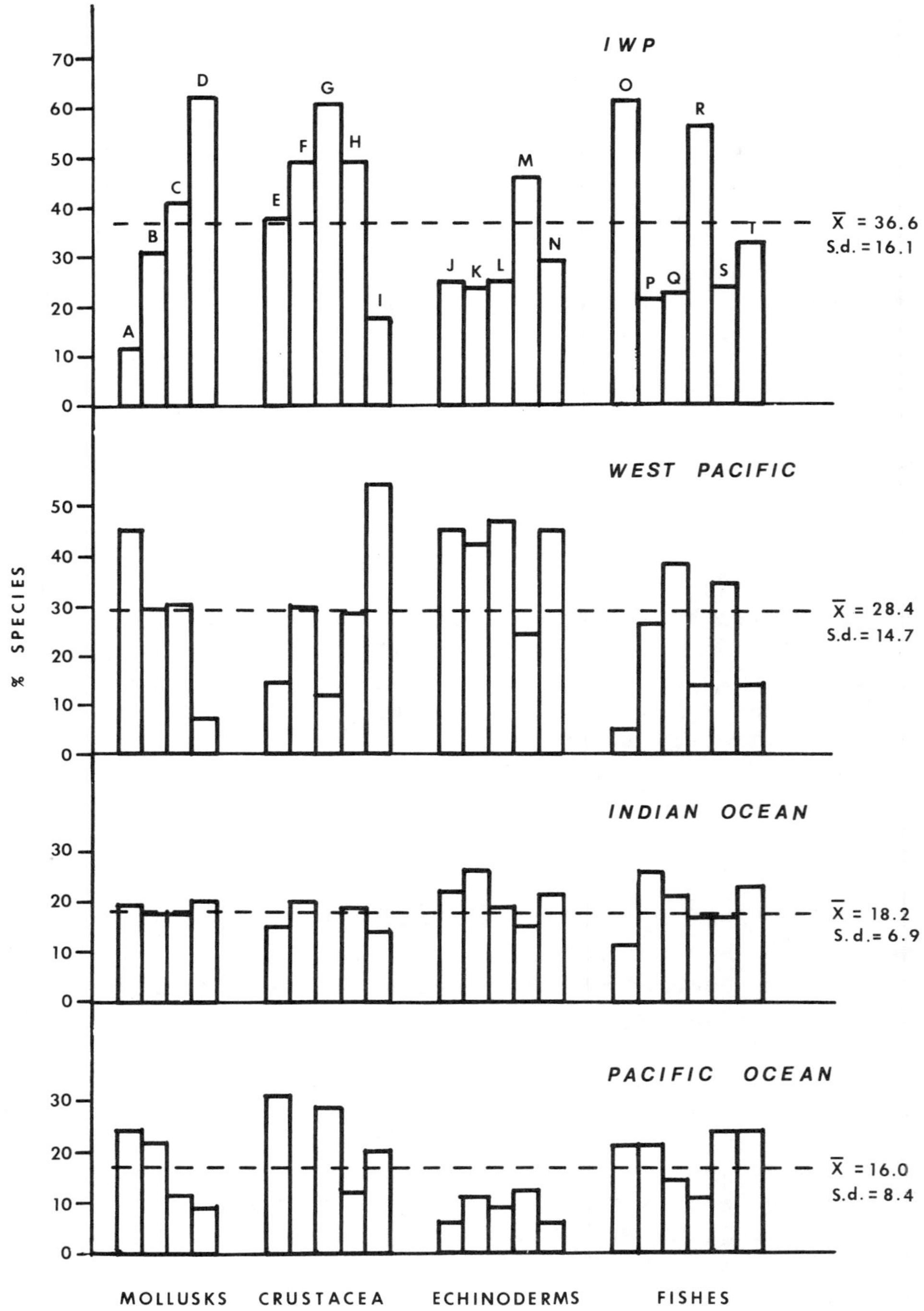

Fig. 1. Percentages of species endemic to the Indo-west Pacific (IWP). Mollusks: A, Archaeogastropoda; B, Mesogastropoda; C, Neogastropoda; D, Bivalvia. Crustaceans: E, Palinura; F, Anomura; G, Dromididae; H, Xanthidae; I, Hapalocarcinidae. Echinoderms: J, Crinoidea; K, Asteroidea; L, Ophiuroidea; M, Echinoidea; N, Holothuroidea. Fishes: O, Holocentridae; P, Serranidae and Chaegodontidae; Q, Pomacentridae and Labridae; R, Acanthuridae; S, Cirrihitidae; T, Tetradontidae.

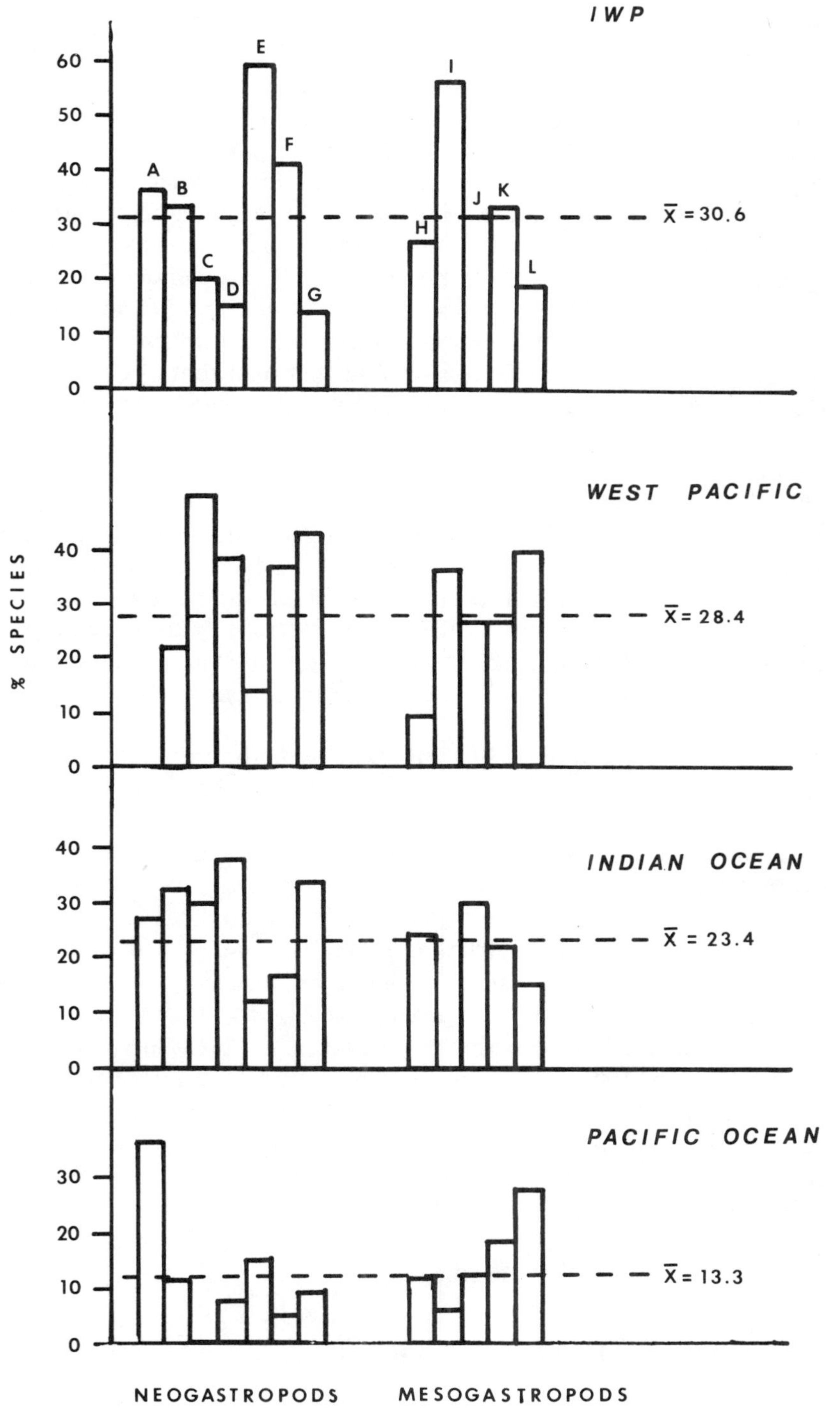

Fig. 2. Percentages of gastropod species endemic to the Indo-west Pacific (IWP). Neogastropoda: A, Thaidinae; B, Harpidae; C, Buccinidae; D, Vasidae; E, Mitridae; F, Conidae; G, Turridae. Mesogastropoda: H, Littorinidae; I, Cerithiidae; J, Strombidae; K, Cypraeidae; L, Cassidae.

The Meaning of Differences

What do these differences mean? Are they a reflection of the systematic treatment of the groups and thus artifacts of taxonomy? If they reflect evolutionary history or habits, what factors have influenced these strikingly different patterns?

It can be shown fairly clearly that the differences are not entirely artifacts of taxonomy. In the fishes, for example, many of the data are derived from a single author (Randall, 1955–1981 in the appendix), but even with his consistent treatment there are noticeable disparities among the families included in the analysis in the proportions of Indo-west Pacific species and those endemic to the other areas. Abbott (1960, 1968) who revised the molluscan families Strombidae and Cassidae noted a similar phenomenon, finding the concentration of species of Strombidae in the western Pacific less pronounced than that of Cassidae.

Nor do the distribution patterns of the bivalves appear to be attributable to the whims of the systematist: there is in general apparently less endemism in the bivalves than in gastropods. In French Polynesia, for example, Salvat and Rives (1975) list 27 gastropods—7% of their list—as endemic but not a single bivalve is noted as endemic. In the Hawaiian Islands 15% of the well known gastropod families (cowries, cones, miters, etc.) are endemic but less than 6% of such well known bivalves as the mytilids, pectens, pen shells, and the like are endemic (Kay, 1979). Rehder (1980) reports 15% of the gastropods are peculiar to Easter Island but only 2.7% of the bivalves are endemic. (Significant: χ^2 test each location, $P < 0.001$ for null hypothesis = no difference in endemism gastropods vs. bivalves.)

If the differences are not entirely attributable to taxonomic whimsy, then how are they to be explained? Stanley (1979) has suggested that there is a correlation between geographic range and geological longevity: widespread species have not put all their eggs in one basket and can tolerate the kind of local habitat destruction that extirpates endemic forms. Rates of evolution may also be important in affecting the length of some of the bars on the histograms in Fig. 1 and 2. Again Stanley (1979) provides some intriguing suggestions: rates of bivalve evolution are probably less than those of gastropods, and there are differences in evolutionary rates among the different families of gastropods. The habits of the animals themselves may also affect their distribution: *Conus* and miters are well known for their long-lived larval stages; limpets and haliotids have relatively short-lived larval forms. [Corpuz (1981) reports length of pelagic larval life in the Hawaiian limpets *Cellana* as about seven days. Heslinga (1981) reports a planktonic stage of *Trochus niloticus* as from three to eight days.] If larval life has anything to do with dispersal capability [and Scheltema (1971) has provided evidence that it does], then those animals with long-lived planktonic larvae should be the most widely spread in the Indo-west Pacific.

ENDEMISM IN THE PACIFIC BASIN MARINE FAUNA

A First Approximation

The Pacific Basin marine fauna is here defined essentially as that which occurs on the Pacific tectonic plate and its margins, as recognized by Kay (1980) and Springer (1981, 1982). What proportion of the marine organisms on the Pacific Plate is endemic to it? The figures for mollusks, crustaceans, echinoderms, and

Table 2. The speciation pattern on the Pacific Plate.*

	Number of species on Pacific Plate	% Indo-west Pacific species	% Pacific Plate species
Mollusca	202	63	37
Echinoderms	257	60	40
Crustacea	143	68	32
Fishes	200	48	52
Averages		60	40

* Data from sources cited in the Appendix; figures rounded.

fishes dealt with here indicate that Pacific Basin endemics comprise the smaller of the two faunal elements on the Pacific Plate—40%, while 60% of the fauna is widespread in the Indo-west Pacific (Table 2).

The Distribution of Pacific Plate Endemics

Is there any pattern to the distribution of the Pacific Plate endemics? From the tabulations, about half (55.5%) the endemics (Table 3) are widely distributed on the Pacific Plate as is, for example, *Strombus maculatus* (Fig. 3). The other half (44.5%) is localized in two areas: the Hawaiian Islands and southeastern Polynesia (Table 3). The latter area includes the Society Islands, the Tuamotus, the Marquesas, the Austral and Cook islands, and Easter Island although it is on the very small, neighboring Nazca Plate.

The figures receive some confirmation in a series of comparisons of similarity indices for nine families of mollusks. The dendrograms (Fig. 4) derive from similarity matrices computed from the Jaccard coefficient of similarity and analyzed in a Q-mode analysis (Q-mode clusters are based on sites) using unweighted pair groups. In the dendrograms this rather consistent picture appears: in six of the nine groups Guam, the Ryukyus, the Marshall Islands, and Fiji comprise one cluster; Hawaii and southeastern Polynesia comprise a second cluster. In three instances there is a slightly different pattern where Hawaii may be clustered with Guam and the Marshalls and Fiji clusters with southeastern Polynesia. These differences may simply reflect different phylogenetic histories, as pointed out by Diamond (this volume).

The patterns distinguished thus far tell us only that the molluscan faunas of

Table 3. Foci of endemism on the Pacific Plate.*

	Number of Pacific Plate species	% Widespread Pacific Plate endemics	% Hawaii Southeast Polynesia endemics	% Hawaii endemics	% Southeast Polynesia endemics
Mollusca	74	45	55	26	29
Echinoderms	102	69	31	17	14
Crustacea	48	60	40	15	25
Fishes	98	48	52	20	32
Averages		55.5	44.5	19.5	25

* Data from sources cited in the Appendix; figures rounded.

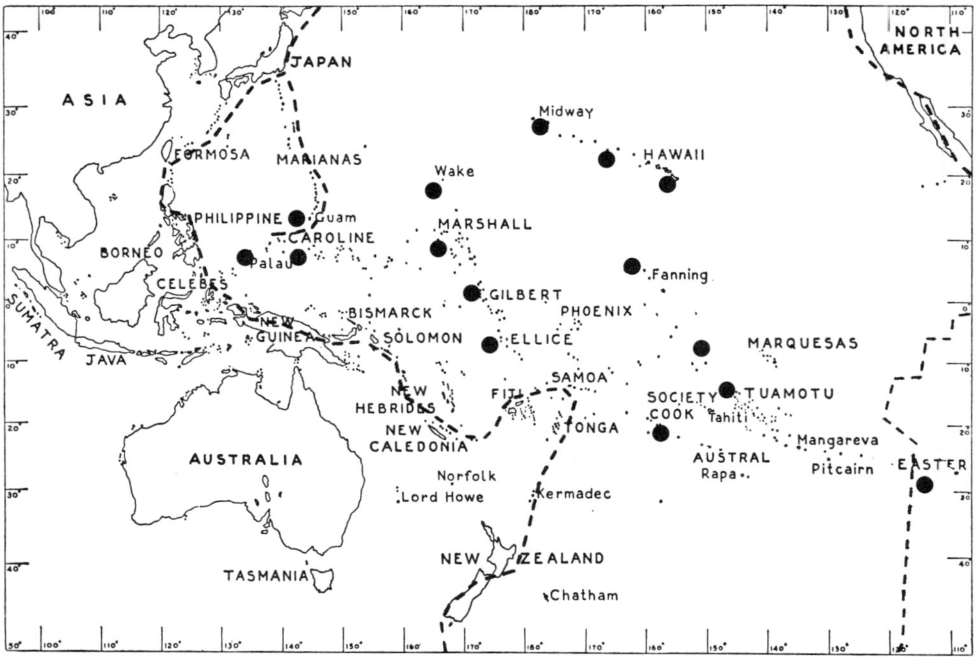

Fig. 3. The distribution of *Strombus maculatus* on the Pacific plate. Dotted lines indicate boundries of the Pacific plate.

Hawaii and southeastern Polynesia are somehow "different" and that there are both island endemics and endemics widespread on the Pacific Plate. For a more detailed assessment of the patterns of speciation, one group of mollusks, the cowries (Family Cypraeidae), is analyzed.

Cowrie Endemism in the Pacific

Although the cowries are conveniently grouped in a single family and often in the single genus *Cypraea* (see Kay, 1960), several species complexes (genera or subgenera depending on one's taxonomic preference) can be distinguished (Kay, 1981). Twenty of these species complexes occur in the Pacific Basin and 12 of the 20 groups show Pacific Plate endemism in the following ways (Table 4): (1) seven species complexes are endemic to either Hawaii or southeastern Polynesia; (2) three species groups have species widespread on the Pacific Plate and species endemic to the Hawaiian Islands or southeastern Polynesia (Table 5); and (3) two species complexes have species widespread on the Pacific Plate but no island endemics, although each one is found in either Hawaii or southeastern Polynesia.

There are two other features of the pattern. (1) Each of the Pacific Plate endemics, whether an island endemic or widespread on the Pacific Plate, has a presumed ancestral species found throughout the Indo-west Pacific. (2) Among the 19 island endemics (that is, species found only in Hawaii or in southeastern Polynesia), in only two instances (*C. semiplota* and *C. granulata* in the subgenus *Nuclearia* in Hawaii and *C. bouteti* and *C. bernardi* in the subgenus *Lyncina* in southeastern Polynesia) are there more than one species in each species complex.

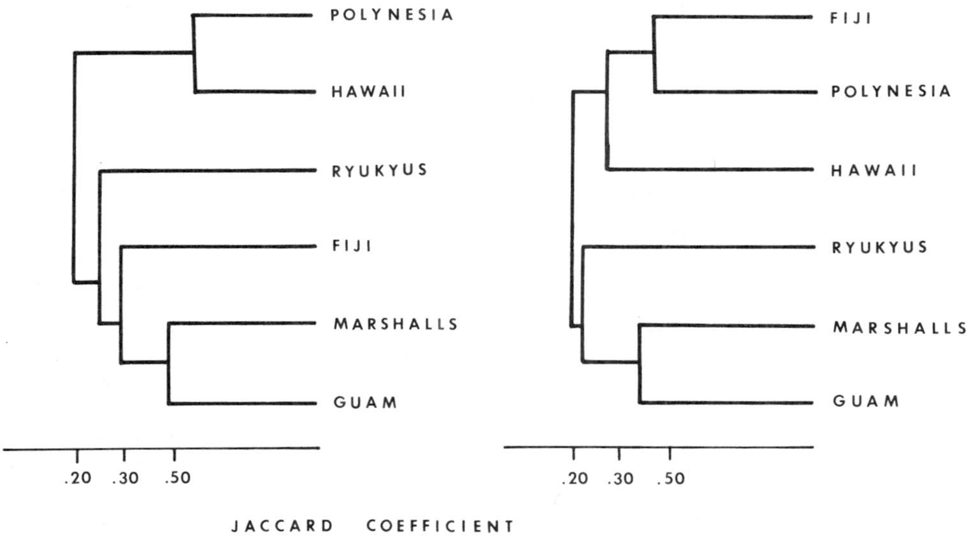

Fig. 4. Island associates in a Q-mode cluster analysis based on the Jaccard coefficient and unweighted pair-group method. A. The pattern for the Cassidae, Strombidae, Cypraeidae, Thaidinae, Vasidae, Harpidae and Bivalvia. B. The pattern for Littorinidae, Cerithiidae and Mitridae.

The low genus or subgenus to species ratio is not limited to the cowries among the marine animals in the Pacific. In 32 genera and 86 species of fishes in the Pomacentridae, Labridae, and Chaetodontidae and two families of crustaceans, the genus/species ratio averages 1:3—a ratio which is to be contrasted with a 1:10 genus/species ratio estimated for insects and plants in the Hawaiian Islands and some islands in southeastern Polynesia. I suggest that Nelson and Platnick's (1981) note that "... one might reasonably expect that most native Hawaiian marine organisms might ultimately be regarded as endemic, as is the case for land plants..." is not correct. The pattern of speciation in marine organisms in the Pacific differs from that in terrestrial organisms on high islands in that there is little or no adaptive radiation noticeable among the endemics (Kay, 1980).

EXPLANATIONS FOR PACIFIC PLATE ENDEMISM

Geological History: The Cretaceous Pacific

Data collected by dredging and the Deep Sea Drilling Project during the last decade indicate that there have been islands and a marine fauna in the Pacific at least since the mid-Cretaceous, 100 m.y.a. The oldest oceanic crust in the Pacific is of Jurassic age, 150 m.y. old (Fischer et al., 1970). During Barremian-Aptian time to late Cretaceous time (115 to 70 m.y.a.), widespread volcanic activity apparently resulted in sea mount and island formation and reef growth in the Central Pacific (Schlanger and Premoli Silva, 1981). Bivalves dredged from the Manihiki Plateau in southeastern Polynesia have been dated from Aptian and Albian time (Kauffman, 1976). Campanian and Maestrichian reefs (78–70 m.y.a.) grew along both the Line and Marshall Islands (Schlanger and Premoli Silva, 1981). Basalts of Meiji in the Emperor Seamounts are dated at 70 m.y. (Dalrymple

Table 4. Cowrie (*Cypraea*) endemism on the Pacific Plate.

Cowrie stock (subgenus)	Hawaiian Islands	Southeast Polynesia	Pacific Plate
(*Palmadusta*)		*serrulifera*	
(*Cribrarula*)	*gaskoini*	*cumingi*	
(*Luria*)	*tessellata*		
(*Monetaria*)		*obvelata*	
(*Ravitrona*)		*caputdraconis*	
(*Nuclearia*)	*semiplota*	*cassiaui*	
	granulata		
(*Talostolida*)	*burgessi*	*subteres*	
	rashleighana		
(*Lyncina*)	*sulcidentata*	*bouteti*	*leviathan*
		bernardi	*propinqua*
			ventriculus
			schilderorum
			aurantium
(*Pustularia*)	*mauiensis*	*dillwyni*	*mariae*
(*Erosaria*)	*ostergaardi*	*englerti*	*irrorata*
			labrolineata
			beckii
(*Mauritia*)			*maculifera*
(*Blasicrura*)			*goodalli*

et al., 1973). A minimum age of 50 m.y. has been estimated for the Tuamotu Islands (Schlanger et al., 1976).

Kauffman's (1976) analysis of the Cretaceous bivalves and those of others for ammonoids (Renz, 1973), rudists (Hamilton, 1956), and foraminiferans (Beckmann, 1976; Premoli Silva and Brusa, 1981) suggest that the fauna of the Cretaceous Pacific was cosmopolitan, with its relatives in western America and the Caribbean as well as to the west. To explain the Caribbean connection, Premoli Silva and Brusa (1981) propose that the Caribbean Plate extended into the Pacific nearly to the Marshall and Line islands in Cretaceous time and shallow island stepping stones would have permitted rapid colonization of the Line and Marshall areas by Caribbean faunas.

Table 5. Distribution of Pacific Plate cowrie endemics.

	Hawaii	Southeast Polynesia
Cypraea maculifera	+	+
C. goodalli		+
C. schilderorum	+	+
C. ventriculus		+
C. aurantium		+
C. leviathan	+	
C. mariae	+	+
C. irrorata		+
C. beckii	+	
C. labrolineata	+	

As interesting as that fauna is, it did not survive. The Palaeocene and Eocene were periods of great marine faunal change worldwide, as rudists were replaced by scleractinian corals and the cephalopod belemnites and ammonites were replaced by bony fishes (Newell, 1952; Taylor et al., 1980). In the Pacific, not only did the cosmopolitan fauna become extinct, but there was a change in faunal relationships. Premoli Silva and Brusa (1981) suggest that rapid ocean floor spreading, resulting in the formation of what is today called the East Pacific barrier, cut affinities with the Caribbean Province, and from the Upper Eocene on the Pacific Plate marine fauna became part of the fauna of the Indo-west Pacific.

The Pacific From the Eocene

The geographical distribution of the Indo-west Pacific fauna in Eocene and Miocene time was markedly different from that today. *Nautilus, Tridacna,* gastropods such as cones and cowries, bony fishes and scleractinian corals are all found in Eocene and Miocene horizons of Europe. That Tertiary fauna was also strongly entrenched from the Eocene onward in India (Davis, 1975), Java (Martin, 1879–80), the Mariana Islands (Ladd, 1972), and Fiji (Ladd, 1934), and, on the Pacific Plate, in the Tuamotus (Cole, 1959) and at Enewetak and Bikini in the Marshall Islands (Ladd, 1966, 1972, 1977). A Miocene fauna with Indo-Pacific affinities has been drilled from Midway in the Hawaiian chain (Ladd et al., 1970).

This Tertiary fauna, often termed a Tethyan fauna, was related to that of the seaway that extended virtually around the world between the continents of Laurasia and Gondwanaland from Jurassic time. Thus, there was a direct marine connection from the present Mediterranean-European province into the present Pacific throughout the Mesozoic and through part of the Tertiary. With time, that seaway changed as major rearrangements of land masses in and around the seaway occurred. India, moving northward from the beginning of the Tertiary, may have collided with Eurasia about 55 m.y.a., in the Eocene (Audley-Charles et al., 1981). The ancestral Mediterranean lost connection with the Indian Ocean in the Miocene. Major tectonic events took place in the western Pacific and the Indo-Malaysian archipelago well into the Pleistocene. These events left three major remnants of the once extensive Tethyan fauna: the Indian Ocean, the western Pacific, and the Pacific Ocean basin; in each, speciation of the marine fauna has proceeded independently.

The geological history of the cowries parallels that of the other members of the Tethyan fauna. The earliest known cowries appear in the late Jurassic in southern Europe and cowries flourished through the Eocene in Europe, India, the Americas and the Caribbean. Eocene fossils are known from Burma, Java, and Fiji in the western Pacific, and from the Miocene of the Marshall Islands (Ladd, 1977). The species complexes that are represented by endemics on the Pacific Plate, for example, trace back to the Eocene, Miocene, and Pliocene.

Explanations for Pacific Plate Endemism

The traditional explanation for speciation in the far-flung reaches of the Indo-west Pacific is that a few larvae were dispersed over great distances and that, given time, as a result of limited gene flow, the populations differentiated to the point that we recognize them today as genera, species, or subspecies. The necessary ingredients for this model, essentially that of geographical speciation as envisioned by Mayr (1942), are here: Hawaii and southeastern Polynesia are at the farthest reaches of the Indo-west Pacific, many cowries (and other animals) apparently have fairly long-lived larval stages such that they may be carried by currents to those far reaches, and there is the element of time, at least 50 to 70 m.y. Some of the islands of Hawaii and those of southeastern Polynesia are among the oldest on the Pacific Plate.

Not only are island endemics accomodated by this model, but so too are the species widespread on the Pacific Plate; each one of them occurs either in Hawaii or Polynesia, if not in both areas. Could they not have speciated in either area and then dispersed westward across the plate as suggested by Ladd (1960)?

Vicariance theory provides another explanation for speciation. As an alternative, Springer (1981) suggests that "At some time in the past. . .each of the Pacific plate endemics was represented by a single widespread taxon that ranged from the continental areas of the Indo-west Pacific onto the islands of the Pacific plate, and that geological events along the western margin of the Pacific plate, resulted in a barrier to dispersal (reproduction) between the continental and Pacific plate populations of each taxon, allowing the populations to vicariate and attain. . .a familial, specific or subspecific level of differentiation." Springer (1982) has enlarged on this point in his review of the distribution of shorefishes, suggesting vicariance scenarios that produced three main types of endemism (widely distributed plate endemics, endemism restricted to a limited region on the plate, and endemism restricted to single islands of island groups).

CONCLUSIONS

The major premises concerning the fauna of the Indo-west Pacific as Ekman (1953) conceived it were that there was a "centre and focus," the Indo-Malayan region, and that subregions or areas of differences in the fauna derived their faunas from that area. The data presented here suggest that there are patterns of speciation within the Indo-west Pacific that indicate three major areas of endemism, the Indian Ocean, the western Pacific, and the Pacific Basin. Within the Pacific today there are two foci of endemism, the Hawaiian Islands and southeastern Polynesia. It is suggested these two foci have not only spawned island endemics but endemics that have dispersed westward (for example, the widespread cowry *Cypraea maculifera* may be a westward spreading species).

The model of patterns of speciation in the Indo-west Pacific presented here has several implications. If the model is plausible, then it would appear that some of the subregions or subprovinces that have been recognized in the past are artifacts, comprised of mixtures of species derived from several different sources of endemism. The western Pacific, Ekman's (1953) center of origin, is especially susceptible to this interpretation. Rather than being a source area, it may instead represent an area of accumulation with three separate sources contributing to its diversity: (1) the accumulation of such animals as *Nautilus* and *Tridacna* the ranges of which were truncated perhaps during the Pliocene and Pleistocene and now appear to virtually terminate in that area, as suggested by Taylor (1971); (2) the accumulation of species endemic to the Pacific Basin, as suggested by Ladd (1960); and (3) the accumulation of species endemic to the area itself.

The recognition of subregions and subprovinces within a faunal province has long been a matter of opinion, as Schmidt (1954) pointed out. It is of interest that although the Hawaiian Islands have long been recognized as having unique marine fauna, few workers have recognized southeastern Polynesia as an entity and focus of endemism. J. D. Dana's (1853) Rarotongan Province, however, essentially encompassed the islands that have been included in what is here considered southeastern Polynesia. More recently, Rehder (1980) has identified the area in terms of its endemism and pointed to its many unique features.

ACKNOWLEDGMENTS

I am grateful to Dr. John D. Taylor and Dr. Alan J. Kohn for criticizing the manuscript, and to Dr. C. G. Adams and Professor Anthony Halleck for discussion of points of geological history.

LITERATURE CITED

Abbott, R. T. 1960. The genus *Strombus* in the Indo-Pacific. *Indo-Pac. Mollusca* 1(2): 33–145.

———. 1968. The helmet shells of the world (Cassidae). Part I. *Indo-Pac. Mollusca* 2(9): 15–209.

Audley-Charles, M. G., A. M. Hurley & A. G. Smith. 1981. Continental movements in the Mesozoic and Cenozoic. *In:* Whitmore, T. C., ed., *Wallace's line and plate tectonics.* Clarendon Press, Oxford.

Beckman, J. P. 1976. Shallow-water foraminfers and associated microfossils from sites 315, 316 and 318, DSDP Leg 33. *Initial Reports of the Deep Sea Drilling Project,* 33: 467–90. U.S. Government Printing Office, Washington, D.C.

Briggs, J. C. 1974. *Marine zoogeography.* McGraw-Hill, New York.

Cole, W. S. 1959. *Asterocyclina* from a Pacific seamount. *Cushman Found. Foraminiferal Res. Contrib.* 10(1): 10–14.

Corpuz, G. C. 1981. Laboratory culture of *Cellana exarata* Reeve (Gastropoda: Prosobranchia, Patellidae). *Aquaculture* 24: 219–31.

Dalrymple, G. G., E. A. Silver & E. D. Jackson. 1973. Origin of the Hawaiian Islands. *Am. Sci.* 61: 294–308.

Dana, J. D. 1853. On an isothermal oceanic chart, illustrating the geographical distribution of marine animals. *Am. J. Sci.* 16: 153–67, 314–27.

Darwin, C. R. 1859. *On the origin of species.* John Murray, London.

Davis, A. M. 1975. *Tertiary faunas.* Vol. 2. George Allen and Unwin, Ltd., London.

Ekman, S. 1935. *Tiergeographie des Meeres.* Akad. Verlagsges, Leipzig.

———. 1953. *Zoogeography of the Sea.* Sidgwick and Jackson, London.

Fischer, A. G., B. C. Heezen, R. E. Boyce, D. Bukry, R. G. Douglas, R. E. Garrison, S. A. Kling, V. Krasheninikov, A. P. Lisitzin & A. C. Pimm. 1970. Geological history of the western North Pacific. *Science* 168: 1210–14.

Hamilton, E. L. 1956. Sunken islands of the Mid-Pacific Mountains. *Geol. Soc. Am. Mem.* 64.

Heslinga, G. A. 1981. Larval development, settlement and metamorphosis of the tropical gastropod *Trochus niloticus. Malacologia* 20(2): 349–57.

Kauffman, E. 1976. Deep sea Cretaceous macrofossils, Hole 317A, Manihiki Plateau. *Initial Reports of the Deep Sea Drilling Project* 33: 503–34. U.S. Government Printing Office, Washington, D.C.

Kay, E. A. 1960. Generic revision of Cypraeidae. *Proc. Malacol. Soc. London* 33: 278–87.

———. 1979. Hawaiian marine shells. Reef and shore fauna of Hawaii. Section 4. *Bernice P. Bishop Museum Special Publ.* 64(4).

———. 1980. *Little worlds of the Pacific. An essay on Pacific Basin biogeography.* University of Hawaii Lyon Arboretum, Lecture No. 9.

———. 1981. The classification and distribution of the cowries; some new directions. *Bull. Am. Malacol. Union* 30.

Knox, G. A. 1958. The distribution of polychaetes within the Indo-Pacific. *Proc. 8th Pac. Sci. Congr.* 3: 403–11.

Ladd, H. S. 1934. Geology of Vitilevu, Fiji. *Bernice P. Bishop Mus. Bull.* 119.
———. 1960. Origin of the Pacific Island molluscan fauna. *Am. J. Sci.* 258-A: 137–50.
———. 1966. Chitons and gastropods (Haliotidae through Adeorbidae) from the western Pacific islands. *U.S. Geol. Surv. Prof. Pap.* 531.
———. 1972. Cenozoic fossil mollusks from western Pacific islands; gastropods (Turritellidae through Strombidae). *U.S. Geol. Surv. Prof. Pap.* 532.
———. 1977. Cenozoic fossil mollusks from Western Pacific Islands; gastropods (Eratoidae through Harpidae). *U.S. Geol. Surv. Prof. Pap.* 533.
Ladd, H. S., J. I. Tracey, Jr. & M. G. Gross. 1970. Deep drilling on Midway Atoll. *U.S. Geol. Surv. Prof. Pap.* 680-A.
Martin, K. 1879–80. *Die Tertiarschicten auf Java.* E. J. Brill, Leiden.
Mayr, E. 1942. *Systematics and the origin of species.* Columbia University Press, New York.
Nelson, G. & N. Platnick. 1981. *Systematics and biogeography. Cladistics and vicariance.* Columbia University Press, New York.
Newell, N. D. 1952. Periodicity in invertebrate evolution, *J. Paleontol.* 26: 371–85.
Powell, A. W. B. 1958. Marine provinces of the Indo-west Pacific. *Proc. 8th Pac. Sci. Congr.* 3: 359–62.
Premoli Silva, I. & C. Brusa. 1981. Shallow-water skeletal debris and larger foraminifers from deep sea drilling project site 462, Nauru Basin, Western Equatorial Pacific. *Initial Reports of the Deep Sea Drilling Project* 61: 439–53. U.S. Government Printing Office, Washington, D. C.
Randall, J. E. 1955. Revision of the surgeon fish genera *Zebrasoma* and *Paracanthurus. Pac. Sci.* 9: 396–412.
———. 1964. A revision of the file fish genera *Amanses* and *Cantherhines. Copeia:* 331–61.
———. 1980. Revision of the fish genus *Plectranthias* (Serrandidae: Anthiinae) with descriptions of 13 new species. *Micronesica* 16: 101–87.
Rehder, H. A. 1980. The marine mollusks of Easter Island (Isla de Pascua) and Sala y Gomez. *Smithson. Contrib. Zool.* 289.
Renz, O. 1973. Two Lamellaptychi (Ammonoidea) from the Magellan Rise in the central Pacific. *Initial Reports of the Deep Sea Drilling Project* 17: 895–97. U.S. Government Printing Office, Washington, D.C.
Richardson, J. 1845. Reports on the ichthyology of the seas of China and Japan. *Rep. Br. Assoc. Adv. Sci.:* 187–320.
Salvat, B. & C. Rives. 1975. *Coquillages de Polynesie.* Les editions du Pacifique, Tahiti.
Scheltema, R. S. 1971. Larval dispersal as a means of genetic exchange between geographically separated populations of benthic marine gastropods. *Biol. Bull.* 140: 284–322.
Schilder, F. A. 1961. The geographical distribution of cowries (Mollusca: Gastropoda). *Veliger* 7: 171–83.
Schlanger, S. O. & I. Premoli Silva. 1981. Tectonic, volcanic and paleo-geographic implications of redeposited reef faunas of Late Cretaceous and Tertiary age from the Nauru Basin and the Line Islands. *Initial Reports of the Deep Sea Drilling Project* 61: 817–27. U.S. Government Printing Office, Washington, D.C.
Schlanger, S. O., E. D. Jackson, A. Kaneps & S. F. Serocki. 1976. Introduction and explanatory note. *Initial Reports Deep Sea Drilling Project* 33: 5–11. U.S. Government Printing Office, Washington, D.C.
Schmidt, K. P. 1954. Faunal realms, regions and provinces. *Q. Rev. Biol.* 29: 322–31.
Springer, V. G. 1981. Comments on Solem's land-snail biogeography with an hypothesized explanation of the distribution of the Endodontidae. *In:* Nelson, G., & D. E. Rosen, eds., *Vicariance biogeography.* Columbia University Press, New York.
———. 1982. Pacific plate biogeography, with special reference to shore fishes. *Smithson. Contrib. Zool.* No. 367.
Stanley, S. M. 1979. *Macroevolution. Pattern and process.* W. H. Freeman, San Francisco.
Stehli, F. G. & J. W. Wells. 1971. Diversity and age patterns in hermatypic corals. *Syst. Zool.* 20: 115–26.
Taylor, J. D. 1971. Reef associated molluscan assemblages in the Western Indian Ocean. *In:* Stoddart, D. R. & M. Yonge, eds., *Regional variation in Indian Ocean coral reefs.* Academic Press, London.
Taylor, J. D., N. J. Morris & C. N. Taylor. 1980. Food specialization and the evolution of predatory gastropods. *Palaeontology* 23: 375–409.

APPENDIX

Data sources from which the lists of Mollusca, Arthropoda, Echinodermata, and fishes were compiled are given.

Mollusca

Abbott, R. T. 1959. The family Vasidae in the Indo-Pacific. *Indo-Pac. Mollusca* 1(1): 15–32.

———. 1960. The genus *Strombus* in the Indo-Pacific. *Indo-Pac. Mollusca* 1(1 and 2): 33–146.

———. 1961. The genus *Lambis* in the Indo-Pacific. *Indo-Pac. Mollusca* 1(3): 142–74.

———. 1968. The helmet shells of the world (Cassidae). Part I. *Indo-Pac. Mollusca* 2(9): 15–201.

Boss, K. J. 1969. The subfamily Tellinidae in South African waters (Bivalvia, Mollusca). *Bull. Mus. Comp. Zool.* 138(4): 1–162.

Burgess, C. M. 1970. *The living cowries.* A. S. Barnes and Co., Inc., South Brunswick.

Cate, C. N. 1973. A systematic revision of the recent cypraeid family Ovulidae (Mollusca: Gastropoda). *Veliger* 15. Suppl.

Cernohorsky, W. O. 1976. The Mitridae of the world. Part I. The subfamily Mitrinae. *Indo-Pac. Mollusca* 3(17): 273–528.

Emerson, W. K. & W. O. Cernohorsky. 1973. The genus *Drupa* in the Indo-west Pacific. *Indo-Pac. Mollusca* 3(13): 1–35.

Houbrick, R. S. 1978. The family Cerithiidae in the Indo-Pacific. Part I. The genera *Rhinoclavis, Pseudovertagus* and *Clavocerithium. Monogr. Mar. Mollusca* 1: 1–130.

Powell, A. W. B. 1964. The family Turridae in the Indo-Pacific. Part I. The subfamily Turrinae. *Indo-Pac. Mollusca* 1(5): 227–345.

———. 1967. The family Turridae in the Indo-Pacific. Part Ia. The subfamily Turrinae concluded. *Indo-Pac. Mollusca* 1(7): 409–43.

———. 1967. Additions and corrections to *Gemmula* and *Xenuroturris. Indo-Pac. Mollusca* 1(7): 433–43.

———. 1969. The family Turridae in the Indo-Pacific. Part II. The subfamily Turriculinae. *Indo-Pac. Mollusca* 2(10): 215–415.

———. 1973. The patellid limpets of the world (Patellidae). *Indo-Pac. Mollusca* 3(15): 75–203.

Ranson, G. 1961. Les especès d'huitres perliers du genre *Pinctada* (biologie de quelques unes d'entre elles). *Mem. Inst. Sci. Nat. Belg.* 67: 1–95.

Rehder, H. A. 1973. The family Harpidae of the world. *Indo-Pac. Mollusca* 3(16): 207–74.

Rosewater, J. 1961. The family Pinnidae in the Indo-Pacific. *Indo-Pac. Mollusca* 1(4): 175–226.

———. 1965. The family Tridacnidae in the Indo-Pacific. *Indo-Pac. Mollusca* 1(6): 347–96.

———. 1970. The family Littorinidae in the Indo-Pacific. Part I. The subfamily Littorininae. *Indo-Pac. Mollusca* 2(11): 417–506.

———. 1972. The family Littorinidae in the Indo-Pacific. Part II. The subfamilies Tectariinae and Echiniinae. *Indo-Pac. Mollusca* 2(12): 507–34.

Talmadge, R. R. 1963. Insular haliotids in the western Pacific (Mollusca; Gastropoda). *Veliger* 5: 129–39.

———. 1974. Some notes on the distribution and taxonomy of haliotids. *Veliger* 16: 411–12.

Waller, T. R. 1972. The Pectinidae (Mollusca, Bivalvia) of Enewetok Atoll, Marshall Islands. *Veliger* 14: 221–64.

Walls, J. G. 1979. *Cone shells. A synopsis of living species.* T. F. H. Publications.

Crustacea

Bright, D. B. & C. L. Hogue. 1972. A synopsis of the burrowing land crabs of the world and list of their arthropod symbionts and burrow-associates. *Contrib. in Sci. Nat. Hist. Mus. Los Ang. Cty.* 220.

George, R. W. 1969. Natural distribution and speciation of marine animals. *J. R. Soc. Aust.* 52: 33–40.

Holthuis, L. B. 1966. On spiny lobsters of the genera *Palinurellus, Linuparus and Puernulus* (Crustacea Decapoda, Palinuridae). Symposium on Crustacea. *Proc. Mar. Biol. Assoc. India.* Part I: 260–78.

Serene, R. 1966. Note sur la taxonomie et la distribution geographique des Haplocarcinidae (Decapoda: Brachyura). *Proc. Mar. Biol. Assoc. India.* Part I: 395–98.

———. 1972. Observations on the Indo-Pacific species of *Kraussia* Dana, 1852 (Decapoda: Brachyura). *J. R. Soc. W. Aust.* 55(2): 41–62.

Stephenson, W. 1972. An annotated checklist and key to the Indo-west Pacific swimming crabs (Crustacea: Decapoda: Portunidae). *Bull. R. Soc. N. Z.* 10.

Yaldwyn, J. E. & K. Wodzicki. 1979. Systematics and ecology of the land crabs (Decapoda: Coenobitidae, Grapsidae and Geocarcinidae) of the Tokelau islands, Central Pacific. *Atoll Res. Bull.* 235.

Echinoderms

Clark, A. M. 1976. Echinoderms of coral reefs. *In:* Jones, O. A. & R. Endean, eds., *Biology and geology of coral reefs.* Vol. 3: Biology 2.

Clark, A. M. & F. W. E. Rowe. 1971. Shallow water Indo-west Pacific echinoderms. Br. Mus. Nat. Hist.

Fishes

Allen, G. R. 1975. *Damselfishes of the South Seas.* T. F. H. Publications.

Allen, G. R. & J. E. Randall. 1977. Review of the sharpnose puffer-fishes (subfamily Canthigasterinae) of the Indo-Pacific. *Rec. Aust. Mus.* 30: 495–517.

Burgess, W. E. 1978. *Butterflyfishes of the world.* T. F. H. Publications.

Greenfield, D. W. 1974. A revision of the squirrelfish genus *Myripristis* Cuvier (Pisces: Holocentridae). *Bull. Nat. Hist. Los Ang. Cty.* 19: 1–54.

Hoese, D. F. 1975. A revision of the gobiid fish genus *Kelloggella. Rec. Aust. Mus.* 29: 474–84.

Randall, H. A. & G. R. Allen. 1977. A revision of the damselfish genus *Dascyllus* (Pomacentridae) with the description of a new species. *Rec. Aust. Mus.* 31: 349–85.

Randall, J. E. 1955. A revision of the surgeon fish genera *Zebrasoma* and *Paracanthurus. Pac. Sci.* 9: 396–412.

———. 1956. A revision of the surgeon fish genus *Acanthurus. Pac. Sci.* 10: 159–235.

———. 1963. Review of the hawkfishes (Family Cirrhitidae). *Proc. U.S. Natl. Mus.* 114: 389–451.

———. 1964. A revision of the file fish genera *Amanses* and *Cantherhines. Copeia* 64:331–61.

———. 1972. The Hawaiian trunkfishes of the genus *Ostracion. Copeia* 72:756–68.

———. 1972. A revision of the labrid fish genus *Anampses. Micronesica* 8: 151–95.

———. 1975a. *Ostracion trachys,* a new species of trunkfish from Mauritius (Ostraciontidae). *Matsya* 1: 59–62.

———. 1975b. A revision of the Indo-Pacific angelfish genus *Genicanthus,* with descriptions of three new species. *Bull. Mar. Sci.* 25: 393–421.

———. 1978. A revision of the Indo-Pacific labrid fish genus *Macropharyngodon,* with descriptions of five new species. *Bull. Mar. Sci.* 28: 742–70.

———. 1980. Revision of the fish genus *Plectranthias* (Serranidae: Anthiinae) with descriptions of 13 new species. *Micronesica* 16: 101–87.

———. 1981. Revision of the labrid fish genus *Labropsis* with descriptions of five new species. *Micronesica* 17: 125–55.

Randall, J. E. & J. K. Dooley. 1974. Revision of the Indo-Pacific branchiostegid fish genus *Hoplolatilus,* with descriptions of two new species. *Copeia* 74: 457–71.

Randall, J. E. & C. J. Ferraris, Jr. 1981. A revision of the Indo-Pacific labrid fish genus *Leptojulis* with descriptions of two new species. *Rev. Franc. Aquariol.* 8: 89–96.

Randall, J. E. & P. Gueze. 1981. The holocentrid fishes of the genus *Myripristis* of the Red Sea, with clarification of the *Muridjan* and *Hexagonus* complexes. *Contrib. Sci. Nat. Hist. Mus. Los Ang. Cty.* 334: 1–16.

Randall, J. E. & M. L. Harmelin-Vivien. 1977. A review of the labrid fishes of the genus *Paracheilinus* with description of two new species from the western Indian Ocean. *Bull. Mus. Natl. d'Hist. Nat.* 3 Ser. 436: 329–42.

Randall, J. E. & J. C. Kay. 1974. *Stethojulis axillarix,* a junior synonym of the Hawaiian labrid fish *Stethojulis balteata,* with a key to the species of the genus. *Pac. Sci.* 28: 101–07.

Randall, J. E. & R. Lubbock. 1981. Labrid fishes of the genus *Paracheilinus,* with descriptions of three new species from the Philippines. *Jpn. J. Ichthyol.* 28: 19–30.

———. 1981. A revision of the serranid fishes of the subgenus *Mirolabrichthys* (Anthiinae: *Anthias),* with descriptions of five new species. *Contrib. Sci. Nat. Hist. Mus. Los Ang. Cty.* 333: 1–27.

———. 1982. Three new labrid fishes of the genus *Cirrhilabrus* from the southwestern Pacific. *Occas. Pap. Bernice P. Bishop Mus.* 25(2): 1–12.

Randall, J. E., K. Matsuura & A. Zama. 1978. A revision of the triggerfish genus *Xanthichthys,* with description of a new species. *Bull. Mar. Sci.* 28: 688–706.

Randall, J. E. & H. A. Randall. 1981. A revision of the labrid fish genus *Pseudojuloides,* with descriptions of five new species. *Pac. Sci.* 35: 51–74.

Randall, J. E. & V. G. Springer. 1973. The monotypic Indo-Pacific labrid fish genera *Labrichthys* and *Diproctacanthus* with description of a new related genus, *Larabicus. Proc. Biol. Soc. Wash.* 86(23): 279–97.

Smith, J. L. B. 1960. Coral fishes of the family Pomacentridae from the western Indian Ocean and the Red Sea. *Ichthyol. Bull.* 19.

Phytogeographic Comparison of Polynesia and Micronesia

F. R. Fosberg
Department of Botany
National Museum of Natural History
Washington, D.C. 20560 USA

INTRODUCTION

Traditionally the Pacific Islands, excluding those close to America and close to Asia and Australia, are divided, originally on the basis of their indigenous aboriginal peoples, into three regions: Polynesia, Micronesia, and Melanesia. This division, with some modifications, serves phytogeography well enough, and continues to be used. The modifications are to exclude New Zealand and the Kermadec Islands as a separate region; to include the Gilbert Islands, Nauru, and Banaba in Polynesia instead of Micronesia; and to include the Volcano and Bonin islands in the eastern Asian region. This leaves the three main Pacific island regions as completely tropical and subtropical, and Polynesia and Micronesia as completely oceanic (in a strict sense). Of Polynesia, the part north of the equator is east of 180° longitude, and the part south of the equator is both east and west of 180°. Micronesia is north of the equator and west of 180°. Melanesia is largely continental (in a strict biogeographic sense) and west of that part of Polynesia south of the equator. Floristically, these regions are definable, but their boundaries are not clear and their floras have many species in common.

The two of these regions to be considered here, Polynesia and Micronesia, are composed entirely of oceanic islands. This term, as I use it, implies that these islands have never been connected to any continental land-mass. Peripherally included as oceanic are islands that may at one time have had continental or larger island connections that have subsequently been completely submerged. Using this definition, we can assume that ancestors of all prehuman indigenous plants arrived by long-distance transport on or over water. The relative importance of the several likely agents of such long-distance transport or dispersal has been the subject of much speculation, with little general agreement among those who have speculated.

These agents, in the broadest sense, are water (ocean currents), wind (ordinary air currents, high altitude "jet streams," and hurricanes or typhoons), and animals (birds and bats). To these must be added man, if we include present day phytogeography in our discussion. Morphological adaptations for dispersal by these agents and physiological adaptations favoring establishment after dispersal are important and extensive subjects for investigation. Far from enough is known about these factors, which are based on inference from what can be observed without experimental tests.

NATURE AND SCOPE OF PHYTOGEOGRAPHY

There seems to be, as yet, no general agreement on the nature and scope of biogeography or, consequently, of its subdivision phytogeography or plant geography. To make clear what I am talking about I will, in the briefest of terms, indicate the sense in which I use the word. For a more comprehensive discussion, I refer my audience to a paper given to the Association of American Geographers (Fosberg, 1976). There are many valid and useful subdivisions of phytogeography, three of which are particularly pertinent here, with a fourth peripheral but also important.

(1) *Floristics or floristic plant geography.* This deals with comparison of floras, the putative geographical relations of their components, their apparent evolutionary complexity and adaptation to their habitats, and the relative representation of the plant families—the so-called "harmony" or "disharmony"—of the floristic composition. The term "flora" is used here to mean the total list of plant taxa known to be or to have been present. This subdivision is classical phytogeography and it is frequently what people mean by the term in a restricted sense. Its subject matter is lists of taxa with information on their nature and affinities.

(2) *Historical plant geography.* This is closely related to the preceding but with a time-dimension added. It deals with the origins, development, and changes that the floras have undergone. Its data are the same as those of floristics, but with the addition of fossil evidence and knowledge of the geological and physiographic history of the regions concerned and of the world. Its contributions are historical inferences on the origins, evolution, and movements of floras. Its conclusions are not to be regarded as facts, but as the nearest their authors have come to understanding, expressed as historical reconstructions.

(3) *Vegetation, or physiological (Schimperian) plant geography, or vegetation science* [by those who want an independent science, or who restrict plant geography to (1) and (2) of this scheme]. Vegetation is the plant cover of the earth, or the mosaic of plant communities covering the area under consideration. It deals with *physiognomy* (visual appearance), *structure* (distribution of biomass in space), *function* (morphological and physiological adaptations to the habitat), and *floristic composition,* and with the interactions of the plant taxa with each other and with animal taxa that influence them.

(4) *Anthropic (or human) plant geography.* This deals with the changes that have resulted from the appearance of man on the scene, or of his presence as part of the biota. Normally, in most continental situations, the results of man's influences and activities are very hard to sort out from prehuman or nonhuman

conditions, as earliest man was a normal member of the biota and did not exert a disproportionate influence. His emergence as an, or rather *the,* ecological dominant was not abrupt but very gradual and his early traces are hard to discern and document, though the later ones are often most obvious (Fosberg, 1959).

On oceanic islands, however, man's arrival was comparatively recent, and he had already reached the stage where he could build boats, make tools and weapons, use fire, domesticate animals and plants, and thus produce much of his own food. His effects, while not always obvious, are discernible to the experienced eye and it is possible to reconstruct in a fairly plausible manner the sequence of events following the advent. I will attempt in this paper to bring in results of man's activities only briefly against a background of my interpretation of the prehuman island situation.

THE STATE OF WHAT WE KNOW

Good accounts of the flora are available for very few of the Pacific islands or island groups. For many of them not even up-to-date lists are available, and of those that exist many are not published. Every significant collecting trip outside the closest and most obvious areas yields species new to the island concerned and often new to science. Obviously, phytogeographic conclusions on island floras must be regarded as most tentative except at the broadest level.

Vegetation descriptions are available for some islands, but they are drawn from various viewpoints and using various methods. For most islands there are no reliable current accounts, and for some nothing at all. Descriptions of pre-European conditions are nearly lacking and what there are must be pieced together from casual remarks and fragments in journals of early voyages. Documented accounts of events in this field in post-European time are rare, sketchy, anecdotal, and incomplete. Hence, attempts at phytogeographic reconstruction are difficult, if not at times exercises in futility. But the subject remains fascinating.

TROPICAL PACIFIC OCEANIC ISLANDS

Both Polynesia and Micronesia can be roughly divided into "high" and "low" islands. The high islands are mostly volcanic (the exception is Yap, which is partly of metamorphic rocks), and most of them, except those with volcanoes only dormant or still active, have been subject to substantial subsidence, although some have been secondarily tectonically uplifted. Thus some few, not yet subsided, have regular coastlines of volcanic rocks. Some others, in cooler waters, have indented coasts of mostly volcanic rocks. Otherwise, almost all are wholly or partly surrounded by sea-level or elevated reefs of coral limestone. Some also have rough or dissected limestone terraces to well above sea level. A few are uplifted limestone platforms, often rough and pitted, and rugged karsts, all of these with little or no volcanic material exposed. Some, such as Guam, are partly volcanic, partly limestone. The volcanic surfaces may be fresh new lava flows, rough or smooth, fresh ash, cinders and scoria, more usually deeply dissected, eroded slopes, some of deeply weathered ancient material eroded down to gently rounded or sharply dissected forms up to a few hundred meters elevation. Of course there is every transition among these conditions.

LOW ISLANDS

Low islands are coral atolls or flat "table reefs," lying practically at sea-level or only a few meters above. The substratum is limestone—consolidated as rock, usually somewhat pitted or rough, or loose, sand to gravel or boulders, flats or beach ridges, storm ridges of cobbles and boulders, or sand dunes, usually small.

The low islands are well-distributed throughout most of both Polynesia and Micronesia, forming a number of large archipelagoes and many scattered more or less isolated islands or small groups.

Because their floras are much more uniform and much more impoverished than those of the high islands, and the topography, substratum, and vegetation show much less diversity, the low islands are here treated separately, as a "dispersed region," so as not to obscure the differences apparent between the two regions of high islands.

This low island "region" comprises a myriad of tiny dots of land, dispersed irregularly over a vast expanse of sea, extending from Helen Reef, southwest of Palau, in the west, to Ducie Atoll between Pitcairn and Easter Island in the east, and from Kure and Marcus in the north to Maria (Austral Is.) and Ducie in the south. The vast majority are atolls, irregular ring-like reefs outlining the tops of submerged mountains, bearing on their peripheries strips of dry land or, usually, one to very many tiny flat islets made up of billions of fragments of skeletons of lime-secreting plants and animals. Their floras are, in general, the strand floras of the high islands, mostly composed of widespread halophytic species. The numbers of such species vary with the degree of isolation and the dryness or wetness of the climate, from 1 (*Tournefortia argentea* on Ducie) to up to 100 or more (including exotics) on some of the wettest atolls in the southern Marshalls and the Carolines. Vascular epiphytes are present only on the wetter atolls, and then not many. Endemic species are very few: *Nesogenes euphrasioides* (Chloanthaceae?), *Digitaria stenotaphrodes* (Poaceae), and *Hedyotis romanzoffiensis* (Rubiaceae) in eastern Polynesian atolls (and similar barrier reef islets); *Digitaria pacifica* (Poaceae), *Eragrostis whitneyi* (Poaceae), *Lepturus pilgerianus* and *L. gasparricensis* (Poaceae, scarcely distinct), and *Lepturopetium* sp. (Poaceae, unpublished), in the central Pacific, Gilbert, and Marshall atolls. Of floristic and ecological interest is the fact that, in the same climatic area, the floras of atolls with areas of even slightly elevated limestone are significantly richer in species. This may be connected with the probability that the islets only 2 m high or less may have been completely submerged during the postglacial rise in sea level (climatic optimum or xerothermic period), eliminating their terrestrial biota. If that explanation is considered untenable, as by some geologists, the larger floras may be due to greater leaching out of salt on the higher areas, permitting less halophytic species to survive.

In accord with the lesser topographic and substratum diversity and smaller floras, the vegetation shows less diversity and simpler patterns, becoming generally more simple toward the dry end of the gradient. Of interest, however, toward the dry end, is the prevalence of single-species, or almost single-species, vegetation types, unusual in the tropics. This may be a response to extreme conditions and a small flora.

The general original vegetation of coral islands was probably broadleaf forest of a number of types: mixed, or pure *Pisonia grandis,* pure *Neisosperma oppositifolia,* pure *Barringtonia asiatica,* as well as scrub of *Scaevola sericea* and *Tournefortia argentea,* rarely pure *Tournefortia,* and savanna of *Lepturus repens,* sometimes with *Heliotropium anomalum,* with or without scattered *Tournefortia* or *Guettarda speciosa,* and other less common types. A mixture of several tree species is perhaps the commonest type. On all but the driest atolls, most of this has now been replaced by coconut (*Cocos nucifera*) plantation or a mixed forest of breadfruit (*Artocarpus altilis*) and coconut palms. This, of course, is the common conception of atoll vegetation, even of general South Sea Island vegetation, but it seems certainly to be strictly a product of human intervention, mostly created since the advent of Europeans. An additional feature of this artificial ecosystem, is the widespread presence of "taro-pits," depressions excavated to below the water table, in which are cultivated taro and other taro-like edible Araceae, and various other nonindigenous food plants. In some islands, these are falling into disuse because of changing human life styles and invasion of persistent exotic weeds, such as *Paspalum distichum.* In many atolls with strong prevailing trade winds, crescent-shaped remnants of native forest or scrub are left as windbreaks to protect the coconut and breadfruit stands from salt spray.

POLYNESIAN HIGH ISLANDS

The high islands of Polynesia are scattered from Rotuma, Samoa, Tonga, Wallis, Nauru, and Banaba on the west, eastward to include Niue, the Cook, Austral, Society, and Marquesas archipelagoes, to Makatea, Mangareva, Pitcairn, Henderson, and Easter Island, and north to the Hawaiian Chain.

They have a flora less rich in species than Fiji and Melanesia, and largely an eastern attenuation of the Fijian-Melanesian assemblage. However, since the oceanic island floras are composed of descendants of random colonists from all directions, there are notable components from the south, west, and even the north—elements of south temperate, tropical American, temperate North American, and even Alaskan and East Asiatic origin. The distribution of these non-Indo-Pacific elements is somewhat random. Additionally, there is what has been called an "Old Pacific" component, the affinities of which are as yet obscure.

Differences resulting from greater or less isolation and uneven presence of the non-Indo-Pacific components make it convenient to discuss Polynesian high islands as three geographic and floristic subregions, each with certain distinctive features of phytogeographic interest. These subregions are Western Polynesia, Eastern Polynesia, and Hawaii.

Western Polynesia

Including the Samoan, Tongan, and Wallis archipelagoes and the scattered islands of Rotuma, Nauru, Banaba, and Niue, Western Polynesia is floristically the least distinct from Fiji and Melanesia of the three Polynesian subregions. Its flora is primarily derived from the immediate west, with few indigenous elements not fitting this pattern. It has few endemic genera but a considerable number of endemic species. The "Old Pacific" element is scarcely represented, though a member of the Hawaiian genus *Phyllostegia* has recently been discovered in Tonga.

Vegetationally it does not have a well-developed sclerophyllous rainshadow vegetation, nor extensive savannas. Tall rain forest is to some extent present, though not like that farther west. Montane-type rain forest and some cloud forest or mossy forest, clothe the mountain slopes. On some newer lava flows and summits of volcanoes a scrub or scrub forest may be seen. No adequate accounts of the vegetation of the volcanoes on Upolu and Savaii, the larger islands, are available. Eastern Samoa is lower, mostly broadleafed wet forest, now largely greatly altered and denuded by human agencies; it has been well described by Whistler (1980).

Tonga is a very diverse group of islands, ranging from very active volcanoes to elevated limestone, one of the latter, Tongatabu, being the principal island and contains the capital of the group. The vegetation of this is mostly cleared away and not much is known of its original nature. The islands of Vavau and Eua have a number of endemic species, belonging mostly to Fijian and Indo-Malayan genera.

Except for a recent probably rather incomplete account of the flora, little is known of the Wallis group botanically speaking. Good collections have been made on Rotuma but not written up. Nauru has recently been very well-collected, but nothing has been published on its flora and vegetation. Its elevated plateau surface is being completely mined for phosphate. Banaba is being similarly mined, and its flora is not at all well known.

Eastern Polynesia

The Cook, Austral, Society, and Marquesas archipelagoes contain some of the most spectacular and beautiful of oceanic islands, culminating in Tahiti, a magnificent, deeply dissected dead volcano 7300 feet high. Its upper slopes are clothed with montane rain forest and cloud forest of many species and a spectacular development of tree ferns. Its flora, thought still predominantly Indo-Pacific in affinities, contains such plants as *Coprosma, Fuchsia, Pittosporum crassifolium, Weinmannia, Oreobolus,* and *Metrosideros* of New Zealand and generally Austral affinities. *Fitchia, Sclerotheca, Meryta,* and *Lepinia* represent the "Old Pacific" flora.

The Austral Islands and Cook Islands, with high peaks on Rapa and Rarotonga, respectively, have floras similar to that of Tahiti but smaller, with even stronger representation of southern elements, especially in Rapa, which has an *Olearia,* a *Hebe,* two species of *Coprosma,* and a *Corokia;* New Zealand genera; and *Fitchia, Meryta,* and *Oparanthus* of the "Old Pacific." Tubuai and Rarotonga share with Hawaii *Charpentiera,* an "Old Pacific" Amaranthaceae. The native vegetation of these islands, largely gone now, is mostly a mixed montane, scrubby, broad-leaf forest with many tree ferns, becoming a stiff scrub, largely *Metrosideros,* a New Zealand-Polynesian genus of dry-fruited Myrtaceae, on high ridges. A large species of *Metrosideros,* related to the one in Tahiti, is common lower down; also a *Santalum* and a *Myoporum,* the latter close to New Zealand species. Remarkable are a thick-stemmed *Plantago,* possibly close to Hawaiian ones, *Exocarpos cupressoides* (Santalaceae) with Hawaiian or Australian affinities, and the "Old Pacific" genus *Nesoluma* (Sapotaceae).

Eastward, in the Tuamotus and beyond, are Makatea, an elevated atoll with large phosphate deposits and a flora with some endemics, and Mangareva, its original vegetation almost completely replaced by a sterile anthropogenous savanna, but with other remnant vegetation almost all consisting of endemics—a *Coprosma,* a remarkable arborescent *Aerva* (Amaranthaceae), a *Gouania* and a *Byttneria,* both apparently of American affinities, and a *Fitchia* (now extinct). East of this is Pitcairn, a tiny island with some indigenous plants left, but mostly under cultivation, with a species of *Homalium* (Flacourtiaceae) as the most important remaining tree. Henderson Island, an elevated atoll, has a still almost completely untouched forest of *Pandanus, Nesoluma, Hernandia, Santalum,* and *Pittosporum*(?). There is said to have been a *Fitchia* there, but the specimen may possibly have come from Tahiti. The most remarkable plant is *Bidens hendersonensis,* probably belonging to the section *Campylotheca,* widespread in Polynesia, but a small tree about 6 meters tall and with a woody trunk. It does not seem to be close to any other species but has a variety on nearby Oeno Atoll. The indigenous flora of Easter Island, the eastern extremity of Polynesia, is almost completely gone but for a sedge of probably American connections and an endemic *Axonopus,* an American genus of grasses. The last woody plant was *Sophora toromiro,* belonging to the circum-antarctic section *Edwardsia.*

North of these islands, near the equator, are the Marquesas, an isolated group of mostly high rugged islands with a limited but highly endemic flora, the easternmost important extension of the Indo-Pacific flora, but with a notable component of "Old Pacific" species, and *Lebronnecia,* an endemic genus of Malvaceae. *Oxalis gagneorum, Abutilon sachetianum, Cheirodendron marquisense, Santalum insulare, Coprosma esulcata, Ixora spathoidea,* a number of *Cyrtandra* and *Psychotria* species, and *Tournefortia marchionica,* this with probable American relatives, form a remarkable group of endemics. The genus *Trimenia* (Monimiaceae) reaches a disjunct eastern limit here, jumping from Fiji. Species of *Sclerotheca* and *Oparanthus* connect with those southward in Rapa and Tahiti. *Cordia lutea* is an American species, probably native here. *Waltheria lophantha* may be closest to a Hawaiian species, or perhaps American. Species of *Pelea* (Rutaceae) provide another Hawaiian connection.

Hawaii

The Hawaiian flora and vegetation are comparatively well-known, and are by far the most diverse of any oceanic island group. The most isolated island group in the world, Hawaii's flora is nearly 97% endemic, including many endemic genera. Its mountains extend above timberline, the highest in the Pacific Basin outside New Guinea. Its eight main islands and 12 much smaller ones lie in a southeast-northwest chain, across the northeast trade wind belt and the Tropic of Cancer. Local climates range from desert to what is said to be the wettest in the world, on Mt. Waialeale, Kauai. Land has been above water in the group for probably at least 30 million years. Substrata include hot, fresh, rough or smooth lava through ash, to deeply weathered soil, to sterile leached bauxite, to rough limestone, to coral sand, to marl and volcanic mud, muck, and peat.

The age and complexity of the floras and vegetation of the individual islands increase northwestward to Kauai, which has a rich and diverse flora, intricate and deeply dissected terrain, a varied and unclear mosaic of vegetation communities, including vast areas of bogs in all stages of development. Beyond Kauai, the islands abruptly are smaller, ending in a chain of half a dozen coral atolls and a number of banks and seamounts, which represent ancient volcanoes long since vanished beneath the sea, but in all probability having played a critical role in the origin and early evolution of the Hawaiian flora.

This situation, with its ecological and geographic diversity and great time-depth, by random colonization from all directions and subsequent evolution, has produced a fantastic diversity of plants unequalled in any other group of oceanic islands in the world. With this great habitat diversity and the relatively limited range of successful colonists (see Fosberg, 1948), nature has set up an unrivaled experiment to demonstrate "adaptive radiation" (Carlquist, 1965, 1974), with familiar genera or families producing bizarre and unusual growth forms. The vegetation occupying the segments of this environmental mosaic is as varied in physiognomy, structure, function, and floristic composition and is a perfect reflection of the complexity of the mosaic.

Space is insufficient here even to begin to describe the vegetation or to give an adequate idea of the flora and its relationships. In 1948, I presented an account of the origin and derivation of the Hawaiian flora to which the audience is referred for more detail (see also Fosberg, 1951). I would suggest very few changes in this, after 35 years, except to account for some new discoveries in the flora, and to correct one or two misjudgements. The flora is, as expected, predominantly Indo-Malayan as represented by 42.7% of original colonists, Austral (12.2%), American (16.2%), boreal (2.1%), East Asian (1.0%).

One more thing needs to be said of the Hawaiian flora and vegetation. The original vegetation has been altered, much of it beyond recognition, by the hand of man, trampling and browsing by his domestic animals, and crowding by his imported plants, which have occupied vast areas, displacing the native flora. According to a recent estimate, based on a careful analysis (Fosberg and Herbst, 1975) of the original flora found here by the Europeans, 273 species and well marked varieties have become extinct and 1690 are rare or endangered. The total indigenous flora is about 2500 species and well marked varieties.

MICRONESIA

Micronesia comprises the oceanic island groups and isolated islands lying in the northwest quadrant of the tropical Pacific. In our definition, as treated in the Flora of Micronesia, it includes the Mariana, Caroline, Marshall, and Gilbert archipelagoes and the more isolated islands of Marcus, Wake, and Mapia. For vegetation purposes we can exclude the Gilberts, Nauru, and Banaba. Of those remaining, the Marshalls, many of the Carolines, Marcus, Wake, and Mapia are low coral atolls and for our present purposes have been included in a separate floristic region. This leaves the Marianas, a northern chain of steep young, mostly active andesitic-basaltic volcanoes and a southern older chain of deeply weathered volcanic islands partly or mostly covered by terraces and caps of old coral limestone, and the Caroline Islands, an enormously long archipelago of mostly older high islands quite diverse in origin, types of rocks, and topography (with a swarm of low atolls not considered in this section). The vegetation and flora of these islands is under active study, but the results are not yet sufficient for a definitive phytogeographic treatment. The latest full list was an enumeration of the Micronesian vascular plants by Kanehira in 1935, which left much to be desired as collecting at that time was sparse indeed. Vastly greater collections are now available, but the taxonomic problems have not all been cleared up. Enough is known to give an idea of the vegetation and of the nature of the flora, with suggestions as to the origins and geographic relationships of some of its components.

The vegetation types, viewed in broad perspective, are much influenced, their distributions controlled, by the chemical and physical nature of the substratum, as well as by the rainfall patterns and seasonality. Since the highest point in these islands in the northern Marianas is only about 1000 m, and the area of land and the geographic extent are much smaller than in Polynesia, the diversity is much less, and the flora itself is much more limited.

The major vegetation formations of Micronesian high islands are moist to very wet broad-leafed evergreen forest, broad-sclerophyll swamp forest (mangrove), and savanna. Other types, while interesting, are of minor extent. Strand vegetation will be considered in the next section. (For a more detailed consideration than what follows, see Fosberg, 1960.)

The forest types on soils derived from volcanic rocks are generally different from those on limestone, in structure and especially in floristic composition. Any original vegetation on volcanic soils in the Marianas has been so degraded and altered as to bear little or no resemblance to what was probably originally there. What has been termed "ravine forest," in the valleys in southern Guam is a mixture of indigenous elements with areca palms and several other exotic trees and shrubs, especially *Triphasia.* A very poorly known forest type exists on rough lava flows in the northern Marianas volcanoes, which somewhat resembles the forest on limestone in southern Marianas. Ravine forest also exists here, but much of the lower slope forest has been replaced by coconut palms.

The forest on the lowest volcanic slopes and valley bottoms of the Carolines was a tall rain forest, which has now been mostly replaced by breadfruit, coconuts, and by large-leafed secondary forest, especially *Hibiscus tiliaceus.* The mid-slopes are still covered by a montane rain forest of many species with several kinds of native palms, which locally become dominant, especially on Ponape, possibly as a result of human interference. Epiphytes are abundant and diverse.

At high elevations, especially on Ponape and Kusaie, is a mossy, scrubby cloud forest with great numbers of bryophytes and ferns, terrestrial and epiphytic. Locally on Ponape this becomes dominated by the endemic *Pandanus patina.* The remarkable genus *Lepinia* (Apocynaceae) exists here in one of its three known areas.

The forests on Yap and the northern part of Palau are more seasonal and with more diverse floristic composition. These forests have been much influenced by logging and other human disturbance. Many genera reach their eastern limits in these forests of the western Carolines.

The forests on limestone are of two main types (in addition to the strand and atoll forests). In the Marianas, they are dominated by *Ficus prolixa, Artocarpus mariannensis, Pandanus, Intsia* and other large tree genera. *Cycas circinalis* and various dicotyledonous shrubs and small trees dominate the undergrowth. Where they are on steep slopes and cliffs their composition changes, and *Mammea* and *Cordia,* and, on windward slopes, *Pemphis,* become important.

On the limestone karst of southern Palau other trees, especially Clusiaceae and palms, become more abundant. There is little soil here and the trees grow on and in pits and crevices of wet, rough limestone.

The Micronesian mangrove forest is strictly similar to that farther west, but with fewer species. Floristically it attenuates both northward to Saipan and eastward to the Marshalls. It occupies most areas of relatively quiet shallow muddy bottoms, especially throughout the Caroline Islands.

Savannas dominated by coarse grasses and ferns exist on all the higher Micronesian islands with volcanic rock exposure, but they are much more developed in western Micronesia where the climate tends to be more seasonal. The larger areas are on Guam, the northern Marianas volcanoes, Yap, and Palau. Except for those in the northern Marianas, they have been greatly modified and expanded by man, but some natural savannas must have existed as there are quite a few endemic savanna plants.

The widespread members of the savanna flora of Micronesia are mostly found in drier areas of Indonesia, New Guinea, and Australia, and in a few cases east to the Society Islands and beyond. Several of the less common grasses–*Ectrosia, Ectrosiopsis,* the sedge *Cyperus tristachya,* and the small herb genus *Stackhousia* are, other than in Palau and Yap, only known from Australia, New Guinea, the Moluccas, to the Philippines. The Micronesian savanna flora has not yet been critically enumerated and distinguished from the rest of the flora or compared with floras of other savanna areas to the south and west, but it is similar to these.

The general floristic pattern in Micronesia consists of an Indo-Malayan flora, related especially to the Molucca-Celebes-Philippine part as might be expected. This is complicated somewhat by the overlap of this Indo-Malayan element with species of Australian-New Guinean affinities that extend to the Moluccas and Philippines. These, except for some obvious ones, are still to be sorted out. A Melanesian component, even less clearly understood, is also involved. Some of its genera, as *Metroxylon* and *Clinostigma,* are clear enough. These are from the central and eastern Carolines.

A few species seem to have eastern Asiatic affinities and perhaps came through the Bonin Islands. *Trema argentea, Fimbristylis dichotoma* var. *boninensis,* and perhaps *Zoysia* are examples. *Myoporum boninense* and *Styphelia marianensis* are puzzling, as they are generally of Australian affinity, but may have come to Micronesia via other areas.

A small Polynesian-Fijian element is discernible, but its species, such as *Peperomia leptostachya* and *Piper latifolium,* are mostly widespread. *Piper guhamense* is an endemic related to *P. latifolium.* Others are similarly uncertain, e.g., *Digitaria* sect. *Solitaria* is definitely a Polynesian-Micronesian connection, but it is uncertain in which direction migration has taken place.

No American or North Pacific relationship has been discerned. An "Old Pacific" element, if it exists, is vague and still to be clarified. *Digitaria* sect. *Solitaria* may be a representative.

CONCLUDING COMPARISONS

A widespread Indo-Pacific and pantropical strand element is prominent in both Polynesia and Micronesia, in both low and high islands. Some of its members, such as *Scaevola sericea* and *Dodonaea viscosa,* belong to families or genera that are mostly Australian, but these very widely distributed species are not likely to be directly derived from the Australian flora. This strand flora has, in all likelihood, been associated with and occupied the strand and coastal habitat for a long time. The same is true of the flora of the mangrove formation, the Micronesian members of which extend westward to India or even East Africa. Contrary to the common belief, many of the strand and even mangrove plants are not monotonously uniform over their vast ranges. Some, such as *Suriana maritima, Ipomoea macrantha, Casuarina equisetifolia, Thuarea involuta, Cassytha filiformis,* or *Paspalum distichum,* indeed, show very little geographical variation, but others, as *Lepturus repens, Scaevola sericea, Pandanus tectorius,* and *Timonius polygama,* show striking differentiation and polymorphism. The strand flora cannot, except for a few species, be considered to be differentiated into a Polynesian and a Micronesian subflora.

One can point out a few species, such as *Hedyotis romanzoffiensis, Nesogenes euphrasioides, Digitaria stenotaphrodes, D. pacifica, Eragrostis whitneyi,* and *Timonius polygama,* that are strictly Polynesian. Strictly Micronesian strand species are scarce, but include *Sporobolus farinosa.* There are some, such as *Excoecaria agallocha, Ximenia americana, Xylocarpus moluccana,* and *Scyphiphora hydrophyllacea* (the last two marginally mangrove species) that have their main distributions farther west but reach Micronesia. Most of the strand flora and mangrove flora are found in both regions, though many species reach only western Polynesia.

Compared with Micronesia, which has rather few plant groups that are not of either Indo-Malesian or Austro-Melanesian distribution, Polynesia, with basically an attenuated Indo-Malesian and Austro-Melanesian flora, shows a much greater diversity of other floristic elements, especially in the Hawaiian and Eastern Polynesian subregions. Hawaii, for example, has a substantial American element, an even larger Austral element, and a small but definite boreal group of species. In addition to these geographically identifiable elements, Eastern Polynesia and Hawaii have a vaguely defined category with uncertain affinities sometimes called the "Old Pacific element." This has been the subject of considerable attention by such eminent phytogeographers as Skottsberg, but definite derivations or relationships are hard to establish. Included are such groups as the Campanulaceae-Lobelioideae; Asteraceae such as *Hesperomannia, Fitchia, Remya, Oparanthus; Lebronnecia, Kokia,* and the *Abutilon sachetianum* group of the Malvaceae; *Oxalis gagneorum; Hillebrandia;* the palms *Pritchardia* and *Pelagodoxa; Geranium* sect. *Neurophyllodes*; Amaranthaceae such as *Charpentiera, Nototrichium,* and *Aerua arborescens,* to name only obvious ones.

Although much effort and study has gone into the systematics of Polynesian and Micronesian plants, one can only conclude that their geographical affinities are in many cases poorly understood. The floras are not completely enough known to make convincing statements on degrees of floristic relationship. With increasing understanding of the geology of the Pacific Basin, the vexed question of ancient land bridges in Polynesia and Micronesia has been fairly well settled in the negative. Dispersal methods are well known, but their relative importance is still very much a matter of speculation. Fossil evidence of the history of the floras of these oceanic islands is scanty, though palynological studies have given us provoking glimpses that make one skeptical even of some of the highly generalized phytogeographical pictures drawn by some of our best plant geographers. An urgent need, from the point of view of phytogeography, is more thorough collecting in poorly known islands and areas difficult of access. Another is more fossil palynological investigation.

Interesting examples of recently discovered important plants are *Scaevola tahitensis,* from the high interior of Tahiti, *Phyllostegia* sp. from Tonga, and *Lebronnecia kokioides, Abutilon sachetianum,* and *Oxalis gagneorum,* all in the Marquesas. Monographs of Pacific genera and modern floras of Pacific Island groups will place plant geography in the Pacific on a much firmer and more convincing basis. The collecting, especially, must be done soon, as the floras are disappearing rapidly.

One must conclude with what to most naturalists who work in the field in the Pacific Islands is only too obvious—if we do not do something soon to protect the remaining vestiges of natural vegetation in the islands, Pacific botany will continue only as herbarium paleobotany.

LITERATURE CITED

Carlquist, S. 1965. *Island life.* Natural History Press, Garden City. 451 p.

———. 1974. *Island biology.* Columbia University Press, New York. 660 p.

Fosberg, F. R. 1948. Derivation of the flora of the Hawaiian Islands. p. 107–19 *in:* Zimmerman, E. C., ed., *Insects of Hawaii.* Vol. 1.

———. 1951. The American element in the Hawaiian flora. *Pac. Sci.* 5: 204–06.

———. 1959. Man as a dispersal agent. *Southwest. Nat.* 3: 1–6.

———. 1960. The vegetation of Micronesia. *Bull. Am. Mus. Nat. Hist.* 119(1): 1–75, pls. 1–40.

———. 1976. Geography, ecology, and biogeography. *Ann. Assoc. Am. Geogr.* 66: 117–28.

Fosberg, F. R. & Herbst, D. 1975. Rare and endangered species of Hawaiian plants. *Allertonia* 1(1): 1–72.

Kanehira, R. 1935. An enumeration of Micronesian plants. *J. Dep. Agric., Kyusha Imp. Univ.* (Fukuoka, Japan) 4: 237–64.

Whistler, W. G. 1980. The vegetation of eastern Samoa. *Allertonia* 2(2): 45–190.

Speciation and the Founder Effect on a New Oceanic Island[1]

Hampton L. Carson
Department of Genetics
University of Hawaii
Honolulu, Hawaii 96822 USA

The "Big Island"—a name used to distinguish Hawaii Island from the state or island chain—is the newest in the Hawaiian Archipelago. Both magnetic declination and potassium-argon measurements indicate an age of not more than 500,000 years for Kahala, the oldest of the five volcanoes (McDougall and Swanson, 1972). The other four volcanoes are much younger, with two of them, Mauna Loa and Kilauea, currently active.

The Big Island is not only new in time but also in space. This is a matter of extreme importance when interpreting the evolutionary past of the organisms present. The evidence indicates that this island, like the others which are older, has been formed by the upwelling of a fixed plume of mantle that perforates the moving Pacific tectonic plate. Thus, unlike what is apparently the case for certain new volcanoes in Mexico, Galapagos and Africa, each new Hawaiian volcano is somewhat spacially separated, in a linear fashion, from the one just preceding it in age.

When considering the evolution of life on oceanic archipelagos, attempts are often made to start with data on the maximal age of the islands concerned and then to identify the putative primitive forms that were originally introduced. Speculations on the evolutionary events that have led to present biogeographic conditions are then advanced. Although some work of this sort has been done with the Hawaiian drosophilids, much of the recent research has had a different emphasis (for a recent review, see Carson, 1982a).

As a population geneticist, I approach the problems of evolution from the point of view of the most recent events and then extrapolate back to older events. Thus, our work on Hawaiian *Drosophila* has stressed the genetics of natural populations from the newest parts of the newest island. Sequential biogeography is studied in order to find indications as to which of the populations and species present are

[1]This work has been supported by Grant DEB 79-26692 from the National Science Foundation.

the very newest. Following this, data relating to their origin from populations from slightly older areas are scrutinized. This "reverse" approach to evolutionary change has the advantage of dealing first with individuals in demes and then expanding this to the recognition of various population-to-population genetic differences. One then approaches the formation of species through attempts to recognize population events and processes that may be judged to be contributing to species incipience. At this point, the microevolutionary and macroevolutionary approaches to the study of evolution have a possibility of being integrated.

The term speciation is used here to refer to the sum total of the population processes judged to be responsible for incipience. This is very different from the way the term is often applied in a macroevolutionary sense, namely to refer descriptively to the estimation of the degree of proliferation of species without regard to actual mechanisms or modes of speciation as a set of processes. As one traces populations from the present back through recent time on a single island like the Big Island, one must eventually come to the question of how this new island or volcano was colonized shortly after its emergence. The understanding of such processes might be easier were the entire Big Island somewhat newer. Nevertheless, the arrival of saprophytic insects like drosophilids must of necessity follow plant arrivals. Accordingly, the time of colonization of the ancestors of the *Drosophila* species endemic to the Big Island is less than the age of the oldest lava flows.

For reasons of relative ease of laboratory culture and the existence of favorable salivary gland chromosomes, research efforts have been concentrated on the so-called picture-winged *Drosophila* species currently found on the six major high Hawaiian islands. Of the approximately 110 species known from these islands, 102 have had their polytene chromosome banding sequences mapped in relation to a set of arbitrarily chosen standard sequences. These data are useful in a number of ways but their most striking use is in grouping the species systematically and biogeographically (see Carson et al., 1970 and Carson and Yoon, 1982).

Here I intend to concentrate on the 26 species of picture-winged *Drosophila* known from the Big Island, trying to reconstruct their recent history. On morphological, cytological, and genetic grounds, all but one of these species (*D. orthofascia;* see Carson, 1980) have been judged to be unique to the Big Island. Since these species are found there, and only there, we may infer that the cladistic events whereby they arose must not only have been geologically very recent but also indeed to have occurred *in situ* on the Big Island volcanoes to which their gene pools are endemic.

When the chromosome sequences of these 26 species are compared with each other and with the sequences of the 76 other picture-wings from the older islands, we may make inferences about the mode of origin of these species. It should be realized that the inversions concerned represent indelible genetic markers of ancestry; they are unaffected by time or environmental factors.

The Big Island is separated from neighboring Maui by a deep channel 46 km wide; clearly the two islands were never connected by a land bridge. The biogeographical facts suggest the hypothesis, now to be developed, that most Big Island species have each separately arisen from one or more founder individuals arriving by chance dispersal from an older island. Whether such events for the sexually

Table 1. Nineteen hypothetical founder events to explain the origin of the 26 picture-winged *Drosophila* species of the Big Island of Hawaii. For details, see text.

	Species	Key chromosomal markers carried by founder		Inferred island of origin
—	*setosifrons*	3e 4hij 5e		Kauai
—	*assita*	$Xk^3 1^3$		Oahu
—	*formella*	2b;2AA		Oahu
—	*orthofascia*	XE		Maui
—	*psilotarsalis*	$Xp^2 2j$		Maui
—	*hawaiiensis*	Xa^2 4u;2AA		Maui
—	*macrothrix*	3i 5d		Maui
—	*digressa*	5d		Maui
—	*ciliaticrus*	Xg 3o		Maui
—	*engyochracea*	Xg 3	‡	Maui
—	*murphyi*	Xg 3/3o		Maui
—	*pullipes*	X 2 3 4 5	*	Lanai
—	*sproati*	X 2 3 4 5		Maui
—	*alsophila* *conspicua*	4b		Maui
—	*heedi* *silvarentis*	$2A^+A^+$		Maui
—	*ochrobasis* *setosimentum*	Xm^2 31 $4b^2$		Maui
—	*claytonae* *ochracea*	5a		Molokai
—	*heteroneura* *silvestris*	Xr		Molokai
—	*basisetae* *paucipuncta* *prolaticilia*	Xef 3z 4efg		Oahu

‡ Although cytological parsimony suggests a single founder for these three species, conditions on Maui and Oahu indicate that separate founders for each of the three are involved (see Carson, 1980).
* Cytologically indistinguishable. Separate founders have been inferred on morphological and behavioral grounds (see Carson, 1981).

reproducing, cross-fertilizing diploid species of *Drosophila* can be so drastic as to involve just a single fertilized female may be questioned. I believe that the evidence indicates such single founders are indeed involved. I will return to this point later, after the presentation of the following examples.

For the origin of the 26 species of the picture-wings on the Big Island, 19 introductions from outside are inferred (Table 1). These inferences are based mostly on the sharing of precise chromosomal markers or combinations of markers. In a few instances, however, other biological criteria such as morphology and behavior have been invoked where the cytological markers do not provide crucial information (see Table 1).

Looking at the details on Table 1, each of the first 13 species listed is postulated to have arisen separately, each from a single founder. In the first 8, inferences are based on unique cytological markers. In the case of the next three (*ciliaticrus,*

RECIPIENT ISLAND

DONOR ISLAND	Kauai	Oahu	Maui Complex	Hawaii	Total founder donations
Kauai	12	5	2	1	8
Oahu	1	29	10	3	14
Maui Complex	1	7	40	15	23
Hawaii	0	0	0	26	0
Total founder receipts	2	12	12	19	45

Fig. 1. Numbers of picture-winged *Drosophila* species on four Hawaiian islands or island groups (boxes on diagonal) and the numbers of hypothetical founders donated (rows) or received (columns) by each island. (From Carson, 1983.)

engyochracea, and *murphyi*), the founders are very similar and at one time (Carson et al., 1970) a single founder ancestral to these three species was considered to be the parsimonious solution. Discovery of new species on Maui and Oahu, however, has led to the inference that separate founders were involved (Carson, 1980).

In five instances (Table 1), pairs of Big Island species appear to owe their origin to a single founder. In one case, a proliferation of three species appears to have followed the founder event. It should be emphasized that the rule of parsimony has necessarily been applied in these cases and new data may make further hypothetical founders necessary.

Because of the detailed knowledge of the chromosomes and distribution of the picture-winged species from the older islands, it is usually possible to make another inference, namely, the island origins of the founders. These latter are listed at the right in Table 1. *D. setosifrons,* for example, appears to be descended from a founder arriving directly from the distant island of Kauai, since its unique chromosomal condition, consisting of a combination of five fixed inversions, is known only in *D. picticornis,* a species endemic to that island and having no other known close relatives.

The procedure described above for the Big Island has also been employed to

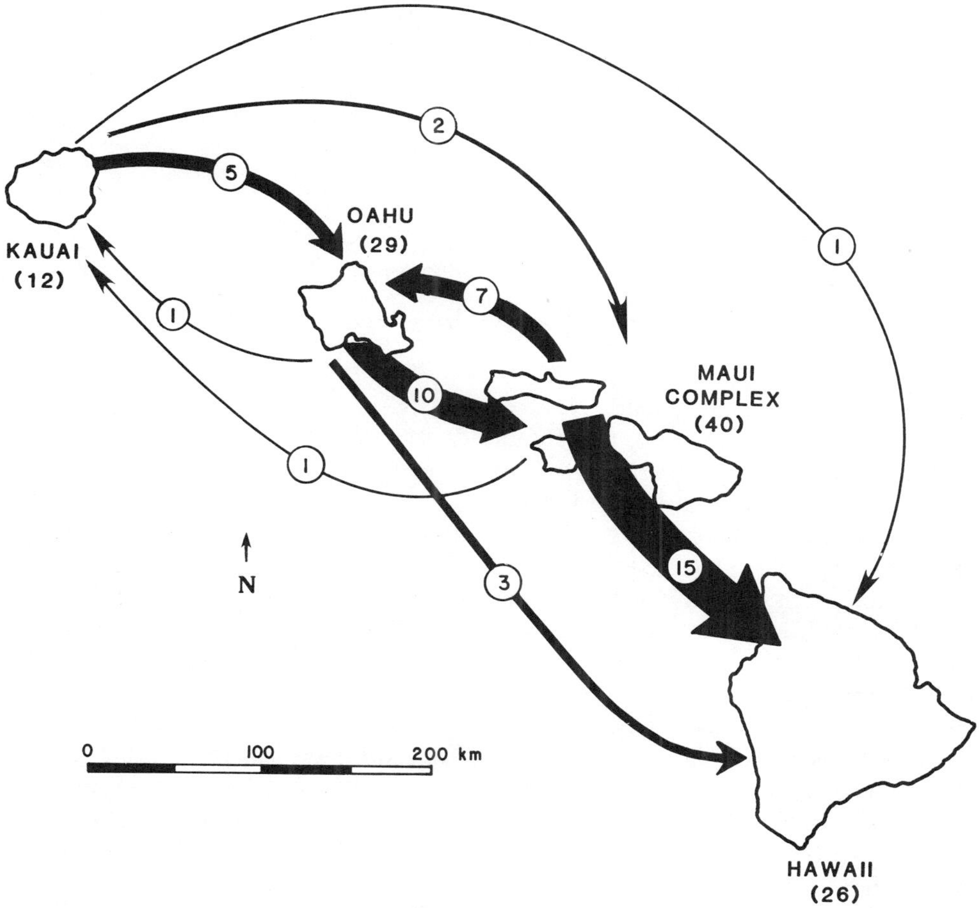

Fig. 2. Geographical summary of the founder events invoked to explain the origin of the picture-winged *Drosophila* species of each Hawaiian island (or island complex). The width of the arrows is proportional to the number of proposed founders. The number of species found on each island is given in parentheses. (From Carson, 1983.)

infer founder events involving the other Hawaiian islands. For this purpose, the three high islands at the middle of the archipelago (Maui, Molokai and Lanai) have been treated as a single complex of islands, the "Maui complex," inasmuch as geological evidence indicates that they were at one time joined into a single land mass. This latter was termed "Maui Nui" by Macdonald and Abbott (1970).

Figures 1 and 2 summarize the data on interisland founders. The details have been presented in a more extended form elsewhere (Carson, 1983). The column headed "Hawaii" in Fig. 1 reviews the material already presented in preceding paragraphs. Of the 26 species present, 15 are traceable to founders arriving from the Maui complex and only four to founders from the older islands. This result is not unexpected in view of the geographical proximity of the Maui complex to the Big Island (Fig. 2). The row labelled "Hawaii" (Fig. 1) reveals that the data do not make it necessary to assume any "back migration" of founders from the

Table 2. Number of species of picture-winged *Drosophila* relative to the age and area of the islands.

	Age of oldest lava flows (years × 10^6)	Size of island (km^2)	No. of endemic species	No. of species per 1,000 km^2	No. of founders received	No. of species per founder
Kauai	5.6	1432	12	8.4	—	—
Oahu	3.3	1575	29	18.4	12	2.4
Maui complex	1.8	2927	40	13.7	12	3.3
Hawaii	0.5	10,458	26	2.5	19	1.4

Big Island to an older island. Nevertheless, it should be noted that the Maui complex apparently gave rise to eight founders going "back" to older islands. This may be ascribed to exuberant population developments and sets of speciation events occurring on Maui and Molokai.

The relation of species diversity to age and area is recorded in Table 2. Here, Hawaii stands out as a large young island with a relatively small number of species. Thus, there are only 2.5 picture-winged species per 1000 km^2, compared with 13.7 for the Maui complex. This is a substantial difference even allowing for the large areas of new bare lava and xeric lowlands on the Big Island.

Another aspect which may be considered is the number of species per founder formed on the recipient island. When the law of parsimony is applied, the difference between the Big Island and the others is quite pronounced (Table 2, right hand column). Whereas on both Oahu and the Maui complex the mean number of species per inferred founder is substantial (2.4 and 3.3, respectively), the number for the Big Island is only 1.4. This low figure correlates with the small number of species per square kilometer.

Finally, we may consider the number of founders that the Island of Hawaii has received. Although this appears to be large (19), it is in fact proportionately low when the size of the island is taken into account by calculating the numbers of founders per thousand square kilometers. This figure is 1.8 for the Big Island but is 5.3 for the Maui complex and Oahu.

Accordingly, we may suggest that not only have successful founders into the Big Island been few but also the proliferative speciation success after arrival has been less than that of the other islands. All these facts appear to support one major conclusion. This is that the newness of all the volcanoes of the Big Island is basically the cause of the biological properties just described. In short, as historic times were approached, the fauna of the island appears not to have reached an age/area equilibrium. Severe degradation of the environment on the Big Island has occurred since the arrival of the first human populations about the 12th century A.D. or earlier. This effectively prevents any further study of the age/area/species problem on this island. This may be used as partial justification for the admittedly speculative nature of this discussion, since it is unlikely that any new information will become available in the future.

Clearly, the picture-winged *Drosophila* species of the Big Island are new species in the geological sense. The fact that a great preponderance of them may be traced to ancestors on the Maui complex provides heightened interest. The question arises as to why the species of the Big Island are, with one exception, clearly recognizable, morphologically and genetically, as full species separate from their founder progenitors. The situation appears to be somewhat different among birds. For example, of 16 species of the family Drepanididae found on the Big Island, there are (or were) ten that have been accorded full specific rank. The remaining six have been relegated to subspecies of forms found on the older islands.

That a considerable number of these drepanidids are considered subspecies appears to represent a difference from the case of the drosophilids. Recognition of subspecific status in drosphilids is a relatively easy matter and it is unlikely that systematic judgments are erroneous to this degree, either for the honeycreepers or for the drosophilids.

I believe the difference is simply due to the low vagility of most of these large drosophilids. They tend to be rather closely restricted to their host plants and there is little evidence of easy dispersability either by active or passive transport. The strongly implied conclusion is that a successful interisland founder event, in the sense that the transported propagule actually reproduces on the new island, is exceedingly rare and is likely to involve an extreme reduction in population size, in fact, very possibly to a single gravid female specimen.

I have argued extensively elsewhere (e.g. Carson, 1971, 1982b) that such a drastic populational event may lead to a rapid recombinational disorganization of the balanced elements of the ancestral gene pool from which the founder comes. If the founder population is to persist in the new site following gene pool disorganization, it is understood that a period of intense selection ensues. I have proposed that during this phase natural selection is forced to seek new polygenic balanced states throughout the genome since the old balanced states have been destroyed or disorganized. This disorganization-by-chance/reorganization-by-selection cycle is considered a powerful force in the origin of species and would be conducive to the formation of a new population ultimately recognizable as a new species.

Templeton (1980) and Lande (1981) have extended this theory. The former has demonstrated that such populational events do not require a drastic loss of genetic variability with concomitant fixation of alleles by classical random drift. A more powerful effect is an accumulation of inbreeding and an induction of gametic disequilibrium. Such changes, as Lande has argued, can be especially manifested in polygenic characters under sexual selection. This can initiate or contribute to rapid speciation. Beyond this, of course, the reorganization phase may also make possible, or indeed require, a shift in adaptation to the ambient environment, although this phase might follow at a later time and at a slower pace.

Accordingly, we may suggest that the situation in the ten species of drepanidids and the 25 species of drosophilids involved very similar founder effects. The six subspecies of drepanidids and *Drosophila orthofascia* may have suffered less drastic founder effects, possibly due to greater vagility and the participation of more than just a single original founder.

Founder effects are not necessarily restricted to interisland events. As has been stated above, considerable speciation has occurred within the confines of a single island. It seems likely that the Big Island species pairs and one trio listed in Table

1 have arisen allopatrically according to reorganization theory on the Big Island volcanoes in a manner similar to that just described for interisland events.

The growth of the Big Island and its five shield volcanoes provides many opportunities for such founder events related to the repeated cycles of plant colonization of lava flows, destruction of these by new flows and colonization of the new flows from isolated remnants of vegetation (kipukas).

A pair of Big Island species that have been most intensively studied are the two giant picture-wings, *D. heteroneura* and *D. silvestris* (for a recent review, see Carson, 1982a). These two species, which are sympatric over large parts of the island, show sharp differences in morphology, behavior, chromosomal polymorphism, and DNA reassociation. They are strongly isolated sexually. All these criteria would appear to label them as good ordinary species. Nevertheless, they are unusual in being electrophoretically virtually indistinguishable and in that they produce fertile hybrids of both sexes in the laboratory and occasionally in nature. Breakdown of their integrity in nature, however, does not occur. They also show a shared chromosomal polymorphism, a circumstance that is rare for full species. Another point of similarity is that where sympatric they occupy extremely similar, if not identical, breeding niches, the decaying branches of a cluster of large *Clermontia* species related to *C. hawaiiensis.* The patchy distribution of the insects over the lava flows of the Big Island volcanoes appears to relate directly to the patchy distribution and rarity of their principal host plants.

The upshot of this is that these two species are judged to have originated from a common ancestor somewhere on the Big Island quite recently; the similarities between them appear to reflect this fact.

A final cautionary note about the evolution of species via the founder effect on oceanic islands may be made. The scheme outlined in the previous pages has clear application only to sexually reproducing, cross-fertilizing, diploid species. In contrast, many organisms like those adapted to strand ecosystems are preadapted to long-distance dispersal. Such forms tend to have natural strongly inbred genetic systems that do not suffer from either inbreeding depression or stochastic reorganization effects when passing through a single individual. Organisms that employ such genetic systems display colonization results that are similar to those of organisms having essentially clonal or apomictic reproduction. Even if an open recombination system prevails, moreover, repeated successful colonizations will serve merely to accomplish transfer of the old gene pool, without reorganization, to a new place without the formation of a new species. Certain Hawaiian plants, notably *Acacia koa* and *Metrosideros collina,* appear to have genetic systems of this latter type.

Conditions on the island of Hawaii appear to serve as an extremely useful paradigm for the evolution of terrestrial life in general. In a more restricted sense, the facts observed may provide a framework for the interpretation of insular evolution on other somewhat older isolated islands in the Pacific Basin. The large size, biological diversity, and extreme youth of the island of Hawaii are virtually unique properties which provide perspectives that can only be dimly perceived elsewhere.

SUMMARY

The "Big Island" of Hawaii is the newest in the archipelago; it shows no lava flows older than 500,000 years. Most of the island is much younger, with two of its five volcanoes currently active. Twenty-six species of large *Drosophila* flies ("picture-wings"), found in the island forests, have been genetically compared with populations of similar forms from older islands. Virtually all are judged to be unique species by almost any criterion. They are neither subspecies nor simple variant populations of species found on adjacent islands. Nevertheless, their affinities to, and origin from, ancestral populations on older Hawaiian islands can be traced by unique genetic markers such as chromosome inversions.

In view of these facts, it is hypothesized that the colonization of Hawaii Island has been accomplished through a finite number of interisland founder events. A single gravid female from a population on an adjacent island is supposed to have founded a new population that then undergoes radical genetic change and differentiates into one or more new species. When compared with the older islands, the Big Island is both young and large. Judging from the picture-wings, its biological properties include the following. There are relatively few species per 1000 km^2. The number of founders is high both relative to the number of species occurring and in absolute terms compared to the other islands (see Figs. 1 and 2). All of these properties relate to the newness of the island which appears (at least historically) not to be at biogeographical equilibrium.

Colonizers that can undergo such evolutionary change are characterized by open systems of genetic recombination. Some organisms, which are preadapted to long-distance dispersal, have closed (e.g. strongly inbred) genetic systems and may colonize without significant genetic change. Even if an open recombination system prevails, repeated successful colonizations serve merely to transfer the old gene pool of a species to a new place without the formation of a new species. These *Drosophila* species appear to have genetic systems especially prone to speciation following a founder event.

LITERATURE CITED

Carson, H. L. 1971. Speciation and the founder principle. *Stadler Symp.* 3: 51–70.

———. 1980. Chromosomes and evolution in some relatives of *Drosophila grimshawi* from Hawaii. *In: Insect Cytogenetics. Symp. R. Entomol. Soc. London* 10: 195–205.

———. 1981. Homosequential species of Hawaiian *Drosophila.* p. 150–64 *In: Chromosomes today.* Vol. 7. George Allen & Unwin, London.

———. 1982a. Evolution of *Drosophila* on the newer Hawaiian volcanoes. *Heredity* 48: 3–25.

———. 1982b. Speciation as a major reorganization of polygenic balances. p. 411–33 *In:* Baragozzi, C., ed., *Mechanisms of Speciation.* Alan R. Liss, New York.

———. 1983. Chromosomal sequences and interisland colonizations in Hawaiian *Drosophila. Genetics* 103: 465–82.

Carson, H. L., D. E. Hardy, H. T. Spieth & W. S. Stone. 1970. The evolutionary biology of the Hawaiian Drosophilidae. p. 437–543 *In:* Hecht, M. K. & W. C. Steere, eds., *Essays in evolution and genetics in honor of Theodosius Dobzhansky.* Appleton Century-Crofts, New York.

Carson, H. L. & J. S. Yoon. 1982. Genetics and evolution of Hawaiian *Drosophila.* p. 297–344 *In:* Ashburner, M., H. L. Carson & J. N. Thompson, Jr., eds., *The genetics and biology of* Drosophila. Vol. 3b. Academic Press, London.

Lande, R. 1981. Models of speciation by sexual selection on polygenic traits. *Proc. Natl. Acad. Sci. USA* 78: 3721–25.

Macdonald, G. A. & A. T. Abbott. 1970. *Volcanoes in the sea.* Univ. of Hawaii Press, Honolulu. 441 p.

McDougall, I. & D. A. Swanson. 1972. Potassium-argon ages of lavas from the Hawi and Pololu volcanic series, Kohala Volcano, Hawaii. *Geol. Soc. Am. Bull.* 83: 3731–38.

Templeton, A. R. 1980. The theory of speciation via the founder principle. *Genetics* 94: 1011–38.

Phytogeography and Evolution of the Flora of the Juan Fernandez Islands: a Progress Report

Tod F. Stuessy
Department of Botany
The Ohio State University
Columbus, Ohio 43210 U.S.A.

Roger W. Sanders
Fairchild Tropical Garden
10935 Old Cutler Road
Miami, Florida 33156 U.S.A.

Mario Silva
Departamento de Botánica
Universidad de Concepción
Concepción, Chile

INTRODUCTION

Some of the most fascinating natural laboratories in which to investigate the process of evolution are oceanic islands. These small land masses, remote from continental areas, contain unique plants that are often quite different from mainland relatives. Such obvious evolutionary divergence within a geographically restricted area offers unique opportunities for studying evolutionary phenomena. A number of island taxa have already been investigated from different evolutionary perspectives, particularly in the Galapagos Islands, Hawaii, and the Canary Islands (e.g. Carlquist, 1965, 1970, 1974; Stern, 1971; Gillett, 1972; Bramwell, 1975, 1979; Kunkel, 1976; Gardner, 1977; Sohmer, 1978; Aldridge, 1979; Gorman, 1979; Carr and Kyhos, 1981).

One set of oceanic islands, the Juan Fernandez Islands, is remarkable for the high degree of endemism of the native flora, especially in view of the small size and limited number of the land masses. These islands are located 600 km west of Valparaiso, Chile at approximately 33° S latitude, and they consist of three separate islands (Fig. 1): Masatierra (= Isla Robinson Crusoe); Santa Clara (just off the coast of Masatierra); and Masafuera (= Isla Alejandro Selkirk). Masatierra comprises about 78 km^2 and Masafuera 59 km^2; Santa Clara is about 3 km^2. The

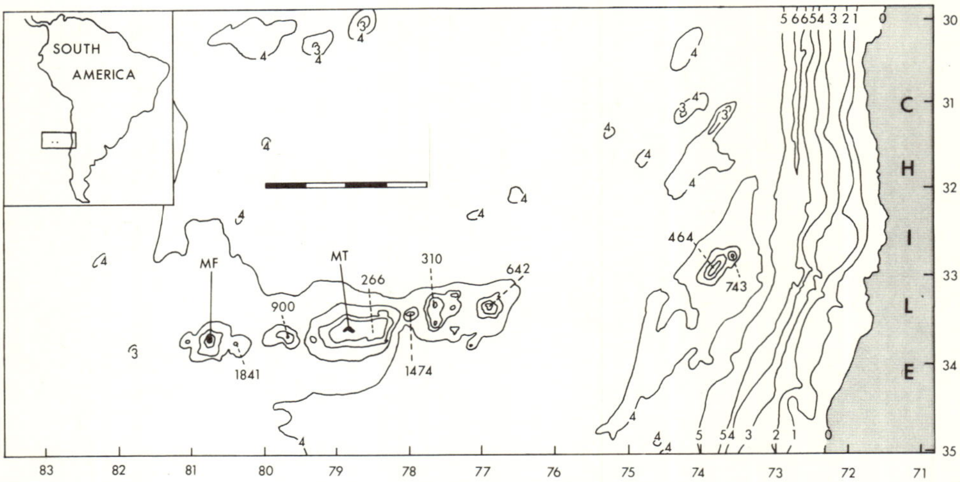

Fig. 1. Bathymetric map of the South Pacific Ocean surrounding the Juan Fernandez Islands. Single-digit depths in thousands of meters. MF = Masafuera, MT = Masatierra. Data from Mammerickx and Smith (1978) and Prince et al. (1980). (Scale = 200 km.)

two major islands are separated by 154 km of ocean. On these islands a flora of 147 indigenous angiosperm species has evolved of which 69% of the species and 19% of the genera are endemic (Skottsberg, 1956). Intensive floristic studies have been done on the islands, principally by Skottsberg (1922, 1951, 1953, 1956), but no biosystematic studies have been completed. A few chromosome counts had been made (Stebbins et al., 1953; Carlquist, 1967) previous to our studies, but no chemical studies of any kind had been initiated.

During the past several years, investigations in laboratories of The Ohio State University and the Universidad de Concepción, Chile, have focused on the evolution of the endemic flora of the Juan Fernandez Islands, with special interest on the tree-Compositae. The objectives of these studies have been to (1) determine the absolute ages of the islands and reconstruct their geological histories; (2) assess the degree of loss of native vegetation in the 60 years since Skottsberg's surveys; (3) investigate the chromosome numbers of the endemic species to learn the role of polyploidy in the evolution of the flora; and (4) understand the phylogeny and modes of speciation of the tree-Compositae, which comprise the largest family and which have undergone more speciation in the islands than any other plant group.

Two expeditions to the Juan Fernandez Islands under a cooperative program have been completed. The first was from 22 January to 23 February, 1980, with the following personnel (research specialties shown in parentheses): J. Arriagada (angiosperms); C. Marticorena (angiosperms and ferns); O. Parra (algae and bryophytes); R. Rodríguez (ferns); R. Sanders (angiosperms); T. Stuessy (angiosperms); and E. Ugarte (ecology of vascular plants). Most of the collecting was done on the larger island, Masatierra (including Santa Clara), with only four days on the smaller and more remote island, Masafuera. The second expedition was from 18 November to 2 December, 1980, and included O. Matthei (vascular plants, especially grasses); R. Sanders (angiosperms); T. Stuessy (angiosperms); and H.

Valdebenito (phytochemistry). All of this collecting was done on Masatierra. The historical details of both expeditions are contained in Marticorena et al. (in preparation).

The purpose of the present paper is to review progress to date on the entire project, including information on geological reconstruction of the islands, chromosome numbers in the native flora, phylogeny and modes of speciation in the tree-Compositae, disappearance of the native flora in recent times, and future research plans.

GEOLOGICAL RECONSTRUCTION OF THE ISLANDS

To determine patterns of phylogeny and speciation in the flora of the Juan Fernandez Islands, the geological setting was investigated. Rock samples were collected from both islands (7 from Masatierra and 3 from Masafuera) and absolute dates obtained in collaboration with Drs. John Sutter and Kenneth Foland of the K-Ar Laboratory of the Department of Geology at Ohio State University. Two dates from each island have been obtained (no previous dates for either island were known; cf. Shields, 1976): 3.8 and 4.2 million years old for Masatierra and 1.0 and 2.4 million years old for Masafuera. One uncertain date was published earlier by P. Baker (in Valenzuela, 1978) of 5.0 million years old for Masatierra, but Dr. Baker has some skepticism about the results (in litt.). It seems clear that the islands are geologically very young—much younger than suggested earlier by Brüggen (in Skottsberg, 1956). This is an interesting point, because the endemic fern genus, *Thyrsopteris,* is apparently very old, having affinities with relatives in the Upper Cretaceous from Japan (Nishida & Nishida, 1979). Also, the endemic family Lactoridaceae is regarded as morphologically primitive with ties to the Magnoliidae (Carlquist, 1964; Cronquist, 1981). The exact affinities of this family are being examined further. The main point is that the radiometric dates suggest that the island flora is young and any apparently "relictual" taxa must have come in by dispersal within the past few million years. Extinction on the mainland, perhaps during the climatic changes of the Pleistocene (Vuilleumier, 1971; Simpson, 1974, 1975), might have caused loss in other places with the only remaining localities being now in these islands.

The Juan Fernandez Islands are entirely volcanic in origin and apparently have never been connected to the mainland (Fig. 1; this is also contrary to the opinion of Brüggen, in Skottsberg, 1956). The Nazcz Plate is moving from west to east and apparently new islands are produced as the plate moves eastward over the "hot spot" (Stearns, 1966; McElhinny, 1973; Jackson, 1976). The absolute ages of ca. 4 million years for Masatierra and 1–2.4 million years for Masafuera correlate well with this viewpoint. The known spreading rates for the margin of the Nazca Plate (8.8 cm/year) agree well with the observed distance separating Masatierra and Masafuera andtheir known ages (150 km/1.7 × 10^6 years). Some easterly seamounts could have been above the surface of the sea at some point in the past, but all but one would most likely have been submerged before Masatierra emerged (Fig. 1). The only possible island could have been 134 km to the east, now some 310 m below the surface.

Based on bathymetric information (Mammerickx and Smith, 1978; Prince et al., 1980) plus general knowledge of changes in sea level during the Pleistocene

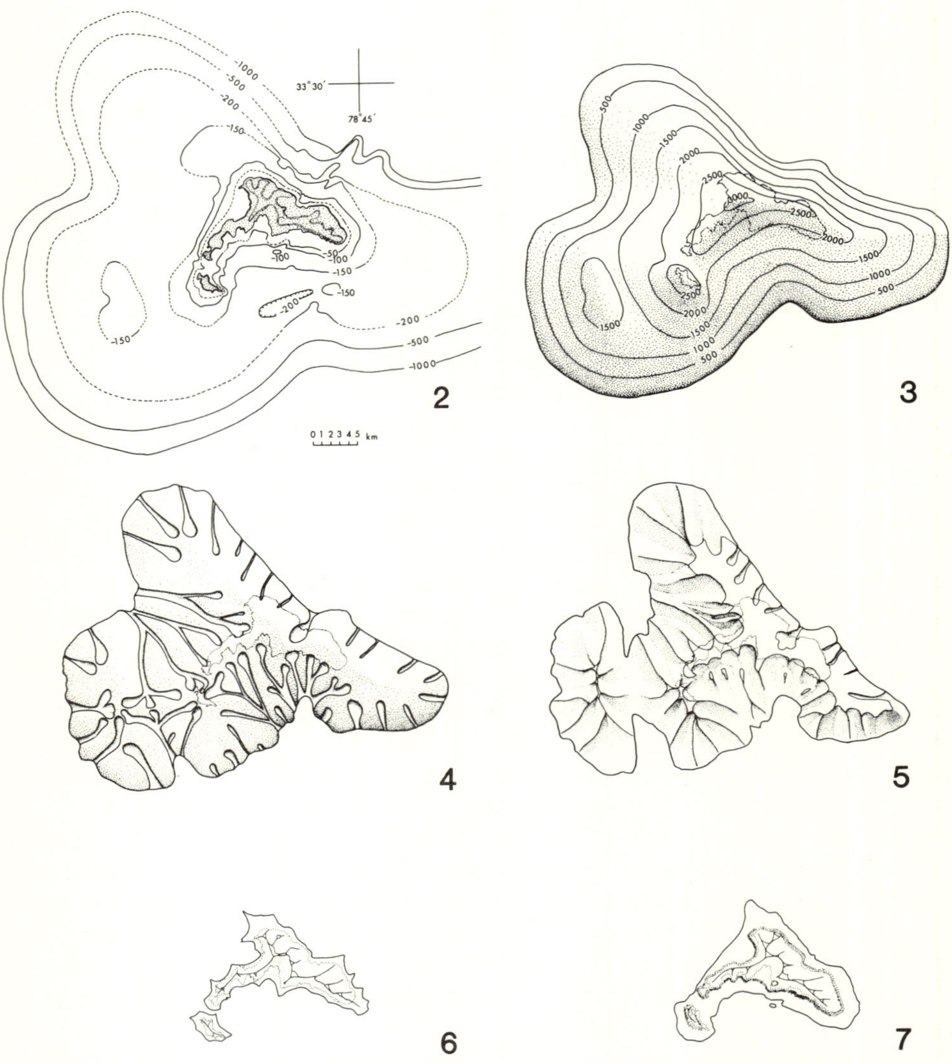

Fig. 2–7. Reconstruction of geological history of Masatierra (and Santa Clara): 2, present configuration of island showing bathymetric contours to -1000 m (data from Mammerickx and Smith, 1978; Prince et al., 1980; dashed line represents interpolated levels between known data points); 3, shape of original island, 4 m.y.b.p. (dashed lines here and in Fig. 4–7 show present island outline); 4, erosional patterns showing amphitheater valleys (cf. Fig. 8 of Masafuera at comparable erosional stage), 3 m.y.b.p.; 5, further erosion showing coalescence of adjacent valleys, 2 m.y.b.p.; 6, further erosion and subsidence, 1 m.y.b.p.; 7, lowering of sea level during Pleistocene glaciation, 10,000 y.b.p.

in the Pacific and erosional rates on other volcanic islands (using Hawaii as an analog; Stearns, 1966), the geological history of both Masatierra and Masafuera can be reconstructed (Fig. 2–9). These data show that Masatierra is now much smaller than it was originally and that the distribution of a particular species in the flora now will not necessarily reflect its place of origin directly nor its mode

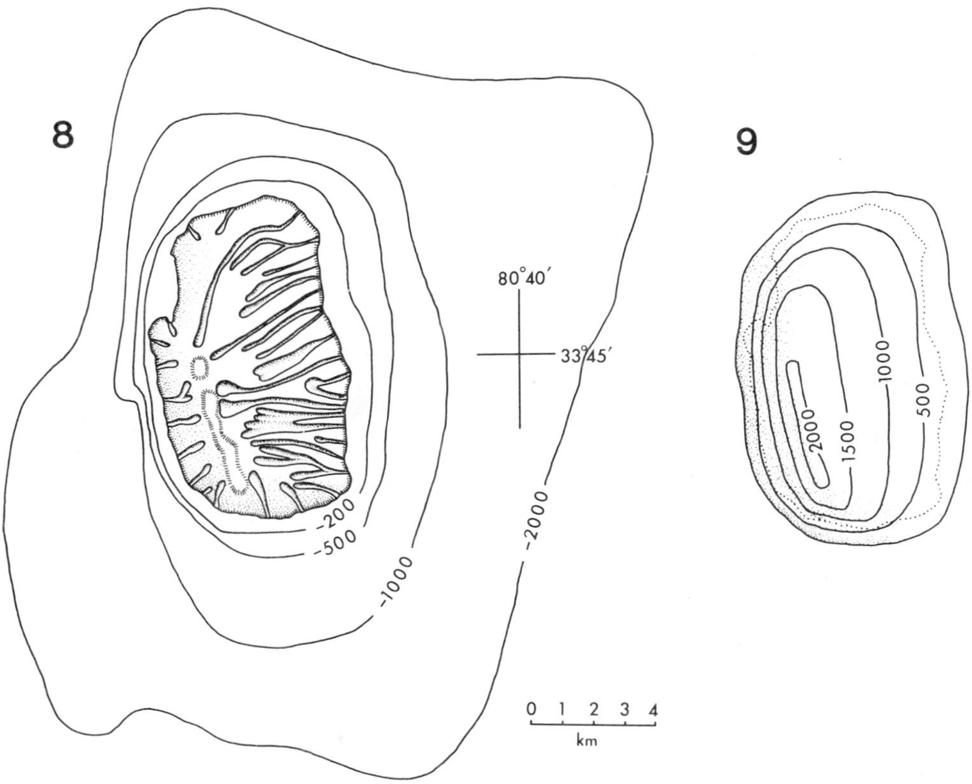

Fig. 8–9. Reconstruction of geological history of Masafuera: 8, present configuration of island showing bathymetric contours to -2000 m (data from Mammerickx and Smith, 1978; Prince et al., 1980); 9, shape of original island, 1–2 m.y.b.p. (dotted line shows present island outline).

of speciation. Masafuera, on the other hand, is younger and only very slightly eroded.

CHROMOSOME NUMBERS IN THE NATIVE FLORA

We have attempted to gain as much cytological information as possible about the native flora to allow for interpretations of evolutionary mechanisms. A major effort during the field work, therefore, has been placed on obtaining bud material for determination of chromosome number. These results have already been reported (Sanders, Stuessy, and Rodriguez, 1983), but a brief summary will be provided here.

Forty-one chromosome counts were obtained for 29 species, bringing the total of chromosomally known native species to 31. Counts are now known from *Azara* (*n*=9, *A. serrata,* Flacourtiaceae), *Dysopsis* (*n*=13, *D. hirsuta,* Euphorbiaceae), *Raphithamnus* (*n*=ca.18, *R. venustus,* Verbenaceae), *Robinsonia (n*=18, *R. gayana* and *R. gracilis,* Compositae), and *Ugni (n*=22, *U. selkirkii,* Myrtaceae). New species counts were also obtained in *Dendroseris* and *Erigeron* (Compositae), *Eryngium* (Apiaceae), *Escallonia* (Saxifragaceae), *Haloragis* (Haloragaceae), *Lib-*

10

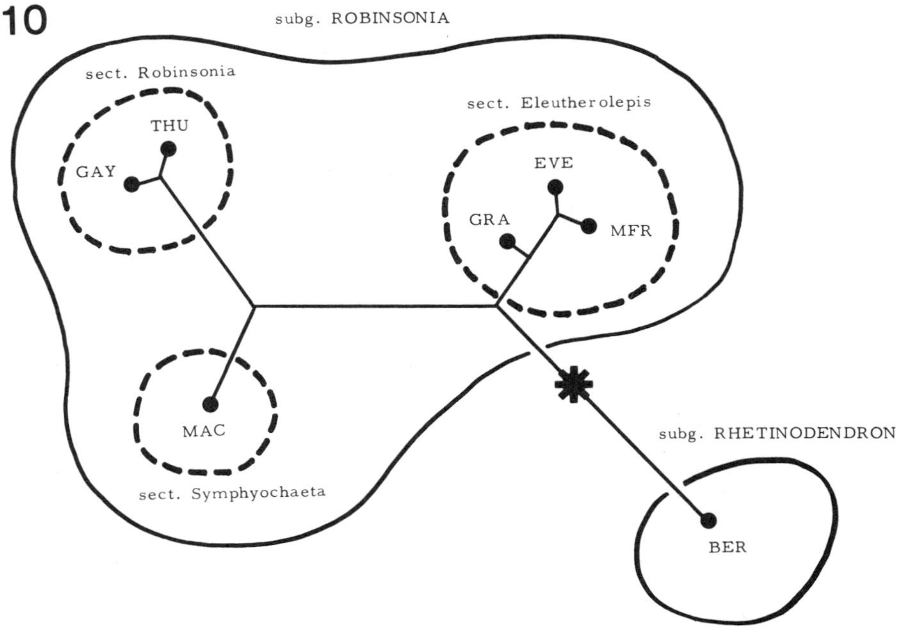

11

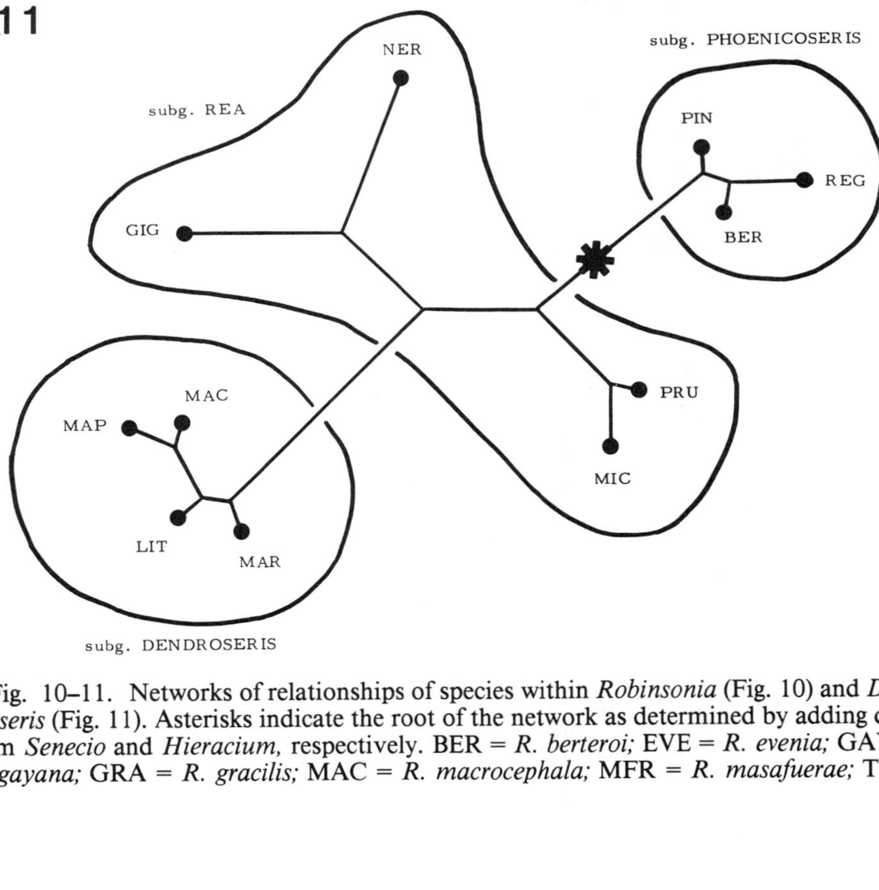

Fig. 10–11. Networks of relationships of species within *Robinsonia* (Fig. 10) and *Dendroseris* (Fig. 11). Asterisks indicate the root of the network as determined by adding data from *Senecio* and *Hieracium*, respectively. BER = *R. berteroi;* EVE = *R. evenia;* GAY = *R. gayana;* GRA = *R. gracilis;* MAC = *R. macrocephala;* MFR = *R. masafuerae;* THU

ertia (Iridaceae), *Myrceugenia* (Myrtaceae), *Pernettya* (Ericaceae), *Peperomia* (Piperaceae), *Spergularia* (Caryophyllaceae), and *Wahlenbergia* (Campanulaceae). Too few endemic species have been examined cytologically (21% known) to provide a complete understanding of the evolutionary role of chromosome number alterations in the flora, but a few generalizations seem in order at this time.

The level of recent polyploidy in species of the Juan Fernandez flora known cytologically is 6%. This is a very low value and suggests that the immigration of species to the archipelago and subsequent initial speciation has not been accompanied by marked chromosomal change. This correlates well with the young ages of the islands. In contrast, the Canary Islands have 28% polyploidy (Borgen, 1979), but these islands are geologically older, once connected to the African mainland (Bramwell, 1976), and in part with a relictual flora. Within the taxa of the Juan Fernandez Archipelago that have diversified beyond two species (especially the largest genera *Dendroseris* and *Robinsonia),* chromosomal alterations in number have not occurred. This agrees with Humphries's (1979) perspective that groups adaptively radiating (in his case using the Canary Islands as the example) all have the same chromosome numbers. It is also worth noting the correlation between chromosome number stability and the high level of the woody-perennial habit in the islands (52% of the endemic angiosperm species and 90% of the endemic genera; Skottsberg, 1956).

PHYLOGENY AND SPECIATION IN THE TREE-COMPOSITAE

The conclusions of previous workers on the taxonomy of the tree-Compositae (Skottsberg, 1922, 1951; Stebbins, Jenkins, and Walters, 1953; Carlquist, 1958, 1967, 1974), principally *Robinsonia* (tribe Senecioneae) and *Dendroseris* (tribe Lactuceae), have been little altered by our studies. However, we do include the monotypic *Rhetinodendron* in *Robinsonia* as a subgenus (Fig. 10) and regard all the segregates of *Dendroseris* as belonging to one genus, also as subgenera (Fig. 11). Each tribal group apparently has had only one introduction to the islands.

The phylogenies of *Robinsonia* and *Dendroseris* have been determined through cladistic analysis using a Wagner network method (Farris, 1970; Nelson and Van Horn, 1975; Fig. 10 and 11). Because ties to the mainland relatives are obscure, it was believed best to develop networks of relationships among the endemic species followed by rooting these networks at different points depending upon which hypothetical ancestor might be suggested. Herbarium studies at the Universidad de Concepción (CONC) and the U.S. National Herbarium (US) suggested that *Robinsonia* is most closely related to *Senecio* of South America and that *Dendroseris* relates most closely to *Hieracium,* also abundant in mainland South America. A full discussion of these methods and results will appear elsewhere (Sanders, Stuessy, and Marticorena, in prep.). The main point is that we now have a framework of relationships for species within both genera based on morphology that can be tested by the accumulation of other types of data, especially phyto-

---

←

= *R. thurifera.* BER = *D. berteriana;* GIG = *D. gigantea;* LIT = *D. litoralis;* MAC = *D. macrantha;* MAP = *D. macrophylla;* MAR = *D. marginata;* MIC = *D. micrantha;* NER = *D. neriifolia;* PIN = *D. pinnata;* PRU = *D. pruinata;* REG = *D. regia.*

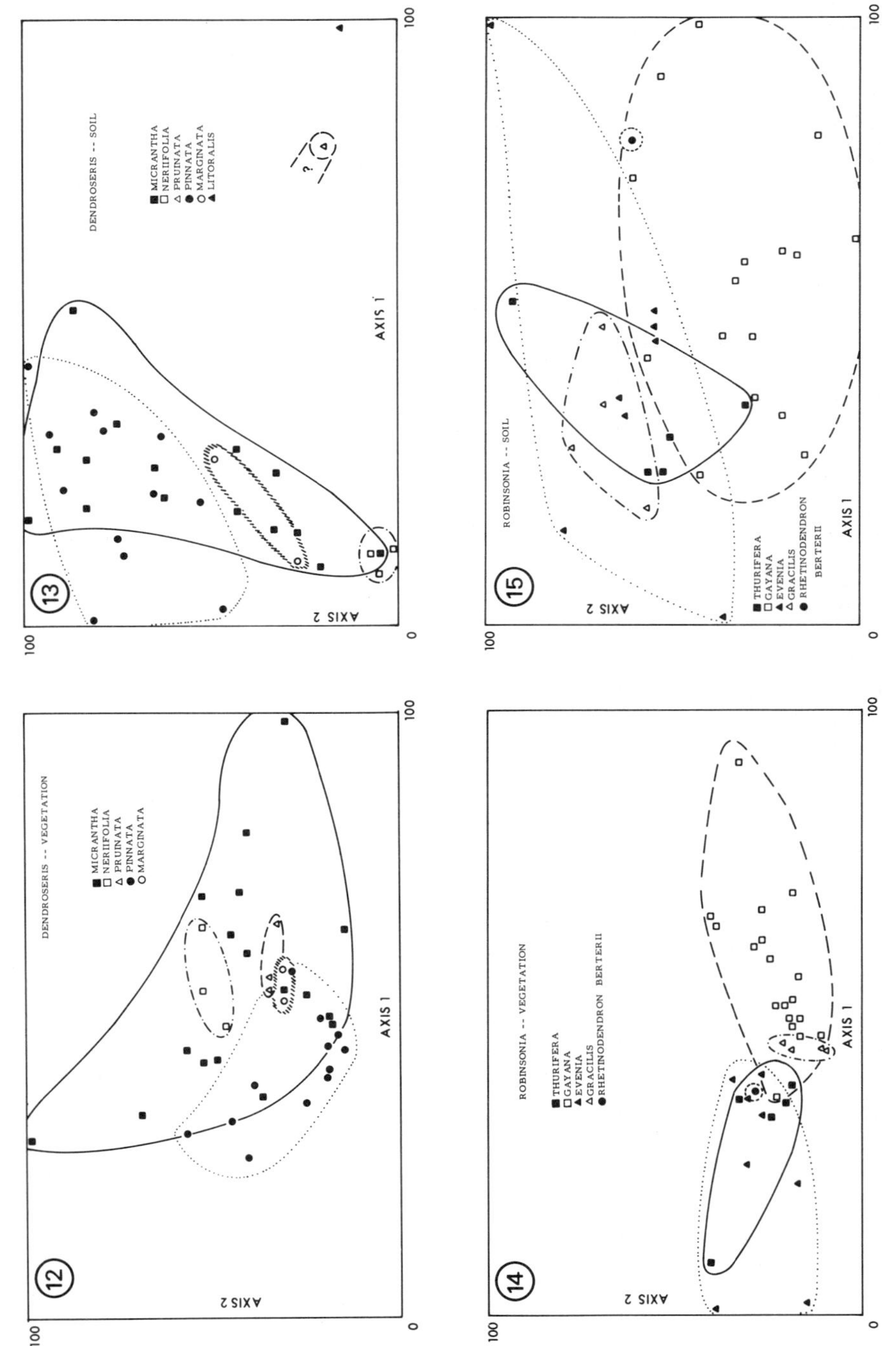

Fig. 12–15. Plots of the first two principal components axes of the species of *Dendroseris* (Fig. 12 and 13) and *Robinsonia* (14 and 15) using vegetational (12 and 14) and soil (13 and 15) factors.

chemistry, and that can be used to suggest patterns of speciation. For example, in *Dendroseris* (Fig. 11), the three species endemic to Masafuera (*D. gigantea, D. macrophylla,* and *D. regia*) occur in each of the major phyletic lines, which suggests that their immediate ancestors dispersed from Masatierra independently and evolved anagenetically in parallel on Masafuera within the past 1–2.4 million years. Without this phyletic insight, it might appear that the three species on Masafuera evolved as a group, for which a search for speciation mechanisms would be fruitless.

Striking patterns of adaptive radiation in *Robinsonia* and *Dendroseris* stressed by Carlquist (1974) have not been borne out by the present studies. Over 80 collections of species of both genera were made throughout the zones of natural vegetation on Masatierra. Two types of data were collected on the ecology of the sites of each collection: soils (P, K, Fe, Ca, Mg, organic matter, pH, and cation exchange capacity, as well as elevation and slope), and associated species (all plants within a 3 m radius, adjusted for cover importance). These data for each collection were subjected to principal components analysis using the Ordiflex program (Gauch, 1977; Gauch and Stone, 1979). The quantitative results (Fig. 12–15) show some ecological differentiation, but not as dramatic a picture as was described earlier (Carlquist, 1974). It is difficult to know what the role of adaptive radiation in the evolution of the tree-Compositae might have been. Strong ecological factors could have been influential initially in speciation with evidence now lacking due to changes in climate, vegetation, and geomorphology of the islands. Nonetheless, there certainly are no strong ecological zones such as can now be seen clearly on Oahu in the Hawaiian Islands (Carlquist, 1970; pers. observ.). Geographical isolation resulting from separation of populations into deep canyons, therefore, seems also a possible factor in speciation in the islands. Full details of this work will appear in Sanders, Stuessy, and Marticorena (in prep.).

Because the flora of the islands (especially Masatierra) has been compacted over geological time, many species occur together in similar ecological settings but without hybridization. For example, in one valley (Quebrada Villagra) three species each of *Dendroseris* and *Robinsonia* occur within approximately one km^2. No hybrids have been found among these taxa even though all but one (*R. thurifera*) flower at the same time. In fact, few hybrids are known in the entire flora. This observation correlates well with the notion that as Masatierra was reduced in area, the species either became extinct, hybridized with other taxa out of existence, or survived if isolation from relatives was sufficient to prohibit swamping through hybridization.

DISAPPEARANCE OF THE NATIVE FLORA

The general collecting done on both expeditions revealed clearly that the native flora has declined in number of species since Skottsberg's observations (made in 1917, published in 1922; also Christensen and Skottsberg, 1920). We have quantified these levels of disappearance (Table 1) from seven of the localities described in detail by Skottsberg and visited again by us, and from 9 to 67% of the native species have been lost in these areas. The main reason for the decline in the endemic flora is the destruction by domesticated and feral animals including goats, sheep, cattle, coatimundis, rabbits, and horses. We have had discussions with

Table 1. Comparison of the numbers of native and introduced species of vascular plants in 1917 and 1980 in seven localities on the Juan Fernandez Islands, Chile.*

Locality	Number of native species		% Decrease of native species	Number of introduced species		% Increase of introduced species
	1917	1980		1917	1980	
MASATIERRA						
Portezuelo, Cumberland side	39	30	23	1	5	400
Portezuelo, Villagra side	32	29	9	0	2	c. 200
Salsipuedes	18	16	11	2	4	100
Damajuana	16	12	25	0	3	c. 300
SANTA CLARA						
Morro Spartan	5	4	20	5	5	0
MASAFUERA						
Quebrada Casas	30	20	33	0	0	0
Quebrada Vacas	6	2	67	2	2	0

* From Sanders et al. (1982).

officials of CONAF (Corporaci6n Nacional Forestal, the equivalent of the U. S. Department of Interior) about this problem. They are aware of the difficulties and have a plan for protection (Anonymous, 1976) but lack funds for its implementation.

In recognition of the need for conservation of the native flora of the islands, the following resolution was passed unanimously at the conclusion of this symposium on Pacific Basin biogeography:

Whereas the flora of the Juan Fernandez Islands contains plant species of great biological and evolutionary interest; whereas the feral goats on Masafuera are a great and immediate danger to the survival of the rare and endemic species on that island; whereas the feral rabbits on Masatierra are placing an equally severe pressure on the rare and endemic flora of that island; whereas these goats and rabbits are only of the most limited systematic and evolutionary significance in their own right;

Be it resolved that the Federal Government of Chile be asked, through its Corporación Nacional Forestal, to use whatever means appropriate to eradicate the goats and rabbits on both islands in the hopes of eventual rejuvenation of this most unusual, remarkable, and biologically significant native flora.

FUTURE RESEARCH PLANS

Because a good general geological framework exists now for the islands, good collections of many of the taxa of the flora are at hand, and data-gathering methods already used have been effective in obtaining answers concerning phylogeny and speciation in the tree-Compositae (although much here remains to be done), it is believed reasonable to examine other genera of the flora for similar information. Once these data are available, they will allow a synthesis on the evolution of the flora of this oceanic archipelago. This particular set of islands has advantages for such a synthesis because the ages of the islands are now known, the younger island is the most remote one from the continent, the taxa are isolated from mainland relatives, and the native angiosperm flora is small (147 species). Furthermore, there are only 11 genera in the entire archipelago that have apparently speciated by some form of cladogenesis; they include a total of 47 species (30% of all native taxa) of which 18 already have been studied, the tree-Compositae. By studying these speciated groups in depth, maximum insights on the evolution of the flora within the islands can be anticipated. Single endemics can give information about evolutionary trends from mainland relatives and can contribute to an understanding of the phytogeography of the entire native flora (this will also be summarized in a later phase of the project,which will also deal with the concepts of island biogeography), but they cannot give much insight on the modes of phylogeny and speciation within the archipelago.

We hope to determine patterns of phylogeny and speciation for the following genera of the islands (family and numbers of species in parentheses): *Chenopodium* (Chenopodiaceae, 3); *Cuminia* (Lamiaceae, 2); *Erigeron* (Asteraceae, 4); *Gunnera* (Haloragaceae, 3); *Haloragis* (Haloragaceae, 2); *Peperomia* (Piperaceae, 3); *Solanum* (Solanaceae, 3); *Sophora* (Fabaceae, 2); and *Wahlenbergia* (Campanulaceae, 5). Both herbs and small woody shrubs (*Wahlenbergia* and some species of *Erigeron* and *Chenopodium*) are represented. Investigations of these generic systems representing different families of flowering plants and different habit types should give a good idea of the patterns of evolution of the entire flora.

Another future objective is the determination of genetic divergence between and among endemic taxa. Because the age of the islands is known, a time frame for divergence is provided. Furthermore, as indicated earlier, the number of possible derivations of closely related endemic taxa from one another is limited. Many taxa are isolated, have small population sizes, and in some cases are restricted only to one remaining population. This represents a good situation in which to employ isozymes for looking at genetic divergence between congeneric species and for measuring the amount of genetic variation within populations.

Gottlieb (1977, 1981) has discussed the value of enzyme electrophoresis in systematic and evolutionary studies and summarized available data for plants. The rather meager data indicate that conspecific populations invariably exhibit the same alleles in similar frequencies at the structural genes coding for the soluble enzymes examined electrophoretically. This is often expressed by the genetic identity statistic of Nei (1972) where the values range from 0 (no alleles in common) to 1 (all alleles present in same frequency). Briefly, conspecific populations normally exhibit a mean genetic identity of 0.95 whereas congeneric species exhibit a sharply reduced mean genetic identity of 0.67. The notable exceptions to this are species pairs thought to be recently diverged and related as progenitor-derivative (e.g. Gottlieb, 1973, 1974; Gottlieb and Pilz, 1976; Crawford and Smith, 1982).

As far as we know, there are no allozyme studies of species of plants where the time of their divergence can be measured nearly as well as can be done on the Juan Fernandez Islands. The question of whether congeneric species in different genera exhibit similar levels of genetic divergence is very worthy of investigation. In addition, as indicated above, inferences about progenitor and derivative taxa can be made with reasonable certainty because of the different ages of the two

Fig. 16. Vegetation map of Morro Spartan, Santa Clara, showing location and size of individuals of *Chenopodium sanctae-clarae* (white enclosed areas). Other dominant species are *Avena barbata* (hatched areas) and *Dendroseris pruinata* (black areas). Analysis and map completed by Eduardo Ugarte.

islands. It will also be of interest to examine genetic variation in populations of these endemic taxa and compare the values for populations of continental species. Available data indicate that in outcrossing species most of the genetic variation detected in the species as a whole may be found in one population whereas in self-pollinating taxa there is more interpopulational variation. Whether this pattern will hold for island species of plants remains to be determined.

A particularly good taxon to examine is *Chenopodium sanctae-clarae,* confined to Morro Spartan, a small rock of 400 m^2 off the coast of Santa Clara (Fig. 16). Fifty-eight individuals of this species are known, and it will be interesting to learn the levels of genetic variation in this population. Seeds were collected from one plant on the first trip, they have germinated, and preliminary banding studies show that the techniques can be easily worked out in this group to yield good results. Progeny of ten or more individuals of this population must be sampled and their isozymes analyzed. One would predict a low level of variation in the population if results are comparable to those for other species of *Chenopodium* (Crawford, 1979; Crawford and Wilson, 1977, 1979), but it will be interesting to test this hypothesis.

Another system that might be examined in a similar fashion is *Juania australis,* an endemic species and genus of palms that has about 1000 individuals scattered throughout the high ridges of the islands (Stuessy, Sanders, and Matthei, 1983). Seedlings are growing in the field plot of CONAF headquarters in San Juan Bautista (at Cumberland Bay) and therefore are easily germinable. CONAF collectors routinely bring back kilos of seeds from trees in different parts of Masatierra. Obtaining germinable seeds from different localities of the island, therefore, should be no difficulty. This study will result in determination of the levels of genetic variation in a morphologically uniform and endemic taxon with widely scattered individuals in an area of about 30 km^2. Comparing the results with a small and more restricted population system as in *Chenopodium sanctae-clarae* will be instructive. Many other generic systems offer potential for study in this fashion, but they are dependent upon abundant seed sources. The success of future field efforts will determine the specific direction this part of the research will take.

ACKNOWLEDGMENTS

It is a pleasure to give thanks to CONAF of Chile for permission to collect on the Juan Fernandez Islands; Universidad de Concepción, Vicerectoria de Investigaciones, CONICYT of Chile, and NSF Grant INT 77-21637 for financial support of the field expeditions; our collecting associates J. Arriagada, C. Marticorena, O. Parra, R. Rodríguez, and E. Ugarte for help and support on the first expedition; Oscar Matthei and Hugo Valdebenito for their unflagging aid on the second expedition; Bernardo Ackermann, Chief of the Juan Fernandez National Park, whose support with guides, pack animals, lodging and storage facilities, and work space was essential for successful completion of the research; the numerous CONAF guides who cheerfully contributed extra hours in the collecting efforts, in particular Oscar Chamorro, Alvis González, Domingo Retamal, and Ramon Schiller; Edwardo Ugarte for undertaking the vegetational survey of Morro Spartan and for permission to include his unpublished map (Fig. 16); and Daniel Crawford for his help in outlining future research plans involving isozyme studies.

LITERATURE CITED

Aldridge, A. E. 1979. Evolution within a single genus: *Sonchus* in Macaronesia. p. 279–91 *In:* Bramwell, D., ed., *Plants and islands.* Academic Press, London.

Anonymous. 1976. Plan de Manejo, Parque Nacional Juan Fernandez. CONAF and FAO, Santiago.

Borgen, L. 1979. Karyology of the Canarian flora. p. 329–46 *In:* Bramwell, D., ed., *Plants and islands.* Academic Press, London.

Bramwell, D. 1975. Some morphological aspects of the adaptive radiation of Canary Islands *Echium* species. *Ann. Inst. Bot. A. J. Cavanilles* 32: 241–54.

———. 1976. The endemic flora of the Canary Islands; distribution, relationships and phytogeography. p. 207–40 *In:* Kunkel, G., ed., *Biogeography and ecology in the Canary Islands.* W. Junk, The Hague.

Bramwell, D., ed. 1979. *Plants and islands.* Academic Press, London.

Carlquist, S. 1958. Anatomy and systematic position of *Centaurodendron* and *Yunquea* (Compositae). *Brittonia* 10: 78–93.

———. 1964. Morphology and relationships of Lactoridaceae. *Aliso* 5: 421–35.

———. 1965. *Island life: a natural history of the islands of the world.* Natural History Press, New York.

———. 1967. Anatomy and systematics of *Dendroseris* (sensu lato). *Brittonia* 19: 99–121.

———. 1970. *Hawaii: a natural history.* Natural History Press, New York.

———. 1974. *Island biology.* Columbia University Press, New York.

Carr, G. D. & D. W. Kyhos. 1981. Adaptive radiation in the Hawaiian silversword alliance (Compositae-Madiinae). I. Cytogenetics of spontaneous hybrids. *Evolution* 35: 543–56.

Christensen, C. & C. Skottsberg. 1920. The Pteridophyta of the Juan Fernandez Islands. *Nat. Hist. Juan Fernandez & Easter Is.* 2: 1–46.

Crawford, D. J. 1979. Allozyme studies in *Chenopodium incanum:* infraspecific variation and comparison with *Chenopodium fremontii. Bull. Torrey Bot. Club* 106: 257–61.

Crawford, D. J. & E. B. Smith. 1982. Allozyme variation in *Coreopsis nuecensoides* and *C. nuecensis* (Compositae), a progenitor derivative species pair. *Evolution* 36: 379–86.

Crawford, D. J. & H. D. Wilson. 1977. Allozyme variation in *Chenopodium fremontii. Syst. Bot.* 2: 180–90.

———. 1979. Allozyme variation in several closely related diploid species of *Chenopodium* of the western United States. *Am. J. Bot.* 66: 237–44.

Cronquist, A. 1981. *An integrated system of classification of flowering plants.* Columbia University Press, New York.

Farris, J. S. 1970. Methods for computing Wagner Trees. *Syst. Zool.* 9: 83–92.

Gardner, R. C. 1977. Evolution and adaptive radiation in *Lipochaeta* (Compositae) of the Hawaiian Islands. *Syst. Bot.* 1: 383–90.

Gauch, H. G., Jr. 1977. *ORDIFLEX: a flexible computer program for four ordination techniques, Release B. Ecology and Systematics.* Cornell University, Ithaca.

Gauch, H. G., Jr. & E. L. Stone. 1979. Vegetation and soil pattern in a mesophytic forest at Ithaca, New York. *Am. Midl. Nat.* 102: 332–45.

Gillett, G. W. 1972. The role of hybridization in the evolution of the Hawaiian flora. p. 205–19 *In:* Valentine, D. H., ed., *Taxonomy, phytogeography, and evolution.* Academic Press, New York.

Gorman, M. 1979. *Island ecology.* Chapman and Hall, London.

Gottlieb, L. D. 1973. Genetic differentiation, sympatric speciation and the origin of a diploid species of *Stephanomeria. Am. J. Bot.* 60: 545–53.

———. 1974. Genetic confirmation of the origin of *Clarkia lingulata. Evolution* 28: 244–50.

———. 1977. Electrophoretic evidence and plant systematics. *Ann. Mo. Bot. Gard.* 64: 161–80.

———. 1981. Electrophoretic evidence and plant populations. *Prog. Phytochem.* 7: 1–46.

Gottlieb, L. D. & G. Pilz. 1976. Genetic similarity between *Guara longiflora* and its obligately outcrossing derivative *G. demareei. Syst.. Bot.* 1: 181–87.

Humphries, C. J. 1979. Endemism and evolution in Macaronesia. p. 171–99 *In:* Bramwell, D., ed., *Plants and islands.* Academic Press, London.

Jackson, E. D. 1976. Linear volcanic chains on the Pacific Plate. p. 319–35 *In:* Sutton, G. H., M.

H. Manghnani & R. Moberly, eds., *The geophysics of the Pacific Ocean and its margin.* Am. Geophys. Union, Washington, D.C.

Kunkel, G., ed. 1976. *Biogeography and ecology in the Canary Islands.* W. Junk, The Hague.

Mammerickx, J. & S. M. Smith. 1978. Bathymetry of the Southeast Pacific. *Geol. Soc. Am. Map and Chart Ser.* MC-26.

McElhinny, M. W. 1973. *Palaeomagnetism and plate tectonics.* Cambridge Univ. Press, Cambridge.

Nei, M. 1972. Genetic distance between populations. *Am. Nat.* 106: 283–92.

Nelson, C. H. & G. S. Van Horn. 1975. A new simplified method for constructing Wagner networks and the cladistics of *Pentachaeta* (Compositae, Astereae). *Brittonia* 27: 362–72.

Nishida, H. & M. Nishida. 1979. *Thyrsopterorachis,* gen. nov., a tree fern rachis from the Upper Cretaceous of Hokkaido, Japan. *Bot. Mag. Tokyo* 92: 187–95.

Prince, R. A., W. J. Schweller, G. E. Ness, W. T. Coulbourn, G. L. Shepherd & A. Masias. 1980. Bathymetry of the Peru-Chile Trench and continental margin. *Geol. Soc. Am. Map and Chart Ser.* MC-26, Part I.

Sanders, R. W., T. F. Stuessy & C. Marticorena. 1982. Recent changes in the flora of the Juan Fernandez Islands, Chile. *Taxon* 31: 284–89.

Sanders, R. W., T. F. Stuessy & R. Rodriguez. 1983. Chromosome numbers from the flora of the Juan Fernandez Islands. *Am. J. Bot.* 70: 799-810.

Shields, O. 1976. A summary of the oldest ages for the world's islands. *Pap. Proc. R. Soc. Tasmania* 110: 35–61.

Simpson, B. B. 1974. Glacial migrations of plants; island biogeographical evidence. *Science* 185: 698–700.

———. 1975. Glacial climates in the eastern tropical South Pacific. *Nature* 253: 34–36.

Skottsberg, C. 1922. The phanerogams of the Juan Fernandez Islands. *Nat. Hist. Juan Fernandez & Easter Is.* 2: 95–240.

———. 1951. A supplement to the pteridophytes and phanerogams of the Juan Fernandez Islands. *Nat. Hist. Juan Fernandez & Easter Is.* 2: 763–92

———. 1953. The vegetation of the Juan Fernandez Islands. *Nat. Hist. Juan Fernandez & Easter Is.* 2: 793–960.

———. 1956. Derivation of the flora and fauna of the Juan Fernandez and Easter Islands. *Nat. Hist. Juan Fernandez & Easter Is.* 1: 193–438.

Sohmer, S. H. 1978. Morphological variation and its taxonomic and evolutionary significance in the Hawaiian *Psychotria* (Rubiaceae). *Brittonia* 30: 256–64.

Stearns, H. T. 1966. *Geology of the state of Hawaii.* Pacific Books, Palo Alto.

Stebbins, G. L., Jr., J. A. Jenkins & M. S. Walters. 1953. Chromosomes and phylogeny in the Compositae, tribe Cichorieae. *Univ. Calif. Publ. Bot.* 26: 401–30.

Stern, W. L., ed. 1971. *Adaptive aspects of insular evolution.* Washington State University Press, Pullman.

Stuessy, T. F., R. W. Sanders & O. R. Matthei. 1983. *Juania australis* revisted in the Juan Fernández Islands, Chile. *Principes* 27: 71–74.

Valenzuela, A. E. 1978. Pleistoceno marino en isla Robinson Crusoe. *Comm. Dep. Geol. Univ. Chile* No. 22: 32–35.

Vuilleumier, B. S. 1971. Pleistocene changes in the fauna and flora of South America. *Science* 173: 771–80.

Floristic Relationships of New Caledonian Rain Forest Phanerogams[1]

Ph. Morat
J.-M. Veillon
Centre ORSTOM de Nouméa
New Caledonia

H. S. MacKee
Service des Eaux et Forêts Nouméa
New Caledonia

INTRODUCTION

New Caledonia has long attracted phytogeographical analysis (Balansa, 1873; Brongniart, 1874; Fournier, 1874; Brousmiche, 1884; Bernard, 1895; Guillaumin, 1921, 1924, 1928, 1934, 1954, 1964a; Virot, 1956; Baumann-Bodenheim, 1956; Thorne, 1965, 1969; Balgooy, 1960, 1971). Early botanists working on the flora were quick to recognize its distinctiveness. When it became better known, together with other Pacific floras, the high proportion of endemics was recognized both at specific (76% in Guillaumin, 1921; 90% in Baumann-Bodenheim, 1956; Guillaumin, 1964a) and at generic (1% in Brongniart, 1874; 13.5% in Thorne, 1969; 16% in Balgooy, 1960) levels. Floristic links were demonstrated, mainly with Australia and New Guinea (Thorne, 1965; Balgooy, 1960, 1971). The presence of numerous archaic forms of gymnosperms and angiosperms led to recognition of the individuality of the island, as shown by its designation as a "Région canaque" (Guillaumin, 1928, 1934), "New Caledonian Region" (Good,1964; Balgooy, 1960, 1971), "New Caledonian Sub-region" (Thorne, 1963), and "New Caledonian Subkingdom" (Takhtajan, 1969). All of these terms apply to the Territory of New Caledonia, which is treated here and which includes the Loyalty Islands and the Isle of Pines.

Reconsideration of the biogeographic relationships of the Territory is timely.

[1]This paper was presented under the same title at the *13th International Botanical Congress* (Sydney, August 1981), but published only as an abstract; it is here modified and brought up to date.

The most recent publications (Thorne, 1965; Balgooy, 1971) are in fact based on distinctly earlier data (Guillaumin, 1948; Baumann-Bodenheim, 1956). Later botanical advances in New Caledonia and in the Pacific generally justify a new approach, and that is the principal aim of the present article. Attention is here focused on the rain forest, an approach that greatly reduces uncertainties in distinguishing native species from those of relatively recent introduction. The rain forest has the clear advantage that, for obvious climatic and paleoclimatic reasons, relict forms are better conserved in it than elsewhere. Unless seriously disturbed, the forest resists invasion by outside elements, and the true floristic affinities are consequently more evident. Such forests exist in New Caledonia (Fig. 1) on a wide range of substrates: limestone, basalt, schist, greywacke, and clays. They also occur on the ultrabasic rocks (peridotites and serpentinites), which produce exceptional edaphic conditions affecting the whole mineral nutrition of plants (excess of Mg and the heavy metals Ni, Cr, Mn; deficiency of P and K). These special conditions are often invoked (Virot, 1956; Thorne, 1965; Jaffré, 1980) to explain the abundance and persistence of primitive relict forms. The second objective of this study is to check this hypothesis by a quantitative analysis of the species according to their habitats.

METHODOLOGY

All available data have been reconsidered in listing as completely as possible all indigenous species convincingly recorded for New Caledonia. Doubtful taxa and introduced plants are excluded.

The Rain Forest

We deal with the rain forest as defined and mapped in the "Atlas de la Nouvelle-Calédonie" (Morat et al., 1981), thus covering the following formations in their essentially intact state: dense evergreen forest of low and medium altitudes, dense mountain forest, and the edaphic variant represented by evergreen forest on limestone. All other forest or similar formations are excluded: sclerophyll forest, swamp forest, mangrove, tall shrubs, gallery forest, etc.). The rain forest covers 300,000 ha on the main island plus 100,000 ha in the Loyalties and the Isle of Pines, equivalent to 22% of the total surface of the Territory of New Caledonia.

Species of the Rain Forest

All phanerogams are included whose occurrence in undisturbed rain forest is constant and unequivocal, regardless of their abundance, their bulk, their association with any particular stratum or biological type or their possible presence in other formations. The stipulation of undisturbed forest excludes all "secondary" species that, by colonizing forest edges or openings where a large tree has fallen, penetrate accidentally and temporarily into the forest, unless they also exist (even rarely) within the intact formation. *Rubus moluccanus,* for example, a species very probably native but known in forest only in disturbed and open situations, is omitted. A few heliophilic species are, however, included (e.g. *Duboisia myoporoides,* a shrub or small tree appearing abundantly whenever an opening, track, or clearing is made) if definitely known, though rare, in normal rain forest. Other species characteristic of adjacent plant communities, such as coastal scrub or gallery forest, occur exceptionally in rain forest. Their presence, due to an acci-

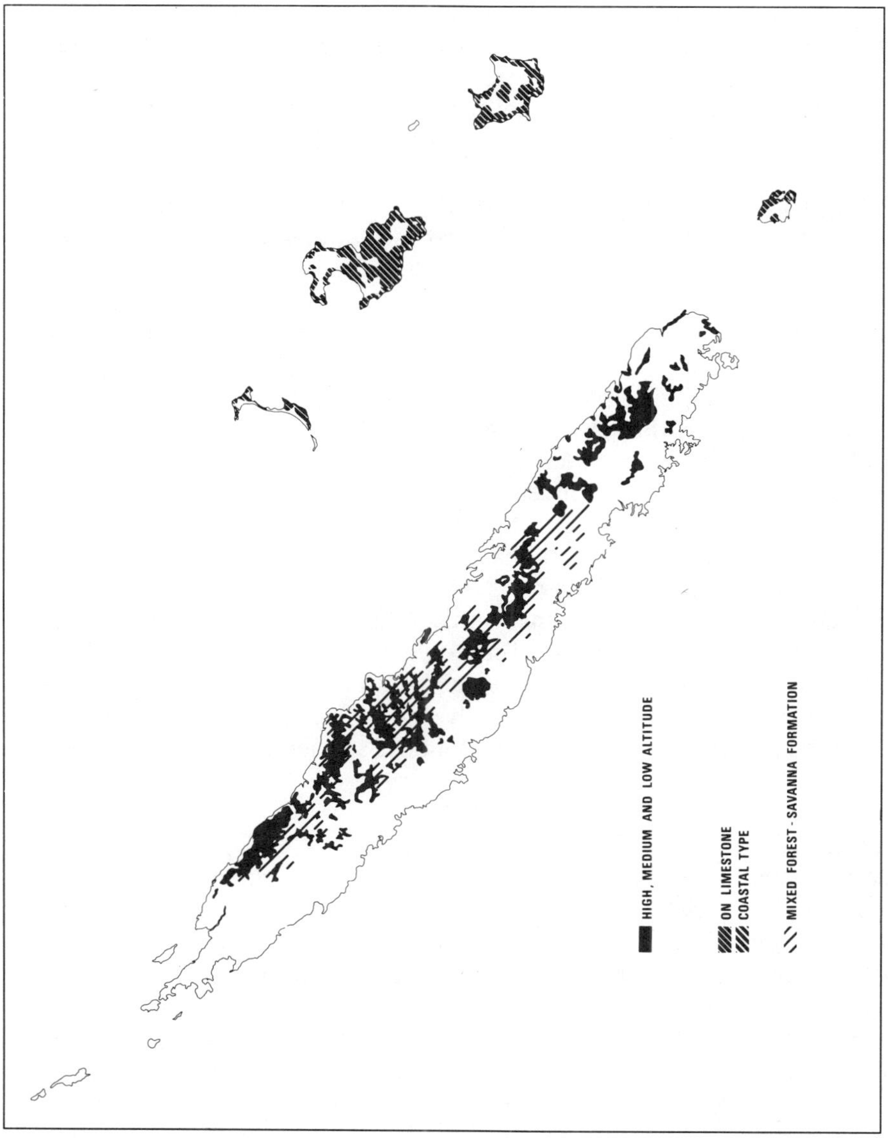

Fig. 1. Rain forest in New Caledonia.

dental introduction or to a transitory situation (edge of a rapidly advancing rain forest), is only sporadic. Such species are excluded, e.g. *Eriaxis rigida* (Orchidaceae), *Gahnia neocaledonica* (Cyperaceae), *Joinvillea* spp. (Flagellariaceae), *Allophylus cobbe* (Sapindaceae), *Malaisia tortuosa* (Moraceae), *Gouania leratii* (Rhamnaceae), *Lindenia vitiensis* (Rubiaceae), *Heliconia* sp. (Musaceae), etc.

Precise delimitation of the rain forest is often difficult, and where it passes progressively into tall shrubbery, gallery forest or secondary forest becomes essentially subjective. Each doubtful taxon has therefore been considered individually and those retained that best meet the given criteria, in the light of existing literature, notes on herbarium sheets, and above all our own observations and impressions.

Rain forest species are listed in Appendix 1. *Forest* species (F) are those found only in rain forest. Species also found in other formations are referred to as *mixed* (M). The following groups are defined to determine the effect of substrates on floristic diversity and on the level of endemism: species growing only on ultrabasic substrates (U), species absent from ultrabasic substrates (A), species growing on both types of substrate (I). Subspecies are not considered. If a species has subspecies of different edaphic status (U or A), the species as a whole is classed as I. The following categories thus exist among strictly forest species and those present also in one or more other formations: FU, FA, FI, MU, MA, MI.

The substrate has little if any influence on the distribution of epiphytes. Possible associations between epiphytes and their hosts are too little known to offer significant correlations with substrates. Parasitic Loranthaceae are, however, much more closely associated with their hosts, whose edaphic status they receive.

With available data, it is sometimes hard to assign a definite edaphic status to particular species. Some parts of the Territory still lack detailed geological maps, and even these cannot show all small-scale variations. Published data and information on herbarium sheets, used critically and with caution, have been further complemented by our own observations. Doubtful species are left without edaphic status.

Genera and Families of the Rain Forest

These are established using the lists of species (see Appendix 1) occurring exclusively (F) or partially (M) in this formation. Certain genera (*Acridocarpus, Erythroxylum, Grevillea, Isachne,* etc.) are excluded as having no forest species in New Caledonia although represented elsewhere in rain forest.

The genus is adopted as the working unit in phytogeographical analyses, for the reasons cited by Balgooy (1971): more homogenous comparisons, relative numerical stability at different dates, definition generally more stable (sounder) than that of species. In some cases, however, the subgenus or section is considered more significant than the genus in establishing floristic relationships: e.g. subgenus *Antholoma* rather than *Sloanea,* section *Scaevola* less its two pantropical species instead of the genus *Scaevola.* Certain genera whose limits seem ill-defined are considered together: e.g. *Caryophyllus* and *Jambosa* with *Syzygium.*

Rain forest genera are listed in Appendix 2. This list does not agree completely with the list of species (Appendix 1), because of nomenclatorial complications arising when authors publish new genera or new synonymy without making all the new combinations that seem to be required at the specific level. Examples include *Gymnostoma* (Casuarinaceae) for some species of *Casuarina* (Johnson, 1980); *Eucarpha* and *Virotia* (Proteaceae) for some species of *Knightia* and *Macadamia* (Johnson and Briggs, 1975); *Cassine* (Celastraceae) for *Elaeodendron* (Ding Hou, 1962); *Anthocarapa* for *Amoora* (Pennington and Styles, 1975); *Pisonia* (Nyctaginaceae) for *Calpidia* (Stemmerik, 1964); and *Lethedon* (Thymeleaceae) for *Microsemma* (Kostermans, 1963). In addition, generic changes are sometimes proposed in unpublished form by the specialists concerned: e.g. transfer to *Medicosma* of *Melicope* p.p. (Rutaceae) (T. G. Hartley, pers. commun.); or revival of *Tristaniopsis,* a genus long reduced to synonymy under *Tristania* (Myrtaceae) (J. W. Dawson, pers. commun.). In neither case is it appropriate for us to make the combinations instead of the authors proposing new generic limits. Such taxa appear, therefore, under their current binomials in the specific list and under the new generic name in the lists of geographical distribution. The quantitative results are not affected, because they are derived separately from each list.

Sources of Information

Sources are too numerous to be cited completely; the most important have already been mentioned or appear in the Literature Cited. For the nomenclatural status of taxa and for their distribution within New Caledonia in relation to edaphic factors, the sources are of three types: the existing literature, the herbarium of Centre ORSTOM (Nouméa), and personal observations. The literature is considerable and scattered through many journals, floras, lists, and compendia. We have first considered the most recent floras and revisions, in particular the *Flore de la Nouvelle-Calédonie et Dépendances* in which 23 families relevant to the present study have appeared (four Gymnosperm families, Sapotaceae, Proteaceae, Lauraceae, Epacridaceae, Orchidaceae, Solanaceae, Acanthaceae, Bignoniaceae, Symplocaceae, Flacourtiaceae, Icacinaceae, Corynocarpaceae, Apocynaceae, Elaeocarpaceae, Monimiaceae, Amborellaceae, Atherospermataceae, Trimeniaceae, Chloranthaceae). Numerous families represented in the rain forest have also been studied recently in whole or in part. These are Anacardiaceae (*Euroschinus*), Araliaceae (*Arthrophyllum*), Balanopaceae, Celastraceae (*Salaciopsis*), Chrysobalanaceae, Cunoniaceae (*Acsmithia*), Cyperaceae (*Baumea, Costularia, Gahnia*), Euphorbiaceae (*Austrobuxus, Baloghia, Bocquillonia*), Fagaceae, Loganiaceae (*Geniostoma*), Meliaceae, Mimosaceae (*Albizia, Pithecellobium, Serianthes*), Moraceae, Myrtaceae–Leptospermoideae, Oncothecaceae, Palmae, Pandanaceae, Pittosporaceae, Rutaceae, Rubiaceae (*Tarenna*), Winteraceae. Recent reliable revisions cover about 65% of the flora here studied. For the rest, we have used the results, modified where necessary, of the Mission Franco-Suisse (Guillaumin, 1957, 1962, 1964b, 1967, 1974). Earlier sources include numerous and varied publications by Schlechter, E. G. Baker, S. Moore, Rendle, Guillaumin, Virot, Thorne, and others. Obvious errors have been corrected, such as the superfluous generic name *Merismostigma* S. Moore based on a species of *Coelospermum* Bl.

The local flora is well represented in the herbarium (50,000 specimens) of Centre ORSTOM (Nouméa). This represents another significant source of information, used critically, on the edaphic status of species. It is also distinctly useful in questions of nomenclature, thanks to many determinations by specialists on families not yet treated in *Flore de la Nouvelle-Calédonie.*

Geographical distributions outside New Caledonia are taken exclusively from the literature. The starting point was the remarkable work of Balgooy (1971), modified and brought up to date in view of new data in later publications already cited and in recent volumes of *Flora Malesiana,* in Pacific Plant Areas, and in the first volume of *Flora Vitiensis Nova.* Finally, the eighth Edition of the *Dictionary of Flowering Plants and Ferns* (Shaw, 1973) and the *Dictionary of Australian Plant Genera* (Burbidge, 1963) served to settle a few outstanding questions.

Geographical Regions

The phytogeographical regions of Balgooy (1971) are modified in relation to our concern with New Caledonia rather than the whole Pacific.

Africa includes the whole continent, plus Madagascar, the Mascarene Islands and the Seychelles; the islands are distinguished by a number (see Appendix II) if a New Caledonian genus in this subdivision is limited to one or more of them.

Asia includes the whole continent (except the Malay Peninsula) plus Japan and Taiwan.

Malesia covers the geographical area of *Flora Malesiana* (including the Philippines), except New Guinea, the Bismarck Archipelago, and the Aru Islands.

New Guinea plus the Bismarck Archipelago and the Aru Islands.

Australia (including Tasmania).

Pacific is divided into nine subregions: (1) *Solomon Islands,* including Bougainville, which geographically belongs here. (2) *New Hebrides,* in the geographical sense, i.e. including the present Vanuatu and the Santa Cruz group. (3) *Lord Howe.* (4) *Norfolk.* (5) *New Zealand* and associated islands (Chathams, Kermadecs, Stewart). (6) *Fiji.* (7) *North Pacific,* an arbitrary unit grouping the following islands: Bonins, Carolines, Marianas, Marshalls, Tuvalu, Phoenix, Tokelau, Hawaii. (8) *Polynesia,* French Polynesia plus the Cook Islands. (9) *Samoa and Tonga* plus Niue, Wallis, and Futuna.

America includes North and South America, which are considered together as being little involved apart from pantropical genera.

The Phoenix, Tokelau and Tuvalu Islands are associated with the North Pacific, on account of their geographical position and because all the relevant New Caledonian genera that occur there are found also in the Carolines or the Mariannas. These genera are *Allophylus, Calophyllum, Calpidia, Epipremnum, Guettarda, Hemigraphis, Hernandia, Intsia, Macaranga, Morinda, Ochrosia, Pandanus, and Planchonella.*

Distributions

The distributional categories defined by Balgooy (1971) and their content are also somewhat modified in the light of recent taxonomic studies and particularly because of the more limited scope of the present work.

Endemic genera (E) are limited to New Caledonia, the Loyalties and the Isle of Pines.

New Caledonian (subendemic) genera (B) are those for which New Caledonia is a primary center of diversification, having the great majority of the species (usually more than 3/4 of the total: *Meryta* 19/25; *Hedycarya* 9/12; *Coronanthera* 9/11; *Dizygotheca* 17/17, one species being also in the New Hebrides). For such genera, New Caledonia usually has a central position (*Balanops, Baloghia, Meryta*) but is sometimes excentric (*Coronanthera, Acianthus*) or even at the limit of their area (*Artia, Delarbrea*). Other genera for which New Caledonia is a secondary center of distribution with many species are not, however, considered "New Caledonian," either because their areas are too wide and scattered for it to be their center of dispersion (*Araucaria,* 13/19; *Soulamea,* 7/14, with one species in the Seychelles, one in Malesia, and one in Polynesia; *Euroschinus,* 4/6; *Agathis,* 5/20; *Campynema,* 1/2; *Arthrophyllum,* 10/31; and also *Dianella, Geniostoma, Phyllanthus, Pittosporum,* etc.) or because their origin may well be elsewhere (*Argophyllum, Austrobuxus, Geissois,* etc.).

Pacific genera (L), are found in one or more island groups in the Pacific but are absent or rare in America, Asia, Australia, Malesia, and New Guinea. Examples are *Cyphosperma* and *Earina.* This type of distribution applies to *Ascarina* [12 species one of which is in Australia and Malesia and another (Jeremie, 1980) in Madagascar]; *Dracophyllum* (50 species, two of which occur in Tasmania and one in continental Australia); *Storckiella* [five species, one (B. P. Hyland, pers. comm.) Australian]; *Astelia* (a few scattered species in Australia, New Guinea, and South America); *Tapeinosperma* (39 species in New Caledonia, 11 in Fiji, 2–3 others in Australia, Malesia, and New Guinea); *Scaevola,* section *Scaevola,* less the two littoral species (extra-Australian but with two species in Malesia and New Guinea). More complex distributions are represented by *Serianthes* [four species in New Caledonia, five in other Pacific islands (Solomons, New Hebrides, Fiji, Polynesia), four in New Guinea or Malesia] and *Austromyrtus* (nine Australian species and 28 others in the Pacific, including 12 in New Caledonia). Their inclusion in this group is justified by the weight of the Pacific species. Similarly, *Campynema* (one species endemic in New Caledonia, one endemic in Tasmania) is better placed here than among the Australian genera.

Subantarctic Pacific genera (J) included here have disjunct distributions in New Caledonia and in the mountains of Australia, New Zealand, and South America. Few such genera occur in New Caledonia, owing to its geographical position and the absence of high mountains. The best examples are *Araucaria* and *Nothofagus;* other examples are *Decussocarpus, Dacrydium,* and *Libocedrus* (three species in New Caledonia, two in New Guinea, two in New Zealand, and one in the Andes).

Subantarctic genera (K) have distributions extending from the Pacific to the south of the Indian Ocean, such as *Cordyline, Dianella* and *Podocarpus,* the last included here in spite of its vast distribution. *Soulamea* (one species in the Seychelles, one in Malesia, one in Polynesia, and seven in New Caledonia) is placed in this group, as is *Cunonia* whose extraordinary distribution includes 20 New Caledonian endemics and a single other species, *C. capensis,* in South Africa. They probably belong to an ancient Gondwanian element.

Australian genera (H) include those occurring mainly in Australia, e.g. *Hibbertia* and *Styphelia;* and even *Duboisia* and *Niemeyera,* each with two species in Australia and one in New Caledonia.

Australian-Papuan genera (I) are centered in Australia and New Guinea, such as *Agathis, Corynocarpus, Euroschinus, Flindersia,* and *Sphenostemon. Cupaniopsis* is placed here; 26 of its 60 described species are New Caledonian, but there are also many in New Guinea and some in Australia.

Malesian-Papuan genera (G) have their center of diversification in Malesia or New Guinea (with the Solomons and Fiji) or covering both these areas, and they are poorly represented in Asia, Australia, and Polynesia. Typical examples are *Bureavella* and *Neuburgia.*

Indo-Malesian genera (F), mainly Asian and Malesian, are little represented in Australia and the Pacific, their distributions often ending in New Caledonia or Fiji. This group, which oddly contains 12 orchid genera, is exemplified by *Dacrycarpus, Procris, Acanthephippium, Appendicula,* and *Coelogyne.*

Indo-Australian genera (D) occur in continental Asia, Malesia, Australia, and the Pacific but are absent or almost so in Africa, such as *Agapetes* (*sensu* Sleumer, 1966) *Desmos, Neisosperma,* and *Pachygone.*

Paleotropical genera (G) are absent only from America.

Pantropical genera (A) occur in all major areas of the tropics, including America.

RESULTS

The New Caledonian rain forest, as defined above, has 1511 species of phanerogams, in 365 genera and 108 families (Appendix 1). A comparison with the total native phanerogamic flora is given in Table 1, showing that the rain forest has approximately 46% of the genera and species and 59% of the families represented in the flora. With almost half the species, it is the richest formation in the Territory.

The level of specific endemism for the entire native phanerogamic flora is about 76%, well below the earlier estimates of Baumann-Bodenheim (1956) and Guillaumin (1964a), whose excessive figure of 90% has often been repeated. Endemicity is highest in the rain forest at all systematic ranks. Indeed, all five endemic families occur there.

Floristic Affinities

The geographic distribution of rain forest genera is shown in Table 2. Genera of widest distribution (A, C, D) represent 45.1% of the rain forest flora, and genera with Malesian distribution total 9.6%. The Australian element (H) is misleadingly small (3%) (see below), but reaches the more significant level of 10.4% when Australian-Papuan genera (I) are added. The complete absence of American genera is notable; this element, though not abundant, does occur in New Caledonia *(Lindenia,* Rubiaceae) but is absent from the rain forest. Southern genera (Pacific-Subantarctic and Subantarctic) are mainly gymnosperms: *Araucaria, Dacrydium, Decussocarpus, Libocedrus, Podocarpus, Prumnopitys.* This old Gondwanian relict element preserved in the rain forest indicates very ancient relationships between the territories where it still exists. Pacific genera (L) are few (4.1%). However, the most striking feature of Table 2 is the large number of endemic genera (83, over 22% of the total). Adding the 15 New Caledonian subendemic genera gives 97 genera (26.8%) confined to or centered in New Caledonia, and these belong to the most primitive families: Amborellaceae, Monimiaceae, Myricaceae, Oncothecaceae, Palmae, Paracryphiaceae, Proteaceae, Winteraceae, etc.

Table 1. Comparison of endemicity of the rain forest flora with the total New Caledonian flora.

	Species			Genera			Families	
	Total	Endemic	% Endemic	Total	Endemic	% Endemic	Total	Endemic
Native flora	3256	2474	76.0	787	108	13.7	182	5
Rain forest flora	1511	1358	89.8	365	82	22.4	108	5
Rain forest flora as % of total flora	46.4			46.4			59.3	

Some endemic genera (*Basselinia, Myodocarpus, Zygogynum*) have a wide range of species, suggesting that their evolution has continued during a long period of isolation. Most, however, are oligotypic, many even monotypic, and appear as relicts at the end of an evolutionary line.

Floristic affinities may be evaluated in several ways. The number of genera shared between New Caledonia and various other regions is shown in Table 3. The number in common is high for New Guinea, Malesia, and Australia, and somewhat less for Fiji, continental Asia, and the New Hebrides. These high figures are inflated by the 68 pantropical genera, which mask the true affinities and whose presence in various regions due to efficient dispersal does not necessarily reflect floristic relationships. The comparison may thus be improved by eliminating the pantropical genera, also shown in Table 3.

There is a general reduction in shared genera when pantropical genera are not considered. This reduction is not, however, uniform (Fig. 2): 18% for America, which drops from 13th to 15th place; 8 to 12% for Africa, Asia, and the North Pacific; 7 to 8% for Malesia, New Guinea, Australia, the Solomons, the New Hebrides, Fiji, and Tonga-Samoa, the first six of these, after some changes in relative position, now taking the lead; 3 to 5% for Lord Howe, Norfolk, and Polynesia; and 1% for New Zealand. The last figures are easily explained as pantropical genera are few in these essentially extratropical areas.

Table 2. Distribution of New Caledonian rain forest genera.

Type	Code	Number of genera	% of total	
Pantropical	A	68	18.6	45.1
Paleotropical	C	51	13.9	
Indo-Australian	D	46	12.6	
Indo-Malesian	F	17	4.7	9.6
Malesian-Papuan	G	18	4.9	
Australian	H	11	3.0	10.4
Australian-Papuan	I	27	7.4	
Pacific-Subantarctic	J	6	1.7	3.8
Subantarctic	K	8	2.2	
Pacific	L	15	4.1	
Endemic	E	83	22.7	26.8
New Caldonian (subendemic)	B	15	4.1	
	Total	365		

With the exclusion of the pantropical element, we see floristic affinities primarily with New Guinea (60%), Australia (57%), Malesia (56%) and then with Fiji (44%), the Solomons (42%), and the New Hebrides (40%). New Zealand (14.5%) is well behind. One could similarly eliminate other widespread elements, for example the paleotropical genera, but this approach is limited as the absence of a genus from the whole American continent is in itself biogeographically significant. It is clear that the fewer the regions containing a taxon, the greater the importance of distribution in these regions in showing their floristic affinity. In the extreme case, a genus found in only two territories implies closer floristic links between them than does a genus common to the same two regions but also occurring in other regions, particularly if these are extensive and geographically distant. Floristic affinities may thus be approached by analyzing the genera shared by New Caledonia and various other phytogeographic regions.

Table 4 presents such an analysis for genera shared by New Caledonia and only one other region. In contrast with the results in Table 3, Australia heads the list with seven genera [*Argophyllum, Canarium* (section *Canariellum*), *Campynema, Duboisia, Medicosma* (T. G. Hartley, pers. commun.); *Niemeyera, Virotia*]; compared with three for New Guinea (*Hunga, Periomphale, Sloanea* subgen. *AnthoFloma*); and two each for Fiji (*Acmopyle, Cyphosperma*) and the New Hebrides (*Cyclophyllum, Dizygotheca*). Africa appears in Table 4 with the genus *Cunonia* which has one species, *C. capensis*, in South Africa, a striking example of relict Gondwanian distribution.

Tables 5, 6, 7, 8, and 9 show the genera shared exclusively by New Caledonia with 2, 3, 4, 5, and 6 other territories, respectively. To go further would be unprofitable, as we then encounter the very widespread genera whose distributions are less and less significant. In each of these tables, Australia, followed fairly closely by New Guinea, has most genera in common with New Caledonia. An attempt may be made to quantify these affinities by assigning to the different territories a correlation coefficient proportional to the number of common genera and inversely proportional to the number of territories in which they occur. For example, five genera (*Anthocarapa, Euroschinus, Eustrephus, Geijera* and *Sphenostemon*) occur, apart from New Caledonia, only in the two territories Australia and New Guinea, each of which is assigned the coefficient 5/2 = 2.5. Coefficients decrease from Table 4 through 9, as the denominator increases with the number of territories. Coefficients are negligible for more than six other territories.

Adding all the coefficients obtained for the territories appearing in Tables 4 to 9 gives the totals shown in Table 10 (converted to percentages in the second

Table 3. New Caledonian rain forest genera present in other regions.

	Africa	Asia	Malesia	New Guinea	Australia	Solomons
All genera (365)	130	187	234	245	232	183
% of New Caledonian	35.6	51.2	64.1	67.1	63.6	50.1
Genera minus pantropical (297)	69	121	166	178	168	124
% of New Caledonian	23.2	40.7	55.9	59.9	56.6	41.8

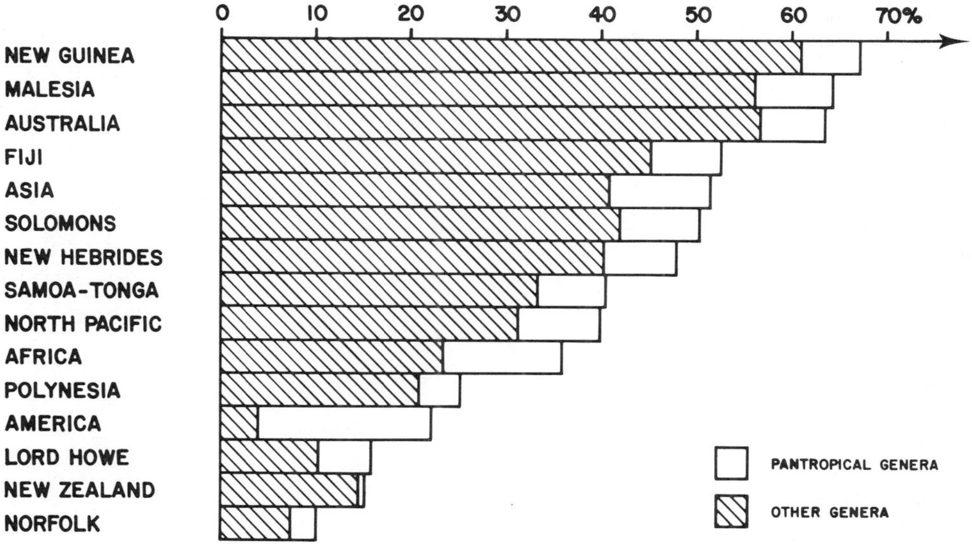

Fig. 2. Graphic representation of Table 3.

column). This confirms the positions of Australia (27%) and New Guinea (20%) as the territories with the closest floristic affinities, followed at some distance by Malesia (11.7%) and then Fiji (9.8%), the New Hebrides (7.6%), and the Solomons (6.8%). New Zealand has only 4.3%. It is no surprise to see thus confirmed and amplified the conclusions reached for the whole New Caledonian flora (Thorne, 1965).

Theoretical objections may be raised that these coefficients depend on the delimitation of the territories chosen. Thus, the exclusive presence of a genus outside New Caledonia in two regions A and B, of very different size (e.g. Australia and Norfolk) gives each the coefficient 1/2 = 0.5. The large region A could be divided into A1, A2, A3 (e.g. Queensland, Tasmania, the rest of Australia). If the genus occurs in all three, they each have the coefficient 1/4 = 0.25, as also now has B, as against a total of 0.75 for A when the coefficients for the three sub-units are added. The system chosen thus seems to favour small territories at the expense of larger areas. This does not in fact reflect the true situation. A genus restricted to a few phytogeographical regions is most unlikely to occur throughout a vast

Table 3. Extended.

New Hebrides	Lord Howe	Norfolk	New Zealand	Fiji	North Pacific	Poly-nesia	Samoa-Tonga	America
174	58	37	56	189	145	92	148	80
47.7	15.9	10.1	15.3	51.8	39.7	25.2	40.5	21.9
119	31	22	43	132	93	62	99	12
40.1	10.4	7.4	14.5	44.4	31.3	20.9	33.3	4.0

Table 4. Genera limited to New Caledonia and one other territory.*

	Africa	Asia	Malesia	New Guinea	Australia	Solomons	New Hebrides	Lord Howe	Norfolk	New Zealand	Fiji	North Pacific	Polynesia	Samoa-Tonga	America	Distribution type	New Caledonia
Periomphale				1												B	2
Campynema					1											L	
Dizygotheca							1									B	2
Canarium																	
(Sect. *Canariellum*)					1											H	2
Hunga				1												B	2
Cunonia	1															K	2
Sloanea																	
(s.g. *Antholoma*)				1												G	
Argophyllum					1											H	2
Cyphosperma											1					L	
Acmopyle											1					L	
Virotia					1											B	2
Cyclophyllum							1									B	2
Medicosma					1											I	
Niemeyera					1											H	
Duboisia					1											H	
Belliolum						1										I	

* Symbols: 1—present in area, 2—species concentration in area; 3, 4, 5 (column headed "Africa"): Madagascar, Mascarenes, Seychelles, respectively. Letters (column headed "Distribution type"): A—Pantropical; B—New Caledonian; C—Paleotropical; D—Indo-Australian; E—Endemic; F—Indo-Malayan; G—Malesian-Papuan; H—Australian; I—Australian-Papuan; J—Pacific Subantarctic; K—Subantarctic; L—Pacific.

territory such as Africa, America, Australia, or New Guinea. If the divisions are accepted, it will probably occur only in A1, A2, or A3, which gives the same results as in the first case.

The distribution area of the great majority of genera, apart from the southern Gondwanian element (Subantarctic and Subantarctic Pacific), is to the west of New Caledonia. Only two genera, *Crossostylis* and *Earina,* are centered to the east and suggest a Pacific Island origin.

Summing up, the results confirm that the rain forest genera come overwhelmingly from the northwest, particular floristic links existing with Australia and New Guinea and at a lower level with Malesia, Fiji, the New Hebrides, and the Solomons; New Zealand is well down in seventh place.

Origin of the New Caledonian Flora

The geological history of New Caledonia must now be evoked in order to relate these results to the origin of its flora. This history, until recently obscure, is now clear in broad outline as a result of recent syntheses (Raven and Axelrod, 1972, 1974; Stevens, 1977; Raven, 1979; and, particularly, Paris, 1981).

During the Permian (corresponding to the oldest dated rocks of New Caledonia) and the whole Mesozoic Era, a series of archipelagoes of varying size occupied

Table 5. Genera limited to New Caledonia and two other territories.*

	Africa	Asia	Malesia	New Guinea	Australia	Solomons	New Hebrides	Lord Howe	Norfolk	New Zealand	Fiji	North Pacific	Polynesia	Samoa-Tonga	America	Distribution type	New Caledonia
Euroschinus				1	1											I	2
Richella			1								1					G	
Storckiella					1						1					L	
Dubouzetia			1	1												G	
Baloghia					1				1							B	2
Coronanthera					1	1										B	2
Anthocarapa				1	1											I	
Piliocalyx							1				1					B	2
Drymoanthus					1					1						L	
Eustrephus				1	1											I	
Falcatifolium			1	1												G	
Geijera				1	1											I	
Sphenostemon				1	1											I	
Lethedon					1		1									B	2

* Symbols as in Table 4.

the site of the present Norfolk Ridge between New Guinea and New Zealand. This arc situated near the edge of the Australian Plate (part of the Gondwanian mass) was connected to it by, perhaps partial, land connections and was, according to paleomagnetic data, in the same relative position to the plate as the present Great Barrier Reef is to Australia. The whole Australian Plate was, however, at much higher latitudes (15–55 degrees further S) than now and thus subjected to

Table 6. Genera limited to New Caledonia and three other territories.*

	Africa	Asia	Malesia	New Guinea	Australia	Solomons	New Hebrides	Lord Howe	Norfolk	New Zealand	Fiji	North Pacific	Polynesia	Samoa-Tonga	America	Distribution type	New Caledonia
Balanops					1		1				1					B	2
Deplanchea			1	1	1											I	
Libocedrus				1						1					1	J	
Dracophyllum					1			1		2						L	
Flindersia			1	1	1											I	
Amylotheca				1	1									1		I	
Austromyrtus					1		1				1					L	
Chamaeanthus		1	1				1									G	
Stenocarpus			1	1	1											I	2
Beccariella			1	1	1											D	
Bubbia				1	1			1								I	

* Symbols as in Table 4.

Table 7. Genera limited to New Caledonia and four other territories.*

	Africa	Asia	Malesia	New Guinea	Australia	Solomons	New Hebrides	Lord Howe	Norfolk	New Zealand	Fiji	North Pacific	Polynesia	Samoa-Tonga	America	Distribution type	New Caledonia
Desmos			1	1	1						1					D	
Pagiantha		1	1			1					1					F	
Araucaria				1	1				1						1	J	2
Corynocarpus				1	1		1			1						I	
Acsmithia			1	1	1						1					I	2
Geissois					1	1	1				1					H	2
Costularia	1		1	1	1											C	
Austrobuxus			1	1	1						1					H	2
Nothofagus				1	1					1					1	J	
Apodytes	1	1	1		1											K	
Arthropodium	3			1	1					1						H	
Tristaniopsis		1	1	1	1											D	
Xanthostemon			1	1	1	1										I	2
Acianthus				1	1	1				1						H	2
Cleisostoma		1	1	1	1											G	
Rhynchophreatia			1	1	1							1				G	
Trachoma		1	1		1		1									D	
Coelospermum		1	1	1	1											D	
Bureavella			1	1	1	1										G	

* Symbols as in Table 4.

a colder climate. The island arc suffered intense tectonic movements in the late Jurassic and early Cretaceous (130 million years ago), when the slow northward drift of the Australian mass provoked the Rangitata Orogeny and the opening of the Tasman Sea. Stevens (1977) considers this period the most favorable for land connections and migration of biota between the New Caledonia-New Zealand area and the Gondwanian mass (New Guinea, Australia, Antarctica, South America). Local dislocations followed by various foldings and metamorphic changes produced first the framework of the Chaîne Centrale and then the rest of New Caledonia in its present form. The Tertiary saw a series of submersions, especially in the Paleocene and the middle Eocene (the date of the first reef formation, which marks a warmer period due to a climatic change or the passage of New Caledonia into the tropics) alternating with general emergences. In spite of geological arguments, these submersions can never have been complete, since floral distributions indicate that considerable surface must have remained above water and served as refuges.

A major event of the upper Eocene was the outpouring of peridotites, which covered almost the whole island to a thickness of about 2000 m. This very slow phenomenon, begun under water and continued in the atmosphere, occurred about 38 million years ago and had very important effects on plant and animal life. Several points may be emphasized.

Table 8. Genera limited to New Caledonia and five other territories.*

	Africa	Asia	Malesia	New Guinea	Australia	Solomons	New Hebrides	Lord Howe	Norfolk	New Zealand	Fiji	North Pacific	Polynesia	Samoa-Tonga	America	Distribution type	New Caledonia
Delarbrea			1	1	1	1	1									B	2
Pandorea			1	1	1	1		1								I	
Hibbertia	1		1	1	1						1					H	
Hedycarya					1		1			1	1			1		B	2
Rhodomyrtus (*Archirhodomyrtus*)		1	1	1	1	1										I	
Olea	2	1	1		1			1								C	
Acanthephippium		1	1	1							1			1		F	
Earina							1			1	1		1	1		L	
Pterostylis			1	1	1			1		1						H	
Decussocarpus			1	1		1					1				1	J	
Prumnopitys			1	1	1					1					1	J	
Crossostylis						1	1				1		1	1		L	
Polyosma		1	1	1	1	1										D	
Quintinia			1	1	1		1			1						D	

* Symbols as in Table 4.

(1) At the end of the lower Cretaceous the bulk of the Australian continent moved northwards faster than its eastern edge, thus isolating the Norfolk Ridge with its associated islands and cutting all possible land connections. This explains the absence of native land mammals except bats and the great poverty in some other zoological groups: fresh-water fish, amphibians, land snakes, and some terrestrial invertebrates. (A land snake of the family Boidae exists in the Loyalty Islands, but its status as a member of the native fauna is uncertain. A small land

Table 9. Genera limited to New Caledonia and six other territories.*

	Africa	Asia	Malesia	New Guinea	Australia	Solomons	New Hebrides	Lord Howe	Norfolk	New Zealand	Fiji	North Pacific	Polynesia	Samoa-Tonga	America	Distribution type	New Caledonia
Agathis			1	1	1		1			1	1					I	2
Gymnostoma (*Casuarina*)			1	1	1	1	1				1					G	2
Codiaeum			1	1	1	1	1				1					I	
Scaevola (Sect. *Scaevola*)			1	1		1					1	1	1			L	
Tapeinosperma			1	1	1	1	1				1					L	
Chrysoglossum		1	1	1		1	1				1					F	

* Symbols as in Table 4.

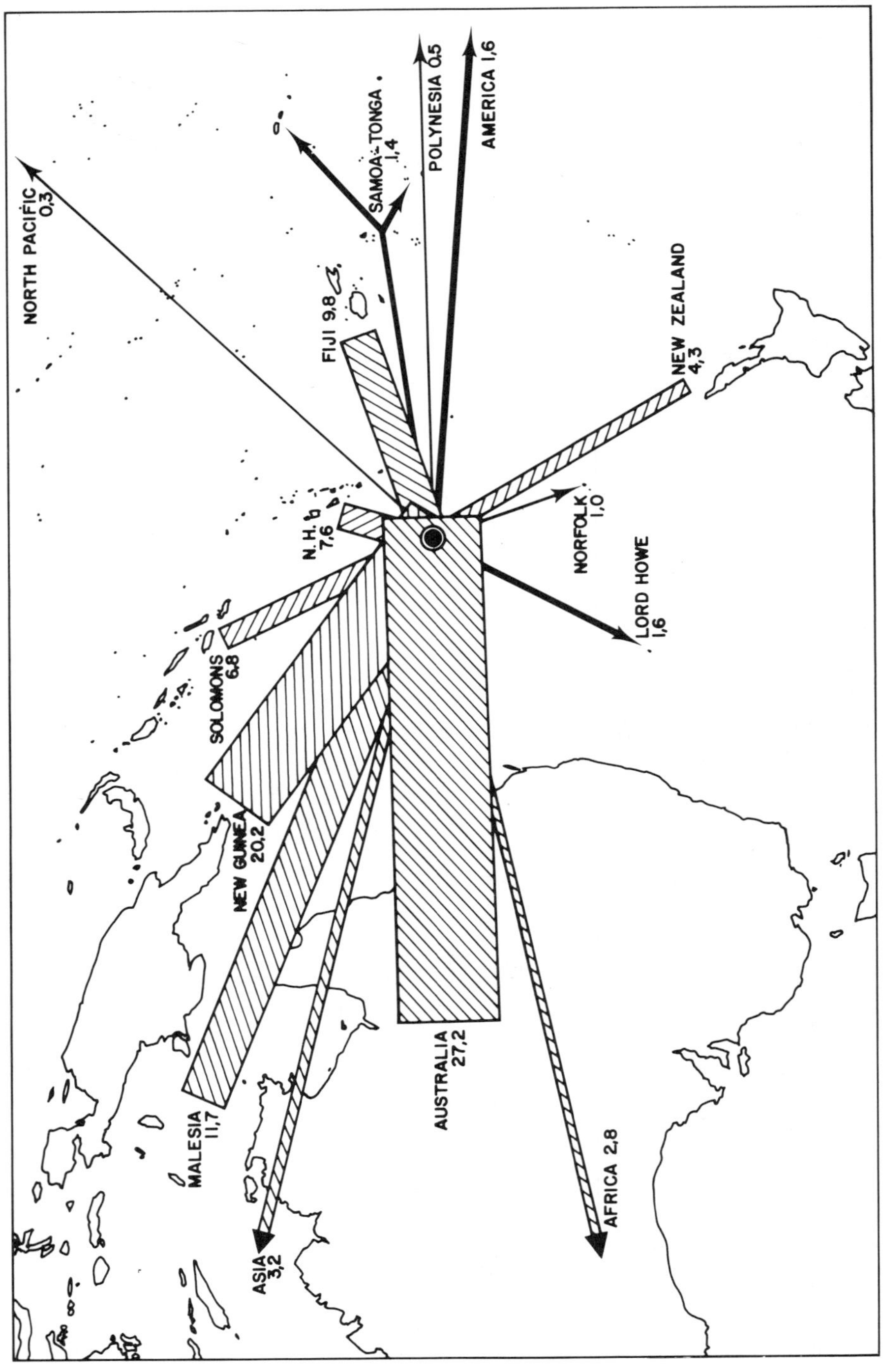

Fig. 3. Floristic relationships of the New Caledonian rain forest.

Table 10. Floristic affinities with New Caledonia as shown by correlation coefficients of regions.

Rank	Region	Coefficient*	Percentage†
1	Australia	21.19	27.2
2	New Guinea	15.77	20.2
3	Malesia	9.13	11.7
4	Fiji	7.63	9.8
5	New Hebrides	5.9	7.6
6	Solomons	5.27	6.8
7	New Zealand	3.22	4.3
8	Asia	2.49	3.2
9	Africa	2.15	2.8
10	Lord Howe	1.26	1.6
11	America	1.23	1.6
12	Samoa-Tonga	1.13	1.4
13	Norfolk	0.75	1.0
14	Polynesia	0.40	0.5
15	North Pacific	0.25	0.3
	Total	77.87	100.0

* Sum of correlation of coefficients for each region as derived from data in Tables 4–9.
† Based on total of summed coefficients for all regions equaling 100%.

snake of the family Typhlopidae is native on the main island of New Caledonia.) Some authors (Gressitt et al., 1963; Raven and Axelrod, 1972, 1974) maintain, however, that during the upper Cretaceous the affinities (gymnosperms, Proteaceae, *Nothofagus*) of the New Caledonian flora with southern lands imply that migration by direct connections or short distance dispersal continued up to the end of the Cretaceous, 60 million years ago. This would explain the continental aspects of the New Caledonian flora. It was long disputed whether migration was from north to south or in the opposite direction, but better knowledge of Antarctic paleoclimates supports a movement from south to north (Cranwell, 1964; Raven and Axelrod, 1974), as at that time New Zealand must have been close to Antarctica and South America.

(2) Migration routes, whether north-south or inversely according to the climatic conditions at different epochs, were reduced and cut during the long isolation following the dislocations and orogenic movements (Cretaceous). This led to a high level of endemism and, in certain genera (*Agathis, Araucaria, Argophyllum, Geissois, Nothofagus, Phyllanthus, Pittosporum, Psychotria, Stenocarpus, Xanthostemon,* etc.), surprisingly active speciation in view of the small surface of the island. This prolonged isolation also explains why many typically Australian taxa—phyllodineous *Acacia, Banksia, Eucalyptus, Macrozamia,* and ten or so forest genera of Proteaceae in the east and particularly the north-east of the continent—are absent or very poorly represented here, as are families common to Australia and New Guinea (Eupomatiaceae, Himantandraceae) or present in both though also represented further afield (Cochlospermaceae, Hamamelidaceae, Myristicaceae, Ochnaceae, Theaceae, etc.).

This ancient flora, diversified by much local evolution, received new elements that from the Cretaceous onwards can only have come by long distance transport.

Table 11. Distribution of rain forest species on different substrates. (Percentage is for 1371 species of known edaphic status.)

	Ultrabasic only (U)		Other substrates only (A)		Ultrabasic and other substrates (I)		Total U + A + I	Epi-phytic species	Total U + A + I + epi-phytes
	Species	%	Species	%	Species	%			
F	290	21.2	526	38.4	273	19.9	1089	76	1165
M	132	9.6	22	1.6	128	9.3	282	1	283
Total F + M	422	30.8	548	40.0	401	29.2	**1371**	77	1448
Non-epiphytes of uncertain edaphic status									63
Total rain forest species									**1511**

(3) In the early Tertiary, the peridotites covering most of the island with a highly selective substrate eliminated much of the ancient flora, discouraged the establishment of certain modern families (e.g. Compositae, Gramineae) well adapted by light, wind-borne diaspores to long distance dispersal, their place being taken by others apparently better adapted such as Cyperaceae, and finally set off a new phase of evolution.

Effect of Substrate on Richness of Species, Endemism, and Conservation of Primitive Forms

The 1511 known rain forest species include 77 epiphytes and 63 others for which edaphic status information is lacking. Of the 1371 species of known edaphic status, 30.8% are limited to ultrabasic rocks (U), 40% to other types of rock (A) and 29.2% occur on both (I) (Table 11). The forest flora is poorer on ultrabasic rocks (U + I) than on other substrates (A + I). This edaphic pattern is enhanced by excluding all species ecologically adaptable enough to occur also in other formations (types MA, MU, and MI), and considering only the strictly forest species (FA, FU, and FI), there are 596 FA species against 290 FU. This relative floristic poverty of the forest on ultrabasic rocks may in part be imputed to the smaller surface area that they cover in New Caledonia. The area of forest on peridotite on the main island is estimated as between 1/3 and 2/5 of the total.

Table 12. Proportion of endemic species in the rain forest. (Percentage based on totals shown in Table 11.)

	Ultrabasic only (U)		Other substrates only (A)		Ultrabasic and other substrates (I)		Total U + A + I	Epi-phytic species	Total U + A + I + epi-phytes
	Species	%	Species	%	Species	%			
F	284	97.9	478	90.8	242	88.6	1004	46	1050
M	131	99.2	12	54.5	112	87.5	255	0	255
Total	415	98.3	490	89.4	354	88.2	**1259**	46	1305
Non-epiphytes of uncertain edaphic status									54
Total endemic rain forest species									**1359**

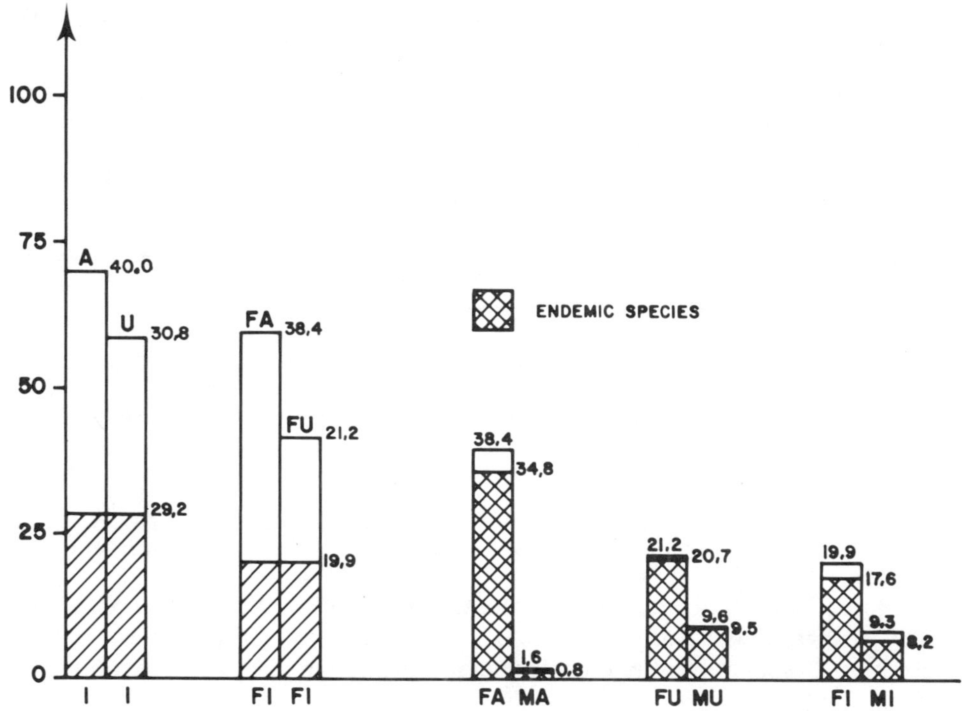

Fig. 4. Relative importance of endemic species based on Tables 11 and 12.

(The Loyalty Islands may be neglected here as their impoverished flora contains mostly species in common with the main island.) This factor cannot however account for so great a difference (little more than half as many strictly forest species), especially as conditions of altitude, topography, and rainfall are equally varied in the two cases. The flora of type A also appears more specialized as it shares only 22 species with other formations such as secondary forest, thickets, savannah, sclerophyll forest, coastal forest, etc. The flora of type U, on the other hand, is less distinctive, with 132 species also found in other formations, mostly maquis. The floristic relationship between maquis and forest on ultrabasic rocks suggests that they are linked, maquis being derived from forest at high altitudes or by disturbance.

The proportion of endemics (89.4%) for the whole rain forest flora (F + M) of type A, although high, is lower than that for type U (98.3%); for exclusively forest species (F), the corresponding figures are 90.8% and 97.9% (Table 12). In absolute numbers, endemic species of type A (478) are not far from twice as many as those of type U (284). Nonendemic forest species are also more numerous in type A (48) than in type U (6).

The 108 families represented in the rain forest may be divided into five groups by their relationship to the substrates as herein defined.

(1) The rain forest species of 30 families are all (underlined) or mostly indifferent (I) to substrate: Agavaceae, Alangiaceae, Alseuosmiaceae, Amaryllidaceae, Ampelidaceae, Anacardiaceae, Aquifoliaceae, Atherospermataceae, Corynocarpaceae, Dilleniaceae, Flindersiaceae, Guttiferae, Hernandiaceae, Icacinaceae, Liliaceae, Linaceae, Menis-

permaceae, Oleaceae, Orchidaceae, Paracryphiaceae, Passifloraceae, Philesiaceae, Podocarpaceae, Rhamnaceae, Rutaceae, Santalaceae, Simarubaceae, Smilacaceae, Solanaceae, Violaceae.

(2) The rain forest species of 48 families are all (underlined) or mostly type A: Acanthaceae, Amborellaceae, Apocynaceae, Araceae, Araliaceae, Asclepiadaceae, Bischofiaceae, Celastraceae, Caesalpiniaceae, Chrysobalanaceae, Commelinaceae, Cyperaceae, Dioscoreaceae, Ebenaceae, Elaeocarpaceae, Ericaceae, Euphorbiaceae, Flacourtiaceae, Gramineae, Hippocrateaceae, Lauraceae, Lecythidaceae, Loganiaceae, Meliaceae, Monimiaceae, Moraceae, Myrsinaceae, Myrtaceae, Nyctaginaceae, Palmae, Pandanaceae, Papilionaceae, Piperaceae, Pittosporaceae, Rhizophoraceae, Rubiaceae, Sapindaceae, Sapotaceae, Saxifragaceae, Sterculiaceae, Symplocaceae, Taxaceae, Tiliaceae, Trimeniaceae, Ulmaceae,Urticaceae, Verbenaceae, Winteraceae.

(3) The rain forest species of 19 families are all (underlined) or mostly type U: Araucariaceae, Balanopaceae, Burseraceae, Casuarinaceae, Cunoniaceae, Cupressaceae, Epacridaceae, Escalloniaceae, Fagaceae, Flagellariaceae, Loranthaceae, Mimosaceae, Myricaceae, Nepenthaceae, Oncothecaceae, Phellinaceae, Sphenostemonaceae, Strasburgeriaceae, Thymeleaceae.

(4) Two families have the same number of rain forest species in type A and type U: Goodeniaceae, Triuridaceae.

(5) Eight families are for various reasons hard to assign definitely to the types A, U, and I: Annonaceae, Balanophoraceae, Bignoniaceae, Caparidaceae, Connaraceae, Cucurbitaceae, Gesneriaceae, Proteaceae.

Three endemic families (Oncothecaceae, Phellinaceae, Strasburgeriaceae) are in the third group but Amborellaceae is in the second and Paracryphiaceae in the first group. Similary, the rain forest species in 22 endemic genera belong completely to type U, compared to 18 of type A and 12 of type I, as follows.

Type A: *Amborella, Alloschmidia, Austrotaxus, Cyphophoenix, Depanthus, Exospermum, Gongrodiscus, Kentiopsis, Kibaropsis, Lavoixia, Leptostylis, Mackeea, Moratia, Pichonia, Pyriluma, Rhopalobrachium, Veillonia.*

Type U: *Actinokentia, Apiopetalum, Arillastrum, Botryomeryta, Campecarpus, Canacomyrica, Cerberiopsis, Clinosperma, Cocconerion, Coilochilus, Daenikera, Dendrophyllanthus, Gastrolepis, Hachettea, Morierina, Neoguillauminia, Oncotheca, Pachyplectron, Pleurocalyptus, Pritchardiopsis, Sebertia, Strasburgeria.*

Type I: *Amphorogyne, Campynemanthe, Chambeyronia, Clematepistephium, Cyphokentia, Nemuaron, Paracryphia, Parasitaxus, Sarcomelicope, Sleumerodendron, Strobilopanax, Zieridium.*

Gymnosperms are fewer in type A (4 species in Araucariaceae, Taxaceae) than in type U (13 species in Araucariaceae, Cupressaceae, Podocarpaceae); they are however well represented in type I (11 species in Araucariaceae, Podocarpaceae). This lack in type A is largely compensated by the great abundance of primitive monocotyledons (Palmae, Pandanaceae) and particularly by the numerous members of archaic families in which the vessels of the wood are absent or primitive (Amborellaceae, Annonaceae, Atherospermataceae, Chloranthaceae, Menispermaceae, Monimiaceae, Piperaceae, Trimeniaceae, Winteraceae). In these families, considered as living fossils (Takhtajan, 1969), type A has 29 rain forest species compared to 18 in type U and 12 in type I.

That the peridotites contributed significantly to the richness of the flora, by covering New Caledonia slowly and massively and inducing a new wave of speciation (species of type U), thus adding to the distinctiveness and endemism of the flora, remains doubtful. First considering the flora, it is not certain that the appearance of taxa associated with this new substrate compensated the inevitable loss of the part of the pre-Eocene flora unable to adapt to the new conditions. The evolutionary capacities of this ancient and already highly diversified flora are shown by its floristic wealth, its endemism, and the abundance of unusual types present today in the forests on non-ultrabasic rocks, in spite of the geological vicissitudes that they have suffered.

On a strictly comparative basis, it is also clear from the floristic analyses that the peridotites, highly selective because of their peculiar chemical composition, provoke a relative impoverishement, not only for modern families (Boraginaceae, Compositae, Gramineae, Labiatae, Scrophulariaceae, etc.) but also for the total number of species, the endemics, and the primitive forms. This casts doubt on the role often attributed to peridotites in conserving archaic ancestral forms (Virot, 1956; Thorne, 1965; Jaffré, 1980). While such an effect may well exist in open formations (maquis), it seems remarkably limited in rain forest.

It appears that the persistence and the abundance of archaic forms in all types of rain forest and irrespective of substrate stem from the extreme isolation of New Caledonia during its geological history.

ACKNOWLEDGEMENTS

Computer input and programming were carried out by M. Hoff, botanist of Centre ORSTOM, Nouméa, whom we thank for this essential collaboration. The following specialists have kindly supplied unpublished data for their respective groups: J. W. Dawson (WELTU)–Myrtaceae (Leptospermoideae); L. L. Forman (K)–Menispermaceae; T. G. Hartley (CANB)–Rutaceae; H. Hürlimann (Z)–*Salacia (Dicarpellum), Salaciopsis;* I. Nielsen (AAU)–Mimosaceae; H. K. Airy Shaw (K)–*Austrobuxus, Baloghia, Bocquillonia, Cocconerion;* and B. C. Stone (KLU)–Pandanaceae.

LITERATURE CITED

Aubréville, A. 1964. Vues d'ensemble sur la géographie et l'écologie des Conifères et Taxacées à propos de l'ouvrage de Rudolf Florin. *Adansonia* 2(4): 4–7.

———. 1965. Les reliques de la flore des Conifères tropicaux en Australie et en Nouvelle-Calédonie. *Adansonia* 2(5): 481–92.

———. 1967. Pt. 1. Sapotacées. p. 3–168 *In: Flore de la Nouvelle-Calédonie et Dépendances.* Mus. Natl. Hist. Nat., Paris.

———. 1973a. Déclin des genres de Conifères tropicaux dans le temps et l'espace. *Adansonia* 2(13): 5–36.

———. 1973b. Distribution des Conifères dans la Pangée. *Adansonia* 2(13): 125–33.

———. 1975. La flore Australo-papoue. Origine et distribution. *Adansonia* 2(15): 159–70.

———. 1976. Centres tertiaires d'origine, radiations et migrations des flores angiospermiques tropicales. *Adansonia* 2(16): 297–354.

Baas, P. 1975. Vegetative anatomy and affinities of Aquifoliaceae, *Sphenostemon, Phelline* and *Oncotheca. Blumea* 22: 311–407.

Balansa, B. 1873. Sur la géographie botanique de l'Océanie et de la Nouvelle-Calédonie. *Bull. Soc. Hist. Nat. Toulouse* 7: 327–32.

Balgooy, M. M. J. van. 1960. Preliminary plant geographical analysis of the Pacific. *Blumea* 10: 385–430.

———. 1969. A study on the diversity of island floras. *Blumea* 17: 139–78.

———. 1971. Plant geography of the Pacific. *Blumea* 6: 1–222.

———. 1975. *Pacific Plant Areas 3 (Z.W.O).* Rijksherbarium, Leiden.

———. 1976. *In* Paijmans, K., ed., *New Guinea vegetation.* C.S.I.R.O. with A.N.U., Canberra.

Barlow, B. A. 1971. Flora of New South Wales. Loranthaceae. Family No. 58: 1–35.

Baumann-Bodenheim, M. B. 1956. Über die Beziehungen der neu-caledonischen Flora zu den tropischen und den süd-hemisphärisch—subtropischen bis—extratropischen Floren und die gürtelmässige Gliederung der Vegetations von Neu-Caledonien. Ber. Geobotan. Inst. Rübel, Zurich.

Bernard, A. 1895. *L'Archipel de la Nouvelle-Calédonie.* Hachette, Paris, p. 216–28.

Bernardi, L. 1979. The New Caledonian genera of Araliaceae and their relationships with those of Oceania and Indonesia. p. 15–325 *In:* Larsen, K., ed., *Tropical botany.* Academic Press, London.

Boiteau, P. 1981. Pt. 10. Apocynacées, p. 5–302. *In: Flore de la Nouvelle-Calédonie et Dépendances.* Mus. Natl. Hist. Nat., Paris.

Boiteau, P. & L. Allorge. 1978. Révision des *Alyxia* de Nouvelle-Calédonie. *Adansonia* 2(18): 443–57.

Boiteau, P., L. Allorge & T. Sévenet. 1972. Révision des *Ochrosia* de Nouvelle-Calédonie. *Adansonia* 2(12): 625–29.

———. 1975. Les *Melodinus* de Nouvelle-Calédonie. *Adansonia* 2(15): 397–407.

———. 1976a. Révision des *Rauwolfia* de Nouvelle-Calédonie. *Adansonia* 2(16): 51–61.

———. 1976b. Apocynacées de Nouvelle-Calédonie. Révision des *Alstonia. Adansonia* 2(16): 465–85.

Boureau, E. 1952. Homoxylées néocalédoniennes. Colloque international sur l'évolution et la phylogénie chez les végétaux. *Ann. Biol.,* C, 3, 2: 129–44.

Brongniart, A. 1874. Nouveaux documents sur la flore de la Nouvelle-Calédonie. *C. R. Hebd. Séances Acad. Sci.* 79: 1442–47.

Brousmiche, A. 1884. Considérations générales sur la végétation de la Nouvelle-Calédonie. *Arch. Med. Navale* 41: 250–60.

Burbidge N. T. 1963. *Dictionary of Australian plant genera—gymnosperms and angiosperms.* Angus and Robertson, Ltd., Sydney.

Carlquist, S. 1980. Anatomy and systematics of Balanopaceae. *Allertonia* 2: 191–246.

Conn, B. J. 1980. A taxonomic revision of *Geniostoma* subg. *Geniostoma* (Loganiaceae). *Blumea* 26: 245–364.

Corner, E. J. H. 1967. *Ficus* in the Solomon Islands. *Phil. Trans. R. Soc.* B 253: 23–159.

———. 1970. *Ficus* subgen. *Pharmacosycea* with reference to the species of New Caledonia. *Phil. Trans. R. Soc.* B 259: 383–433.

———. 1975. *Ficus* in the New Hebrides. *Phil. Trans. R. Soc.* B 272: 343–67.

Cranwell, L. M. 1964. Antarctica: cradle or grave for its *Nothofagus*? p. 87–93 *In:* Cranwell, L. M., ed., *Ancient Pacific floras: the pollen story.* University of Hawaii Press, Honolulu.

Däniker, A. U. 1931. Ergebnisse der Reise von Dr. A.U. Däniker nach Neu-Caledonien und den Loyalitäts-Inseln (1924/25). 3. Die Loyalitäts-Inseln und ihre Vegetation. *Vierteljahrsschr. Naturforsch. Ges. Zür.* 76: 170–213.

———. 1932–33. Ergebnisse der Reise von Dr. A.U. Däniker nach New-Caledonien und den Loyalty-Inseln (1924/6). Katalog der Pteridophyta und Embryophyta siphonogama. *Vierteljahrsschr. Naturforsch. Ges. Zür.* 77: 1–235; 78: 237–395, 397–507.

Dawson, J. W. 1963. New Caledonia and New Zealand—a botanical comparison. *Tuatara* 11: 178–93.

———. 1970. Pacific capsular Myrtaceae 1. Reproductive morphology of *Arillastrum gummiferum* Panch. ex Baillon (New Caledonia). *Blumea* 18: 431–40.

———. 1976. Pacific capsular Myrtaceae 11. Redefinition of *Metrosideros* Banks ex Gaertn. and definition of infrageneric categories. *Blumea* 23: 7–11.

Dickison, W. C. & P. Baas. 1977. The morphology and relationships of *Paracryphia* (Paracryphiaceae). *Blumea* 23: 417–38.

Ding Hou. 1962. Celastraceae. p. 227–291 *In: Flora Malesiana.* Ser. I., Vol. 6. Groningen.

———. 1978. Anacardiaceae. p. 395–548 *In: Flora Malesiana.* Ser. I., Vol. 8. Groningen.

Florin, R. 1963. The distribution of conifer and taxad genera in time and space. *Acta Horti Bergiani 20, Uppsala*—additions and corrections 20, N: 6 (1966).

Fournier, E. 1874. Sur la dispersion géographique des Fougères de la Nouvelle-Calédonie. *Ann. Sci. Nat. Bot. 5ème Sér.* 19: 287–99.

Fosberg, F. R., P. Boiteau & M. H. Sachet. 1977. Nomenclature of the Ochrosinae: synonymy of *Ochrosia* Juss. and *Neisosperma* Raf. *Adansonia* 2(17): 23–33.

Gardner, R. O. 1978. Systematic notes on the Alseuosmiaceae. *Blumea* 24: 138–42.

Good, R. 1955. Madagascar and New Caledonia, a problem in plant geography. *Blumea* 6: 470–74.

———. 1960. On the geographical relationships of the Angiosperm flora of New Guinea. *Bull. Br. Mus. Nat. Hist.* 2(8): 205–26.

———. 1964. *The geography of the flowering plants,* 3rd ed. Longmans, London.

Green, P. S. 1979. Observations on the phytogeography of the New Hebrides, Lord Howe Island and Norfolk Island. *In:* Bramwell, D., ed., *Plants and islands.* Academic Press. New York.

Gressitt, J. L., ed. 1963. *Pacific Basin biogeography—a symposium.* Bishop Museum Press, Honolulu.

Griffith, J. R. 1975. New Zealand and the southwest Pacific margin of Gondwanaland. p. 619–37 *In:* Campbell, K. S. W., ed., *Gondwana geology.* Australian National University Press, Canberra.

Guillaumin, A. 1911. Catalogue des plantes phanérogames de la Nouvelle-Calédonie et Dépendances, Ile des Pins et Loyauté. *Ann. Mus. Colon. Marseille* 2: 77–290.

———. 1921. Essai de géographie botanique de la Nouvelle-Calédonie. p. 256–93 *In:* Sarasin, F. & J. Roux, eds., *Nova Caledonia.* 1. C. W. Kreidel, Berlin and Wiesbaden.

———. 1924. Le peuplement botanique de la Nouvelle-Calédonie. C. R. 48e Session. Assoc. Franc. Avanc. Sci. Liège: 953–54.

———. 1928. Les régions florales du Pacifique d'après leur endémisme et la répartition de quelques plantes phanérogames. *Proc. 3rd Pan. Pac. Sci. Congr., Tokyo* 1: 920–38.

———. 1934. Les régions florales du Pacifique. *Mém. Soc. Biogéogr.* 4: 255–70.

———. 1948. *Flore analytique et synoptique de la Nouvelle-Calédonie. Phanérogames.* Office de la Recherche Scientifique Coloniale, Paris.

———. 1953. Y-a-t-il réellement des rapports entre la flore calédonienne et la flore malgache? *Mém. Soc. Biogéogr. Nouv.* Sér. 1: 97.

———. 1954. A propos de la répartition de quelques Phanérogames de Nouvelle- Calédonie et des Nouvelles-Hébrides. *C. R. Somm. Soc. Biogéogr.* 31: 38–40.

———. 1957. Résultats scientifiques de la mission franco-suisse de botanique en Nouvelle- Calédonie (1950–1952). *Mém. Mus. Natl. Hist. Nat.* Sér. B, 8: 1–120.

———. 1962. Résultats scientifiques de la mission franco-suisse de botanique en Nouvelle- Calédonie (1950–1952). II. *Mém. Mus. Natl. Hist. Nat.* Sér. B, 8: 193–330.

———. 1964a. L'endémisme en Nouvelle-Calédonie. *C. R. Soc. Biogéogr.* 38: 67–75.

———. 1964b. Résultats scientifiques de la mission franco-suisse de botanique en Nouvelle- Calédonie (1950–1952). III. *Mém. Mus. Natl. Hist. Nat.* Sér. B, 15: 1–96.

———. 1967. Résultats scientifiques de la mission franco-suisse de botanique en Nouvelle- Calédonie (1950–1952). IV. *Mém. Mus. Natl. Hist. Nat.* Sér. B, 15: 97–132.

———. 1974. Résultats scientifiques de la mission franco-suisse de botanique en Nouvelle- Calédonie (1950–1952). V. *Mém. Mus. Natl. Hist. Nat.* Sér. B, 22: 1–36.

Haas, J. E. 1977. The Pacific species of *Pittosporum* Banks ex Gaertn. *Allertonia* 1: 73–167.

Hallé, N. 1977. Pt. 8. Orchidacées. p. 4–565 *In: Flore de la Nouvelle-Calédonie et Dépendances.* Mus. Natl. Hist. Nat., Paris.

Hansen, B. 1976. Balanophoraceae. p. 783–806 *In: Flora Malesiana.* Ser. I. Vol. 7. Croningen.

Heine, H. 1976. Pt. 7. Acanthacées, Bignoniacées, Boraginacées et Solanacées. p. 3–212 *In: Flore de la Nouvelle-Calédonie et dépendances.* Mus. Natl. Hist. Nat., Paris.

Hoogland, R. D. 1979. Studies in the Cunoniaceae II. The genera *Caldcluvia, Pullea, Acsmithia* and *Spiraeanthemum. Blumea* 25: 492–505.

Hürlimann, H. 1955. Celastracées nouvelles ou critiques de la Nouvelle-Calédonie. *Candollea* : 69–78.

Jaffré, T. 1974. La végétation et la flore d'un massif de roches ultrabasiques de Nouvelle-Calédonie: Le Koniambo. *Candollea* 29: 427–56.

———. 1980. Étude écologique du peuplement végétal des sols dérivés de roches ultrabasiques en Nouvelle-Calédonie. *Coll. Trav. Doc. O.R.S.T.O.M.* No. 124. 274 p.

Jérémie, J. 1974. A propos du genre *Tarenna* (Rubiaceae—Gardeniae) en Nouvelle-Calédonie. *Adansonia* 2(14): 473–80.

———. 1977. Étude des Monimiaceae: le genre *Kibaropsis. Adansonia* 2(17): 79–87.

———. 1978. Étude des Monimiaceae: révision du genre *Hedycaria. Adansonia* 2(18): 25–53.

———. 1980. Notes sur le genre *Ascarina* (Chloranthaceae) en Nouvelle-Calédonie et à Madagascar. *Adansonia* 2(20): 273–85.

———. 1982. Pt. 11. Monimiacées, Amborellacées, Athérospermatacées, Triméniacées, Chloranthacées. p. 125–79 *In: Flore de la Nouvelle-Calédonie et Dépendances.* Mus. Natl. Hist. Nat., Paris.

Johnson, L. A. S. 1980. Notes on Casuarinaceae. *Telopea* 2: 83–84.

Johnson, L. A. S. & B. G. Briggs. 1975. On the Proteaceae—the evolution and classification of a southern family. *J. Linn. Soc. Lond., Bot.* 70:83–182.

Kanis, A. 1979. The Malesian species of *Serianthes* Bentham (Fabaceae—Mimosoideae). *Brunonia* 2: 289–320.

Kiew, R. 1979. Florae Malesianae Praecursores LX. The Oleaceae of Malesia II. The Genus *Olea. Blumea* 25: 305–13.

Kostermans, A. J. G. H. 1963. The identity of *Lethedon* Spreng. (Thymeleaceae). *J. Bot., Moscow,* 48(6): 830–33.

———. 1974. Pt. 5. Lauracées. p. 3–123 *In: Flore de la Nouvelle-Calédonie et Dépendances.* Mus. Natl. Hist. Nat., Paris.

———. 1977. Notes on Asiatic, Pacific and Australian *Diospyros* (Ebenaceae). *Blumea* 23: 449–74.

Laubenfels, D. J. de. 1972. Pt. 4. Gymnospermes. p. 1–168 *In: Flore de la Nouvelle-Calédonie et Dépendances.* Mus. Natl. Hist. Nat., Paris.

———. 1978. The genus *Prumnopitys* (Podocarpaceae) in Malesia. *Blumea* 24: 189–90.

Leenhouts, P. W. 1955. The genus *Canarium* in the Pacific. Bernice P. Bishop Mus. Bull. 216: 1–53.

Leroy, J. F. 1957. Sur deux amentifères remarquables de la flore asiatico-pacifique et pacifique. *Proc. 8th Pac. Sci. Congr.* 4: 459–64.

Lescot, M. 1980. Pt. 9. Flacourtiacées. p. 3–134 *In: Flore de la Nouvelle-Calédonie et Dépendances.* Mus. Natl. Hist. Nat., Paris.

Lobreau-Callen, D. 1975. Deux genres de Celastraceae *Cassine* L. et *Maytenus* Mol., revus à la lumière de la palynologie. *Adansonia* 2(15): 215–23.

Marais, W. 1978. Pt. 182. Dioscoréacées. p. 1–5 *In: Flore des Mascareignes.* Port-Louis.

Marais, W. & M. J. E. Coode. 1978. Pt. 183. Liliacées. p. 1–41 *In: Flore des Mascareignes.* Port-Louis.

Markgraf, F. 1979. Florae Malesianae Praecursores LIX. Apocynaceae V. *Ochrosia, Neisosperma. Blumea* 25: 233–47.

Martin, H. A. 1977. The history of *Ilex* (Aquifoliaceae) with special reference to Australia: evidence from pollen. *Aust. Bot. J.* 25: 655–73.

Moldenke, H. N. 1971. Vol. 1: 1–483. Vol. 2: 489–974. *In: A fifth summary of the Verbenaceae, Avicenniacae, Stilbaceae, Dicrastylidaceae, Symphoremataceae, Nyctanthaceae and Eriocaulaceae of the world as to valid taxa, geographic distribution, and synonymy.* Braun–Brumfield, Ann Arbor.

Moore, Jr., H. E. 1978. New genera and species of Palmae from New Caledonia. *Gentes Herb.* 11: 291–309.

———. 1980. New genera and species of Palmae from New Caledonia. *Gentes Herb.* 12: 17–24.

Morat, Ph., T. Jaffré, J. M. Veillon & H. S. MacKee. 1981. Les formations végétales. Map 15, *Atlas de la Nouvelle-Calédonie.* O.R.S.T.O.M., Paris.

Morat, Ph. & H. S. MacKee. 1977. Quelques précisions sur le *Trimenia neocaledonia* Bak. f. et la famille des Triméniacees en Nouvelle-Calédonie. *Adansonia* 2(17): 204–13.

Nooteboom, H. P. 1980. Pt. 9. Symplocacées. p. 135–58 *In: Flore de la Nouvelle-Calédonie et Dépendances.* Mus. Natl. Hist. Nat., Paris.

Paris, J. P. 1981. *Géologie de la Nouvelle-Calédonie. Un essai de synthèse.* Orléans. B. R. G. M. 278 p.

Pennington, T. D. & B. T. Styles. 1975. A generic monograph of the Meliaceae. *Blumea* 22: 419–540.

Philipson, W. R. 1977. The identity of *Arthrophyllum* and *Eremopanax* (Araliaceae). *Adansonia* 2(17): 329–33.

Prance, G. T. 1979. New genera and species of Chrysobalanaceae from Malesia and Oceania. *Brittonia* 31: 79–88.

Raven, P. H. 1979. Plate tectonics and Southern Hemisphere biogeography. p. 1–24 *In:* Larsen, K., ed., *Tropical botany.* Academic Press, London.

Raven, P. H. & D. I. Axelrod. 1972. Plate tectonics and Australasian biogeography. *Science* 176: 1379–86.

———. 1974. Angiosperm biogeography and past continental movements. *Ann. Mo. Bot. Gard.* 61: 539–673.

Raynal, J. 1973. Notes cypérologiques: 20. Un *Baumea* néo-calédonien nouveau. *Adansonia* 2(13): 467–69.

———. 1974. Notes cypérologiques–22. Les *Costularia* de Nouvelle-Calédonie. *Adansonia* 2(14): 339–68.

———. 1975a. Les Cypéracées des Nouvelles-Hébrides–Résultats de l'expédition de la Royal Society aux Nouvelles-Hébrides en 1971. *Adansonia* 2(15): 99–119.

———. 1975b. Notes cypérologiques: 23. Un nouveau *Gahnia* néo-calédonien. *Adansonia* 15: 189–91.

Rendle, A. B., E. G. Baker & S. Moore. 1921. A systematic account of the plants collected in New Caledonia and the Isle of Pines by Prof. R. H. Compten, M.A., in 1914. Part 1. Flowering plants (Angiosperms) *Proc. Linn. Soc., Bot.* 45: 245–417.

Rodenburg, W. F. 1971. A revision of the genus *Trimenia* (Trimeniaceae). *Blumea* 19: 3–15.

St. John, H. 1978. Revision of *Joinvillea* (Joinvilleaceae). Pacific plant studies 37. *Phytologia* 40: 369–74.

Schlechter, R. 1905. Pflanzengeograpische Gliederung der Insel Neu-Kaledonien. *Bot. Jahrb.* 36: 1-41.

———. 1907–1908. Beiträge zur Kenntnis der Flora von Neu-Kaledonien. *Bot. Jahrb.* 39: 1–274; 40: 20–45.

Scott, A. J. 1979. A revision of *Xanthomyrtus* (Mytaceae). *Kew Bull.* 33: 461–77.

———. 1980. Notes on Myrtaceae in the Mascarenes with some recombinations for taxa from Aldabra–Malaya–New Caledonia. *Kew Bull.* 34: 473–484–496.

Shaw, H. K. A. 1971. Notes on Malesian and other Asiatic Euphorbiaceae CXXXI—New combinations and new taxa in *Ausrobuxus* Miq. *Kew Bull.* 25: 506.

———. 1972. Notes on Malesian and other Asiatic Euphorbiaceae CLXIV. A misplaced species of *Bocquillonia* Baill. *Kew Bull.* 27: 88.

———. 1973. *In:* Willis, J. C., ed., *A dictionary of the flowering plants and ferns.* 8th ed. University Press, Cambridge.

———. 1974. Notes on Malesian and other Asiatic Euphorbiaceae CLXXV. New species of *Austrobuxus* Miq. with a key to the whole genus. *Kew Bull.* 29: 303.

———. 1978. Notes on Malesian and other Asiatic Euphorbiaceae CCXII. *Austrobuxus. Kew Bull.* 33: 39.

———. 1980. Notes on Euphorbiaceae from Indomalesia, Australia and the Pacific—CCXL *Bocqullonia* Baill. *Kew Bull.* 35: 396–98.

Sleumer, H. 1966. Ericaceae. p. 469–668 *In: Flora Malesiana.* Ser. I, Vol. 6. Groningen.

———. 1974. A concise revision of the Flacourtiaceae of New Caledonia and the Loyalty Islands. *Blumea* 22: 123–47.

Smith, A. C. 1955. Phanerogam genera with distributions terminating in Fiji. *J. Arn. Arb.* 36: 373–92.

———. 1970. The Pacific as a key to flowering plant history. *Harold C. Lyon Arboretum lecture* 1: 1–26.

———. 1976. Studies of Pacific Island plants, XXXIII. The genus *Ascarina* (Chloranthaceae) in the Southern Pacific. *J. Arn. Arb.* 57: 405–25.

———. 1978. A precursor to a new flora of Fiji. *Allertonia* I: 347.

———. 1979. *Flora Vitiensis Nova.* Vol. I. Pacific Tropical Garden, Hawaii.

Soepadmo, E. 1977. Ulmaceae. p. 43–66 *In: Flora Malesiana.* Ser. I, Vol. 8. Groningen.

Steenis, C. G. G. J. van. 1955. Some notes on the flora of New Caledonia and reduction of *Nouhuysia* to *Sphenostemon. Svensk. Bot. Tidskr.* 49: 19–23.

———. 1971a. Revision of *Nothofagus* in New Caledonia. *Adansonia* 2(11): 615–24.

———. 1971b. *Nothofagus*—Key genus of plant geography, in time and space, living and fossil, ecology and phylogeny. *Blumea* 19: 65–98.

———. 1978. The genus *Periomphale* in New Guinea (Caprifoliaceae). Miscellaneous botanical notes XXV. *Blumea* 24: 480–81.

Stemmerik, J. F. 1964. Nyctaginaceae. p. 450–68 *In: Flora Malesiana.* Ser. I, Vol. 8. Groningen.

Stevens, G. R. 1977. Mesozoic biogeography of the South West Pacific and its relationship to plate tectonics. *Int. Symp. Geodyn. South West Pacific, Nouméa 1976.* Technip Ed.: 309–26.

Stone, B. C. 1972. On the genus *Pandanus* (Pandanaceae) in New Caledonia. *Adansonia* 2(12): 409–20.

Takhtajan. A. 1969. *Flowering plants, origin and dispersal.* Oliver and Boyd, Edinburgh.

Thorne, R. F. 1963. Biotic distribution patterns in the tropical Pacific. *In:* Gressit, J. L., ed., *Pacific Basin biogeography.* Bishop Museum Press, Honolulu.

———. 1965. Floristic relationships of New Caledonia. *Univ. Iowa Stud. Nat. Hist.* 20(7): 1–14.

———. 1969. Floristic relationships between New Caledonia and the Solomon Islands. *Phil. Trans. R. Soc.* 255: 595–602.

———. 1972. Major disjunctions in the geographic ranges of seed plants. *Quart. Rev. Biol.* 47: 366–411.

Tirel, C. 1977. A propos du genre *Elaeocarpus* en Nouvelle-Calédonie. *Adansonia* 2(17): 441–54.

———. 1980. Nouvelles espèces de *Sloanea* (Elaeocarpaceae) en Nouvelle-Calédonie. *Adansonia* 2(20): 91–106.

———. 1982. Pt. 11. Eléocarpacées. p. 3–124 *In: Flore de la Nouvelle-Calédonie et Dépendances.* Mus. Natl. Hist. Nat., Paris.

Tirel, C. & J. Raynal. 1980. Recherches bibliographiques sur trois espèces d' *Elaeocarpus* (Elaeocarpacéae). *Adansonia* 2(20): 169–77.

Villiers, J. F. 1980. Pt. 9. Icacinacées–Corynocarpacées. p. 159–78 *In: Flore de la Nouvelle-Calédonie et dépendances.* Mus. Natl. Hist. Natl. Paris.

Vink, W. 1977. The Winteraceae of the Old World II. *Zygogynum*—morphology and taxonomy. *Blumea* 23: 219–50.

Virot, R. 1956. La végétation Canaque. *Mém. Mus. Natl. Hist. Sér. B,* Vol. 7 Bot.: 1–398.

———. 1968. Pt. 2. Protéacées. p. 3–254 *In: Flore de la Nouvelle-Calédonie et Dépendances.* Mus. Natl. Hist. Nat., Paris.

———. 1975. Pt. 6. Epacridacées. p. 3–16 *In: Flore de la Nouvelle-Calédonie et Dépendances.* Mus. Natl. Hist. Nat., Paris.

Whitmore, T. C. 1975. *Tropical rain-forests of the Far East.* University Press, Oxford.

Whitmore, T. C. 1980. A monograph of *Agathis. Plant Syst. Evol.* 135: 41–69. Commonwealth Forestry Institute, Oxford Univeristy.

Whitmore, T. C. & C. N. Page. 1980. Evolutionary implications of the distribution and ecology of the tropical conifer *Agathis. New Phytol.* 84: 407–16.

Wilde, W. J. J. O. de. 1972. The indigenous old world Passifloras. *Blumea* 20: 227–50.

APPENDIX 1
LIST OF RAIN FOREST SPECIES IN NEW CALEDONIA

(Symbols: E–Endemic; P–Native, non-endemic; e–Epiphyte; M–Forest species also in other formations; F–Strictly forest species; U–Species limited to ultrabasic rocks; A–Species limited to other rocks; I–Species indifferent to substrate.)

Acanthaceae		
Graptophyllum		
G. balansae Heine	E	FA
G. pictum (L.) Griffith	P	FA
Hemigraphis		
H. reptans (G. Forster) T. Anderson ex Hemsley	P	MA
Justicia		
J. pinensis S. Moore	E	MA
Pseuderanthemum		
P. comptonii S. Moore	E	MI
P. repandum (G. Forster) Guillaumin	P	MI
Agavaceae		
Cordyline		
C. fruticosa (L. ex Stickman) A. Chevalier	P	FI
C. neocaledonica Linden	E	MU
Alangiaceae		
Alangium		
A. bussyanum (Baillon) Harms	E	FI
Alseuosmiaceae		
Periomphale		
P. balansae Baillon	E	FI
P. gaultherioides (Gilg & Schlechter) Steenis	E	FI
P. neocaledonica (Gilg & Schlechter) Steenis	E	FI
P. pancheri Baillon	E	FI
Amaryllidaceae		
Campynema		
C. neocaledonicum Rendle	E	FI
Campynemanthe		
C. viridiflora Baillon	E	FI
Amborellaceae		
Amborella		
A. trichopoda Baillon	E	FA
Ampelidaceae		
Cayratia		
C. carnosa (Link) Gagnepain	P	FA
C. japonica (Thunberg) Gagnepain	P	FI
Cissus		
C. glaucoramea Planchon	P	MI
Anacardiaceae		
Euroschinus		
E. elegans Engler	E	FI
E. obtusifolius Engler	E	FI
E. verrucosus Engler	E	FI
E. vieillardii Engler	E	FI
Semecarpus		
S. balansae Engler	E	FU
S. neocaledonia Engler	E	FU
Annonaceae		
Desmos		
D. lecardii (Guillaumin) R. E. Fries	E	FU
Polyalthia		
P. nitidissima (Dunal) Bentham	E	FI
Richella		
R. obtusata (Baillon) R. E. Fries	E	FA
Uvaria		
U. baillonii Guillaumin	E	MU
Xylopia		
X. pancheri Baillon	E	MU
X. vieillardii Baillon	E	FA
Apocynaceae		
Alstonia		
A. balansae Guillaumin	E	FI
A. boulindaensis Boiteau	E	FU
A. coriacea Pancher ex S. Moore	E	MU
A. lanceolata Heurck & Mueller Argovie	E	MI
A. lanceolifera S. Moore	E	MI
A. legouixiae Heurck & Mueller Argovie	E	MU
A. lenormandii Heurck & Mueller Argovie	E	MU
A. odontophora Boiteau	E	FU
A. plumosa Labillardière	E	FI
A. quaternata Heurck & Mueller Argovie	E	FA
A. sphaerocapitata Boiteau	E	FI
A. undulata Guillaumin	E	MI
A. vieillardii Heurck & Mueller Argovie	E	MU
Alyxia		
A. baillonii Guillaumin	E	FI

A. celastrinea (Baillon) Schlechter ex Guillaumin	E	MI
A. dolioliflora Guillaumin	E	MU
A. glaucophylla Heurck & Mueller Argovie	E	MU
A. hurlimannii Guillaumin	E	FA
A. integricarpa Boiteau	E	FA
A. laurina Baillon	E	FA
A. leucogyne Heurck & Mueller Argovie	E	MI
A. loeseneriana Schlechter	E	FA
A. longicarpa Boiteau	E	FA
A. margaretae Boiteau	E	FA
A. microcarpa Pancher ex Boiteau	E	MI
A. myrtoides Schlechter	E	MI
A. oppositifolia Boiteau	E	MI
A. oubatchensis (Schlechter) Boiteau	E	FA
A. podocarpa Heurck & Mueller Argovie	E	MI
A. pseudoserpentina Boiteau	E	MI
A. stellata (Forster f.) Roemer & Schultes	P	MA
A. suavis (Baillon) Schlechter	E	MU
A. tisserantii Montrouzier	E	MU
A. torqueata (Baillon) Guillaumin	E	FI
A. vieillardii Boiteau	E	FA
Artia		
A. francii (Guillaumin) Pichon	E	MI
A. lifuana (Baillon) Pichon ex Guillaumin	E	FI
Cerbera		
C. manghas L.	P	MI
Cerberiopsis		
C. candelabra Vieillard	E	FU
C. neriifolia (S. Moore) Boiteau	E	MU
Ervatamia		
E. lifuana Boiteau & Allorge	E	MI
Melodinus		
M. aeneus Baillon	E	FU
M. balansae Baillon	E	MU
M. polyadenus Baillon	E	MI
Neisosperma		
N. brevituba (Boiteau) Boiteau	E	FA
N. lifuana (Guillaumin) Boiteau	E	FA
N. miana (Baillon ex Guillaum.) Boiteau	E	FU
N. sevenetii (Boiteau) Boiteau	E	MU
N. thiollierei (Montrouzier) Boiteau	E	FU
Ochrosia		
O. balansae Baillon & Guillaumin	E	FU
O. grandiflora Boiteau	E	FA
O. silvatica Daeniker	E	FU
Pagiantha		
P. cerifera (Pancher & Sébert) Markgraf	E	MU
Parsonsia		
P. affinis Baillon	E	FA
P. brachiata Baillon	E	MI
P. catalpaecarpa Baillon	E	MI
P. crebriflora Baillon	E	FI
P. edulis (G. Bennett) Guillaumin	E	FA
P. effusa S. Moore	E	MU
P. flexuosa Baillon	E	MU
P. franchetii Baillon	E	MI
P. laxiflora Guillaumin	E	MI
P. longiflora Guillaumin	E	MU
P. macrophylla Pichon	E	FU
P. pachycarpa Guillaumin	E	MI
P. populifolia Baillon	E	MU
P. scabra (Labillardière) Markgraf	E	MI
Rauvolfia		
R. balansae (Baillon) Boiteau	E	FU
R. semperflorens Schlechter	E	MU
R. spathulata Boiteau	E	FU
Aquifoliaceae		
Ilex		
I. sebertii Pancher & Sébert	E	MI
Araceae		
Epipremnum		
E. pinnatum (L.) Engler	P	FA
Araliaceae		
Apiopetalum		
A. arboreum E. G. Baker	E	FU
A. glabratum Baillon	E	
A. pennelii Viguier	E	FU
A. velutinum Baillon	E	MU
Arthrophyllum		
A. angustatum (Baillon) Philipson	E	FU
A. balansae (Baillon) Philipson	E	FA
A. glaberrimum (Baumann-Bodenheim) Philipson	E	FA

A. grandifolium (Guillaumin) Philipson	E	MU
A. hederoides (Baumann-Bodenheim) Philipson	E	FU
A. otopyrenum (Baillon) Philipson	E	FU
A. schlechteri (Harms) Philipson	E	FA
A. vieillardii (Baillon) Philipson	E	FA
Botryomeryta		
B. lecardii Viguier	E	FU
Delarbrea		
D. collina Vieillard	P	FI
D. harmsii Viguier	E	FA
D. longicarpa Viguier	E	FU
D. montana Viguier	E	FA
D. paradoxa Vieillard	E	FA
Dizygotheca		
D. apioidea (Baillon) Viguier	E	FA
D. bailloni Viguier	E	
D. coenosa Viguier	E	
D. elegantissima (Hort.) Viguier & Guillaumin	E	
D. fagueti (Baillon) Viguier	E	FA
D. harmsii Viguier	E	
D. lecardi Viguier	E	FA
D. leptophylla Hemsley	E	
D. ouveana Daeniker	E	FA
D. parvifolia (Pancher & Sébert) Viguier	E	
D. plerandroides Viguier	E	
D. tenuifolia (Pancher) Viguier	E	FU
D. toto (Baillon) Viguier	E	MA
D. vieillardi (Baillon) Viguier	E	FU
Meryta		
M. balansae Baillon	E	FI
M. coriacea Baillon	E	FU
M. denhamii Seemann	E	FA
M. gracilis Guillaumin	E	FA
M. microcarpa Baillon	E	FI
M. oxylaena Baillon	E	FI
M. pandanicarpa Guillaumin	E	FI
M. schlechteri Harms	E	FA
M. sonchifolia Linden & André	E	FI
Myodocarpus		
M. brongniartii Dubard & Viguier	E	FA
M. coronatus Dubard & Viguier	E	FA
M. crassifollius Dubard & Viguier	E	MU
M. elegans Dubard & Viguier	E	FU
M. pinnatus Brongniart & Gris	E	MI
M. simplicifolius Brongniart & Gris	E	MI
M. vieillardii Brongniart & Gris	E	FA
Schefflera		
S. affinis Baillon	E	FA
S. andraeana Baillon	E	FI
S. balansaeana Baillon	E	FU
S. candelabrum Baillon	E	FA
S. combouiensis E. G. Baker	E	MU
S. comptonii E. G. Baker	E	FI
S. crassipes Baillon	E	FU
S. cussoniae Baillon	E	FU
S. elongata Baillon	E	FU
S. emiliana Baillon	E	
S. gabriellae Baillon	E	FI
S. golip Baillon	E	FA
S. lerati Viguier	E	FU
S. marcellana Baillon	E	FA
S. nono Baillon	E	FU
S. pancheri Viguier	E	FU
S. pseudocandelabrum Viguier	E	FA
S. veillonorum Bernardi	E	FA
S. vieillardii Baillon	E	FI
Strobilopanax		
S. macrocarpus (Baillon) Viguier	E	FI
S. macrocephalus (Baillon) Viguier	E	FI
Tieghemopanax		
T. austrocaledonicus Viguier	E	FI
T. balansae Viguier	E	FA
T. bracteatus Viguier	E	FI
T. lecardi Viguier	E	FA
T. microbotrys Viguier	E	FI
T. microcarpus Viguier	E	FA
T. monticola Viguier	E	FA
T. pulchellus Viguier	E	FI
T. reflexus Viguier	E	FI
T. sessiliflorus Viguier	E	MI
T. suborbicularis Viguier	E	MU
Araucariaceae		
Agathis		
A. corbassonii de Laubenfels	E	FA
A. lanceolata Lindley ex Warburg	E	FU
A. montana de Laubenfels	E	FA
A. moorei (Lindley) Masters	E	FI

Araucaria		
A. bernieri Buchholz	E	FU
A. biramulata Buchholz	E	FU
A. columnaris (J. R. & G. Forster) J. D. Hooker	E	FI
A. humboldtensis Buchholz	E	FU
A. laubenfelsii Corbasson	E	MU
A. montana Brongniart & Gris	E	FI
A. muelleri (Carrière) Brongniart & Gris	E	FU
A. schmidii de Laubenfels	E	FA
A. subulata Vieillard	E	FU
Asclepiadaceae		
Hoya		
H. limoniaca S. Moore	E	FA
H. neocaledonica Schlechter	E	FI
Marsdenia		
M. lyonsioides Schlechter	E	FA
M. nigriflora Guillaumin	E	MU
M. oubatchensis Schlechter	E	MI
M. pseudoparsonsia Guillaumin	E	MI
M. raoulii Guillaumin	E	MI
M. speciosa Baillon	E	FA
M. tylophoroides Schlechter	E	FA
Secamone		
S. insularis Schlechter	E	MI
Tylophora		
T. anisotomoides Schlechter	E	FA
T. insulincola S. Moore	E	FA
T. micrantha Guillaumin	E	FA
T. tapeinogyne Schlechter	E	FA
Atherospermataceae		
Nemuaron		
N. vieillardii Baillon	E	FI
Balanophoraceae		
Balanophora		
B. fungosa J. R. & G. Forster	P	MA
Hachettea		
H. austrocaledonica Baillon	E	FU
Balanopaceae		
Balanops		
B. balansae Baillon	E	FU
B. microstachya Baillon	E	FU
B. oliviformis Baillon	E	FU
B. pachyphylla Baillon ex Guillaumin	E	FI
B. pancheri Baillon	E	FU
B. sparsiflora (Schlechter) Hjelmquist	E	FI
B. vieillardii Baillon	E	FI
Bignoniaceae		
Deplanchea		
D. sessilifolia Vieillard ex Steenis	E	MU
D. speciosa Vieillard	E	MI
Pandorea		
P. austrocaledonica (Bureau) Seemann	E	FA
Bischofiaceae		
Bischofia		
B. javanica Blume	P	FA
Burseraceae		
Canarium		
C. balansae Engler	E	FA
C. oleiferum Baillon	E	FU
C. trifoliolatum Engler	E	FI
C. whitei Guillaumin	E	FU
Capparidaceae		
Capparis		
C. sp.	E	MU
Casuarinaceae		
Casuarina		
C. glaucescens Schlechter	E	FU
C. webbiana Miquel	E	FU
Celastraceae		
Cassine		
C. cunninghamii (Montrouzier) Lobreau-Callen	E	FA
Elaeodendron		
E. clusiophyllum Baillon	E	FU
E. curtipendulum Endlicher	P	FA
E. pininsulare Huerlimann	E	FU
E. tuk Guillaumin	E	
Maytenus		
M. fournieri (Pancher & Sébert) Loesener	E	FI
M. hybanthifolia (Guillaumin) Guillaumin	E	FA
Menepetalum		
M. cassinoides Loesener	E	FU
M. pachystimoides Loesener	E	FU
M. salicifolium Loesener	E	FA
M. schlechteri Loesener	E	MI
Salaciopsis		
S. glomerata Huerlimann	E	FA
S. megaphylla (Poisson ex Guillaumin) Loesener	E	FU
S. neocaledonica E. G. Baker	E	FA
S. sparsiflora Huerlimann	E	FI
S. tapeinospermophylla Huerlimann	E	FA

Caesalpiniaceae		
Caesalpinia		
C. rubiginosa Guillaumin	E	FA
C. schlechteri Harms	E	FA
Intsia		
I. bijuga (Colebrooke) Kuntze	P	FA
Mezoneurum		
M. baudouinii Guillaumin	E	FU
M. montrouzieri Guillaumin	E	MU
Storckiella		
S. comptonii E. G. Baker	E	MI
S. pancheri Baillon	E	MI
Chloranthaceae		
Ascarina		
A. rubricaulis Solms	E	FU
A. solmsiana Schlechter	E	FA
Chrysobalanaceae		
Hunga		
H. lifouana (Daeniker) Prance	E	FA
H. rhamnoides (Guillaumin) Prance	E	FI
Commelinaceae		
Aneilema		
A. biflorum R. Brown	P	MA
Connaraceae		
Santaloides		
S. balanseanum (Baillon) Schellenberg	E	MU
S. vieillardii Schellenberg	E	FA
Corynocarpaceae		
Corynocarpus		
C. dissimilis Hemsley	E	FI
Cucurbitaceae		
Diplocyclos		
D. palmatus (L.) C. Jeffrey	P	
Cunoniaceae		
Acsmithia		
A. austrocaledonica (Brongn. & Gris) Hoogland	E	FI
A. laxiflora Hoogland	E	FA
A. pedunculata (Schlechter) Hoogland	E	FU
A. pubescens (Pampanini) Hoogland	E	MU
A. undulata (Vieillard) Hoogland	E	FA
Codia		
C. albifrons Vieillard ex Guillaumin	E	MU
C. arborea Brongniart ex Guillaumin	E	FU
C. incrassata Pampanini	E	FA
C. montana J. R. & G. Forster	E	MI
Cunonia		
C. alticola Guillaumin	E	FU
C. austrocaledonica Brongniart ex Guillaumin	E	FI
C. balansae Brongniart & Gris	E	FU
C. bernieri Guillaumin	E	FU
C. bullata Brongniart & Gris	E	FU
C. montana Schlechter	E	FI
C. ouaiemensis Guillaumin & Virot	E	MA
C. pterophylla Schlechter	E	FU
C. pulchella Brongniart & Gris	E	FI
C. vieillardi Brongniart & Gris	E	FU
Geissois		
G. hippocastaneifolia Guillaumin	E	FU
G. hirsuta Brongniart & Gris	E	FU
G. intermedia Vieillard	E	FI
G. montana Vieillard	E	FA
G. polyphylla Lecard ex Guillaumin	E	FA
G. racemosa Labillardière	E	FI
G. trifoliolata Guillaumin	E	FA
G. velutina Guillaumin	E	FU
Pancheria		
P. brunhesi Pampanini	E	FA
P. calophylla Guillaumin	E	MU
P. engleriana Schlechter	E	FU
P. seberti Guillaumin	E	FI
Weinmannia		
W. dichotoma Brongniart & Gris	E	FU
W. monticola Daeniker	E	FA
W. paitensis Schlechter	E	FU
W. serrata Brongniart & Gris	E	FU
W. thornei Guillaumin	E	FU
Cupressaceae		
Libocedrus		
L. austrocaledonicus Brongniart & Gris	E	FU
L. yateensis Guillaumin	E	MU
Cyperaceae		
Baumea		
B. veillonis Raynal	E	FU
Carex		
C. appressa R. Brown	P	FI

C. brunnea Thunberg	P	MI
C. dietrichiae Boeckeler	P	FI
C. indica L.	P	FA
C. inversonervosa Nelmes	E	FA
C. neurochlamys F. Mueller	P	FA
Costularia		
C. sylvestris Raynal	E	FU
Scleria		
S. ovinux Raynal	E	FA
S. rheophila Raynal	E	FA
Uncinia		
U.uncinata (L. f.) Kuekenthal	E	FI
Dilleniaceae		
Hibbertia		
H. baudouinii Brongniart & Gris	E	MU
H. catargyrea Guillaumin	E	FU
H. comptonii E. G. Baker	E	FA
H. oubatchensis Schlechter	E	FA
H. pancheri (Pancher & Sébert) Briquet	E	FI
H. podocarpifolia Schlechter	E	FI
H. sp.	E	MU
Tetracera		
T. billardieri Martelli	E	FI
Dioscoreaceae		
Dioscorea		
D. bulbifera L.		FA
Ebenaceae		
Diospyros		
D. austrocaledonica Hiern	E	FI
D. balansae Guillaumin	E	FU
D. elliptica (J. R. & G. Forster) P. S. Green	P	
D. fasciculosa (F. Mueller) F. Mueller	P	FA
D. lecardii Guillaumin	E	FA
D. leroyi Kostermans	E	FA
D. macrocarpa Hiern	E	FU
D. olen Hiern	P	FA
D. oubatchensis Kostermans	E	FA
D. pancheri Kostermans	E	
D. parviflora (Schlechter) Bakhuizen	E	FI
D. rufa (Labillardière) Fosberg	P	MI
D. xylopiaefolia Guillaumin	E	FA
Elaeocarpaceae		
Dubouzetia		
Dubouzetia elegans Brongniart & Gris	E	FI
Elaeocarpus		
E. alaternoides Brongniart & Gris	E	MI
E. angustifolius Blume	P	FI
E. baudouinii Brongniart & Gris	E	FU
E. biflorus Tirel	E	FA
E. brachypodus Guillaumin	E	FI
E. bullatus Tirel	E	FA
E. castanaefolius Guillaumin	E	FA
E. colnettianus Guillaumin	E	FA
E. comptonii E. G. Baker	E	FA
E. dognyensis Guillaumin	E	FA
E. geminiflorus Brongniart & Gris	E	FA
E. guillainii Vieillard	E	FA
E. gummatus Guillaumin	E	FU
E. hortensis Guillaumin	P	FI
E. leratii Schlechter	E	FU
E. moratii Tirel	E	FA
E. ovigerus Brongniart & Gris	E	FA
E. pulchellus Brongniart & Gris	E	MI
E. rotundifolius Brongniart & Gris	E	MI
E. seringii Montrouzier	E	MI
E. spathulatus Brongniart & Gris	E	MI
E. speciosus Brongniart & Gris	E	FU
E. toninensis E. G. Baker	E	FI
E. vaccinioides F. Mueller	E	FU
E. vieillardii Brongniart & Gris	E	FA
E. weibelianus Tirel	E	FU
E. yateensis Guillaumin	E	FU
Sloanea		
S. billardieri (Vieillard) A. C. Smith	E	FA
S. haplopoda (Guillaumin) A. C. Smith	E	FU
S. koghiensis Tirel	E	FA
S. lepida Tirel	E	FA
S. magnifolia Tirel	E	FA
S. montana (Labillardière) A. C. Smith	E	FI
S. ramiflora Tirel	E	FA
S. raynaliana Tirel	E	FA
S. suaveolens Tirel	E	FA
Epacridaceae		
Dracophyllum		
D. ramosum Pancher ex Brongniart & Gris	E	MU
D. verticillatum Labillardière	E	MI

Styphelia		
S. balansae Virot	E	MU
S. dammarifolia (Brongniart & Gris) F. Mueller		
S. macrocarpa (Schlechter) Sleumer	E	MU
Ericaceae		
Agapetes		
A. neocaledonica Guillaumin	E	FA
Escalloniaceae		
Argophyllum		
A. brevipetalum Guillaumin	E	FU
A. ellipticum Schlechter	E	FA
A. laxum Schlechter	E	MU
A. nitidum J. R. & G. Forster	E	FA
A. vernicosum Daeniker	E	FU
Euphorbiaceae		
Antidesma		
A. messianianum Guillaumin	E	FA
Austrobuxus		
A. brevipes A. Shaw	E	MU
A. buxoides (Baillon) A. Shaw	E	MU
A. carunculatus (Baillon) A. Shaw	E	MU
A. cuneatus (A. Shaw) A. Shaw	E	MU
A. depauperatus (Baillon) A. Shaw	E	MU
A. eugeniifolius (Guillaumin) A. Shaw	E	MU
A. huerlimannii A. Shaw	E	FU
A. lugubris A. Shaw	E	FU
A. montisdo A. Shaw	E	FU
A. ovalis A. Shaw	E	FU
A. pauciflorus A. Shaw	E	FU
A. rubiginosus (Guillaumin) A. Shaw	E	MU
A. vieillardii (Guillaumin) A. Shaw	E	FA
Baloghia		
B. alternifolia Baillon	E	MI
B. anisomera Guillaumin	E	FU
B. balansae (Baillon) Pax	E	FA
B. bureavii (Baillon) Schlechter	E	MU
B. lucida Endlicher	P	FA
B. montana (Mueller Argovie) Pax	E	FA
B. pininsularis Guillaumin	E	MI
B. pulchella Schlechter ex Pax	E	MU
Bocquillonia		
B. arborea A. Shaw	E	FA
B. brevipes Mueller Argovie	E	FA
B. codonostylis (Baillon) A. Shaw	E	FA
B. grandidens Baillon	E	FA
B. lucidula A. Shaw	E	FI
B. nervosa A. Shaw	E	FI
B. phenacostigma A. Shaw	E	FA
B. rhomboidea (Schlecht.) A. Shaw	E	MU
B. schistophila A. Shaw	E	FA
B. sessiliflora Baillon	E	FU
B. spicata Baillon	E	FU
Claoxylon		
C. insulanum Mueller Argovie	E	FA
Cleidion		
C. claoxyloides Mueller Argovie	E	FU
C. lasiophyllum Pax & Hoffmann	E	FU
C. marcarangoides Guillaumin	E	FA
C. spathulatum Baillon	E	FI
C. verticillatum Baillon	E	MI
C. vieillardii Baillon	E	MI
Cleistanthus		
C. stipitatus Mueller Argovie	E	FI
Cocconerion		
C. balansae Baillon	E	FU
C. minus Baillon	E	MU
Codiaeum		
C. inophyllum (Forst.) Mueller Argovie	P	FI
C. variegatum (L.) Blume	P	MI
Croton		
C. cordatulus A. Shaw	E	FU
Dendrophyllanthus		
D. comptonii S. Moore	E	FU
Drypetes		
D. deplanchei (Brongniart & Gris) Merrill	E	FA
Glochidion		
G. billardieri Baillon	E	FA
G. caledonicum Mueller Argovie	E	FA
G. glaucum (Labillardière) Mueller Argovie		
G. kanalense Baillon	E	FA
G. macrophyllum Mueller Argovie	E	
G. wagapense (Muel. Arg.) Briquet	E	FI

Homalanthus		
H. nutans (J. R. & G. Forster) Guillaumin	P	FI
H. repandus Schlechter	E	FI
H. schlechteri Pax & Hoffmann	E	FI
Macaranga		
M. alchorneoides Pax & Liegelsheim	E	FU
M. coriacea Mueller Argovie	E	MI
M. fulvescens Schlechter	E	FI
M. insularis Schlechter	E	FA
M. oreophila Pax & Hoffman	E	
M. vedeliana Mueller Argovie	E	FI
M. vieillardii Mueller Argovie	E	MU
Mallotus		
M. repandus (Willd.) Mueller Argovie	P	FI
Neoguillauminia		
N. cleopatra (Baillon) Croizat	E	FU
Phyllanthus		
P. amieuensis Guillaumin	E	FA
P. baladensis Baillon	E	FA
P. balansaeanus Guillaumin	E	FU
P. baumannii Guillaumin	E	FU
P. bourgeoisii Baillon	P	FA
P. bupleuroides Baillon	E	MI
P. casearoides S. Moore	E	FA
P. cataractarum Mueller Argovie	E	FU
P. caudatus Mueller Argovie	E	MU
P. chamaecerasus Baillon	E	FA
P. chrysanthus Baillon	E	MA
P. gneissicus S. Moore	E	FA
P. hurlimannii Guillaumin	E	MU
P. induratus S. Moore	E	FU
P. jaubertii Vieillard ex Guillaumin	E	FA
P. koghiensis Guillaumin	E	FU
P. lifuensis Guillaumin	E	FA
P. longeramosus Guillaumin	E	FU
P. loranthoides Baillon	E	MI
P. mareensis Guillaumin	E	FA
P. maytenifolius S. Moore	E	FI
P. ouveanus Daeniker	E	FA
P. platycalyx Mueller Argovie	E	MI
P. rhodocladus S. Moore	E	FA
P. sarasinii Guillaumin	E	FA
P. stenophyllus Guillaumin	E	FA
P. sylvincola S. Moore	E	FA
P. toninensis S. Moore	E	FA
P. torrentium Mueller Argovie	E	FU
P. trichopodus Guillaumin	E	FA
P. umbraecolus Guillaumin	E	FU
P. vespertilio (Mueller Argovie) Baillon	E	MI
P. yahouensis Schlechter	E	FA
Fagaceae		
Nothofagus		
N. aequilateralis (Baumann-Bodenheim) Steenis	E	FI
N. balansae (Baillon) Steenis	E	FU
N. baumanniae (Baumann-Bodenheim) Steenis	E	FU
N. codonandra (Baillon) Steenis	E	Fu
N. discoidea (Baumann-Bodenheim) Steenis	E	FU
Flacourtiaceae		
Casearia		
C. coriifolia Lescot & Sleumer	E	FU
C. lifuana Daeniker	E	FA
C. puberula Guillaumin	E	FU
C. silvana Schlechter	P	MI
Homalium		
H. decurrens (Vieillard) Briquet	E	MI
H. francii Guillaumin	E	FI
H. guillainii (Vieillard) Briquet	E	MI
H. intermedium (Vieillard) Briquet	E	FA
H. polystachyum (Vieillard) Briquet	E	MI
H. rivulare (Vieillard) Briquet	E	FA
H. sleumerianum Lescot	E	FA
Lasiochlamys		
L. cordifolia Sleumer	E	FA
L. coriacea Sleumer	E	FA
L. fasciculata (Guillaumin) Sleumer	E	FA
L. hurlimannii (Guillaumin) Sleumer	E	FU
L. koghiensis (Guillaumin) Sleumer	E	MU
L. mandjeliana Sleumer	E	FA
L. pseudocoriacea Sleumer	E	FU
L. reticulata (Schlechter) Pax & Hoffmann	E	FA

L. rivularis Sleumer	E	FU
L. trichostemona (Guillaumin) Sleumer	E	FU
Xylosma		
X. bernardianum Sleumer	E	FA
X. confusum Guillaumin	E	FU
X. dothioense Guillaumin	E	FU
X. gigantifolium Sleumer	E	FA
X. inaequinervium Sleumer	E	FU
X. lancifolium Sleumer	E	FA
X. lifuanum Guillaumin	E	FA
X. peltatum (Sleumer) Lescot	E	FU
X. tuberculatum Sleumer	E	FU
X. vincentii Guillaumin	E	FI
Flagellariaceae		
Flagellaria		
F. neocaledonica Schlechter	E	MU
Flindersiaceae		
Flindersia		
F. fournieri Pancher & Sébert	E	FI
Gesneriaceae		
Coronanthera		
C. barbata C. B. Clarke	E	FU
C. clarkeana Schlechter	E	FA
C. deltoidifolia Vieillard ex C. B. Clarke	E	FA
C. pancheri C. B. Clarke	E	FU
C. pedunculosa C. B. Clarke	E	MU
C. pinguior C. B. Clarke	E	FA
C. pulchra C. B. Clarke	E	FU
C. sericea C. B. Clarke	E	FU
C. squamata Virot	E	MU
Depanthus		
D. glaber (C. B. Clarke) S. Moore	E	FA
D. pubescens Guillaumin	E	FA
Goodeniaceae		
Scaevola		
S. balansae Guillaumin	E	FU
S. indigofera Schlechter	E	FA
Gramineae		
Centosteca		
C. lappacea (L.) Desvaux	P	MA
Leptaspis		
L. banksii R. Brown	P	FA
Oplismenus		
O. compositus (L.) Palisot de Beauvois	P	MI
O. hirtellus (L.) Palisot de Beauvois	P	MI
Guttifereae		
Calophyllum		
C. caledonicum Vieillard	E	FI
Garcinia		
G. amplexicaulis Vieillard	E	FI
G. balansae Pierre	E	MU
G. densiflora Pierre	E	FA
G. laurina Vieillard ex Guillaumin	E	FA
G. lenormandii Vieillard ex Guillaumin	E	FA
G. neglecta Vieillard	E	FI
G. pedicellata Seemann	E	FI
G. puat Guillaumin	E	FI
G. sessilis Seemann	P	
G. vieillardii Pierre	E	FA
G. virgata Vieillard ex Guillaumin	E	FA
Mammea		
M. neurophylla (Schlechter) Kostermans	E	FI
Montrouziera		
M. cauliflora Planchon & Triana	E	FI
M. gabriellae Baillon	E	FU
M. verticillata Planchon & Triana	E	MU
Hernandiaceae		
Hernandia		
H. cordigera Vieillard	E	FI
Hippocrateaceae		
Dicarpellum		
D. baillonianum (Loes.) A. C. Smith	E	FA
D. dognyensis (Guillaumin) Comb. required	E	FA
D. neocaledonicum (Loes.) A. C. Smith	E	FU
D. pancheri (Baill.) A. C. Smith	E	FI
D. poissonianum (Loes.) A. C. Smith	E	FA
D. pronyense (Guill.) A. C. Smith	E	FU
Icacinaceae		
Apodytes		
A. clusiaefolium (Baillon) Villiers	E	FI
Citronella		
C. macrocarpa Huerlimann	E	FA
C. sarmentosa (Baillon) Howard	E	FI
Gastrolepis		
G. austrocaledonica (Baillon) Howard	E	FU

Lauraceae			
Adenodaphne			
A. macrophylla Kostermans	E		MU
A. spathulata Kostermans	E		MI
Beilschmiedia			
B. neocaledonica Kostermans	E		MI
B. oreophila Schlechter	E		MU
Cryptocarya			
C. aristata Kostermans	E		FA
C. chartacea Kostermans	E		FA
C. elliptica Schlechter	E		FA
C. gracilis Schlechter	E		FI
C. guillauminii Kostermans	E		FU
C. leptospermoides Kostermans	E		MI
C. lifuensis Guillaumin	E		FA
C. mackeei Kostermans	E		MU
C. macrocarpa Guillaumin	E		FI
C. macrodesme Schlechter	E		FA
C. odorata Guillaumin	E		MU
C. oubatchensis Schlechter	E		FA
C. phyllostemon Kostermans	E		FU
C. pluricostata Kostermans	E		MI
C. schmidii Kostermans	E		FA
C. transversa Kostermans	E		FU
C. velutinosa Kostermans	E		FA
Endiandra			
E. baillonii (Pancher & Sébert) Guillaumin	E		MU
E. lecardii Guillaumin	E		FA
E. neocaledonica Kostermans	E		FI
E. polyneura Schlechter	E		FA
E. poueboensis Guillaumin	E		FA
E. sebertii Guillaumin	E		FI
Litsea			
L. lecardii Guillaumin	E		MI
L. neocaledonica S. Moore	E		FI
Lecythidaceae			
Barringtonia			
B. integrifolia (Montrouzier) Schlechter	E		FI
B. neocaledonica Vieillard	E		FA
Liliaceae			
Arthropodium			
A. neocaledonicum E. G. Baker	E		FI
A. paniculatum R. Brown	P		FI
Astelia			
A. neocaledonica Schlechter	E	e	FI
Dianella			
D. ensifolia (L.) A. DC.	P		
D. javanica (Blume) Kunth	P		FI
D. pendula Schlittler	E		MI
Linaceae			
Hugonia			
H. latifolia (Vieillard) Schlechter	E		MI
H. neocaledonica Vieillard	P		FA
H. oreogena Schlechter	E		FI
H. papillosa Guillaumin	E		FI
Loganiaceae			
Fagraea			
F. berteriana A. Gray	P		FI
Geniostoma			
G. balansaeanum Baillon	E		MU
G. celastrineum Baillon	E		MI
G. densiflorum Baillon	E		MU
G. erythrospermum Baillon	E		FA
G. glaucescens Schlechter	E		MA
G. mooreanum Conn	E		FA
G. novaecaledoniae Vieillard ex Baillon	E		FI
G. rupestre J. R. & G. Forster	P		MI
G. vestitum Baillon	E		FA
Neuburgia			
N. neocaledonica (Gilg & Benedict) Comb. required	P		FI
Loranthaceae			
Amyema			
A. artensis (Montrouzier) Danser	P		FU
A. scandens (Tieghem) Danser	E		FU
Amylotheca			
A. dictyophleba (F. Mueller) Tieghem	P		FA
A. pyramidata (Tieghem) Danser	E		FU
Korthalsella			
K. amentacea (Tieghem) Danser	P		
K. dichotoma (Tieghem) Engler	E		FI
K. striata Danser	E		FI
Meliaceae			
Aglaia			
A. elaeagnoidea (Jussieu) Bentham	P		FI
Amoora			
A. balansaeana C. DC.	E		FA
A. vieillardii C. DC.	E		FU
Dysoxylum			
D. albicans Vieillard ex C. DC.	E		MI
D. balansaeanum C. DC.	E		FU

D. comptonii E. G. Baker	E	FA
D. francii C. DC.	E	FA
D. gamosepalum E. G. Baker	E	FA
D. glomeratum Vieillard	E	FU
D. hurlimannii Guillaumin	E	FI
D. kouiriense Virot	E	FA
D. rufescens Vieillard	E	FI
D. vieillardii C. DC.	E	FU
Menispermaceae		
Hypserpa		
H. neocaledonica Diels	P	FI
H. vieillardii Diels	E	MU
Pachygone		
P. loyaltiensis Diels	E	FI
P. tomentella Diels	E	MI
Stephania		
S. forsteri (A. DC.) A. Gray	P	FA
Mimosaceae		
Adenanthera		
A. pavonina L.	P	FA
Albizia		
A. granulosa (Labillardière) Bentham	E	FI
A. paivana Fournier	E	FU
A. streptocarpa Fournier	E	FA
Serianthes		
S. melanesica Fosberg	P	FA
S. sachetae Fosberg	E	FU
Monimiaceae		
Hedycarya		
H. aragoensis Jérémie	E	FA
H. baudouini Baillon	E	FU
H. chrysophylla Perkins	E	FA
H. cupulata Baillon	E	FA
H. engleriana S. Moore	E	FA
H. parvifolia Perkins & Schlechter	E	FU
H. rivularis Guillaumin	E	FA
H. symplocoides S. Moore	E	FA
Kibaropsis		
K. caledonica (Guillaumin) Jérémie	E	FA
Moraceae		
Ficus		
F. asperula Bureau	E	FI
F. austrocaledonica Bureau	E	FI
F. barraui Guillaumin	E	FA
F. cataractarum Vieillard ex Bureau	E	FA
F. dzumacensis Guillaumin	E	FU
F. heteroselis Bureau	E	FA
F. hurlimannii Guillaumin	E	FA
F. leiocarpa (Bureau) Warburg	E	FA
F. lifouensis Corner	E	FA
F. maialis Guillaumin	E	FU
F. mareensis Warburg	E	FA
F. microcarpa L. f.	P	MA
F. microtophora Corner	E	FA
F. mutabilis Bureau	E	MI
F. nitidifolia Bureau	E	FU
F. otophora Corner & Guillaumin	E	FA
F. otophoroides Corner	E	FA
F. pancheriana Bureau	E	FA
F. planchonellaefolia Guillaumin	E	FU
F. prolixa J. R. Forster	P	FA
F. pteroporum Guillaumin	E	FA
F. racemigera Bureau	E	FI
F. schlechteri Warburg	E	MI
F. versicolor Bureau	E	FA
F. vieillardiana Bureau	E	FA
F. virgata Reinwardt	P	MI
F. webbiana (Miquel) Miguel	E	FI
Pseudomorus		
P. brunoniana Bureau	E	FA
Sparattosyce		
S. balansae A. Richter ex Guillaumin	E	FA
S. dioica Bureau	E	FU
Streblus		
S. sclerophyllus Corner	E	FU
Myricaceae		
Canacomyrica		
C. monticola Guillaumin	E	FU
Myrsinaceae		
Maesa		
M. novocaledonica Mez	E	FI
Rapanea		
R. asymmetrica Mez	E	MU
R. citrifolia Mez	E	MI
R. diminuta Mez	E	FU
R. grandifolia S. Moore	E	FU
R. lanceolata (Pancher & Sébert) Mez	E	FI
R. lecardii Mez	E	FA
R. lifuensis Mez	E	FA
R. macrophylla (Pancher & Sébert) Mez	E	FU
R. modesta Mez	E	MA
R. stenophylla Mez	E	FA
R. vieillardii Mez	E	FI
Tapeinosperma		
T. acutangulum Mez	E	FA
T. amosense Guillaumin	E	FA
T. amplexicaule Mez	E	FA
T. aragoense Guillaumin	E	FA
T. clethroides Mez	E	FA

T. colnettianum Guillaumin	E	FA
T. deflexum Mez	E	FU
T. ellipticum Mez	E	FA
T. glandulosum Guillaumin	E	
T. gracile Mez	E	FI
T. grandiflorum Guillaumin	E	
T. koghiense Guillaumin	E	FU
T. laeve Mez	E	FA
T. laurifolium Mez	E	
T. lecardii Mez	E	FA
T. lenormandii J. D. Hooker	E	FU
T. minutum Mez	E	FI
T. multipunctatum Guillaumin	E	FA
T. nectandroides Mez	E	FA
T. nitidum Mez	E	MI
T. oblongifolium Mez	E	FA
T. pancheri Mez	E	FU
T. pauciflorum Mez	E	FA
T. pennelii Guillaumin	E	FA
T. psaladense Mez	E	FA
T. pulchellum Mez	E	
T. robustum Mez	E	FU
T. rubidum Mez	E	FA
T. rubriscapum Guillaumin	E	FA
T. schlechteri Mez	E	FI
T. scrobiculatum (Seem.) Mez	P	
T. squarrosum Mez	E	FA
T. tenue Mez	E	FA
T. vestitum Mez	E	FU
T. vieillardii J. D. Hooker	E	FI
T. wagapense Mez	E	FA
T. whitei Guillaumin	E	FU
Myrtaceae		
Archirhodomyrtus		
A. baladensis (Brongniart & Gris) Burret	E	MU
A. paitensis (Schlechter) Burret	E	MI
A. vieillardii (Brongiart & Gris) Burret	E	FU
Arillastrum		
A. gummiferum Pancher ex Baillon	E	FU
Austromyrtus		
A. ploumensis (Daeniker) Burret	E	FU
A. vieillardii (Brongniart & Gris) Burret	E	FA
Carpolepis		
C. elegans (Montrouzier) Dawson ined.	E	MI
C. laurifolia (Brongniart & Gris) Dawson ined.	E	FI
C. tardiflora Dawson ined.	E	FI
Caryophyllus		
C. amieuensis Guillaumin	E	FA
C. arboreus (E. G. Baker) Guillaumin	E	FA
C. baladensis Brongniart & Gris	E	FA
C. balansae Guillaumin	E	FI
C. deplanchei Guillaumin	E	FI
C. elegans Brongniart & Gris	E	FA
C. ellipticus Labillardière	E	MU
C. garciniifolius Guillaumin	E	FA
C. propinquus Guillaumin	E	FA
C. pterocarpus Vieillard	E	
C. undulatus Guillaumin	E	MI
C. xanthostemifolius Guillaumin	E	MU
Cleistocalyx		
C. brongniartii Merrill & Perry	E	FU
C. pennelii (Guillaumin) Merrill	E	FU
Cupheanthus		
C. comptonii (E. G. Baker) Guillaumin	E	FA
C. neocaledonicus Seemann	E	FA
C. paniensis (E. G. Baker) Guillaumin	E	FA
C. serpentini A. Shaw	E	FU
C. toninensis (E. G. Baker) Guillaumin	E	FA
Eugenia		
E. bullata Pancher	E	FI
E. colnettiana Guillaumin	E	FA
E. crucigera Daeniker	E	FU
E. daaouiensis Guillaumin	E	FA
E. heckelii Pancher & Sébert	E	MU
E. ignambiensis E. G. Baker	E	FA
E. myrtopsidioides Guillaumin	E	FU
E. pronyensis Guillaumin	E	FU
E. pterocarpa Baillon	E	
E. sarasinii Guillaumin	E	FA
Jambosa		
J. acris Pancher	E	FU
J. canalensis Vieillard	E	FI
J. longifolia Brongniart & Gris	E	FA
J. nervosa Vieillard	E	FI

Mearnsia		
M. brevistylis Dawson ined.	E	FI
Metrosideros		
M. dolichandra Schlechter ex Guillaumin	E	FU
M. nitida Brongniart & Gris	E	FI
M. operculata Labillardière	E	MI
M. oreomyrtus Daeniker	E	FI
M. paniensis Dawson ined.	E	FA
M. longipetiolata Dawson ined.	E	MA
M. porphyrea Schlechter	E	FI
Piliocalyx		
P. baudouini Brongniart & Gris	E	MI
P. bullatus Brongniart & Gris	E	FA
P. francii Guillaumin	E	FI
P. micranthus Brongniart & Gris	E	FA
P. robustus Brongniart & Gris	E	FA
P. wagapensis Brongniart & Gris	E	FA
Pleurocalyptus		
P. austrocaledonicus (Guillaumin) Dawson ined.	E	FI
P. pancheri (Brongniart & Gris) Dawson ined.	E	FU
Syzygium		
S. auriculatum Brongniart & Gris	E	MI
S. austrocaledonicum (Seemann) Guillaumin	E	MI
S. conceptionis Guillaumin	E	FA
S. densiflorum Brongniart & Gris	E	FI
S. frutescens Brongniart & Gris	E	FA
S. koghiense Guillaumin	E	FI
S. lateriflorum Brongniart & Gris	E	FA
S. lifuanum Daeniker	E	FA
S. macranthum Brongniart & Gris	E	FU
S. micans Brongniart & Gris	E	FA
S. microsemmifolium Guillaumin	E	
S. mouanum Guillaumin	E	FU
S. neglectum Brongniart & Gris	E	FA
S. pseudopinnatum Daeniker	E	FA
S. quadrangulare Guillaumin	E	FU
S. rivulare Vieillard	E	FA
S. schlechterianum Hochreutiner	E	
S. tenuiflorum Brongniart & Gris	E	MI
S. tripetalum Guillaumin	E	FI
S. verrucosum Daeniker	E	FA
S. wagapense Brongniart & Gris	E	FI
Tristaniopsis		
T. polyandra (Guillaumin) Dawson ined.	E	MU
Xanthostemon		
X. flavum (Brongniart & Gris) Schlechter	E	MU
X. grisei Guillaumin	E	FA
X. sebertii Guillaumin	E	FU
X. vieillardii (Brongniart & Gris) Niedenzu	E	FA
Nepenthaceae		
Nepenthes		
N. vieillardii J. D. Hooker	P	MU
Nyctaginaceae		
Calpidia		
C. gigantocarpa Heimerl	E	FI
C. pancheriana Heimerl	E	FA
Pisonia		
P. aculeata L.	P	FA
Oleaceae		
Jasminum		
J. didymum J. R. & G. Forster	P	FI
J. elatum Pancher ex Guillaumin	E	Fi
J. neocaledonicum Schlechter	E	FI
Linociera		
L. brachystachys (Schlechter) P. S. Green	E	FI
Olea		
O. paniculata R. Brown	P	FI
Osmanthus		
O. austrocaledonicus (Vieillard) Knoblauch	E	MU
Oncothecaceae		
Oncotheca		
O. balansae Baillon	E	FU
O. macrocarpa McPherson Morat & Veillon	E	FU
Orchidaceae		
Acanthephippium		
A. vitiense L. O. Williams	P	FI

Acianthus			
A. atepalus H. Reichenbach	E		FU
A. bracteatus Rendle	E		FU
A. confusus Guillaumin	E		FU
A. corniculatus Rendle	E		FA
A. cymbalariifolius F. Mueller & Kraenzlin	E		FI
A. elegans H. Reichenbach	E		FU
A. grandiflorus Schlechter	E		MU
A. heptadactylus Kraenzlin	E		FI
A. oxyglossus Schlechter	E		FU
A. tenuilabris Schlechter	E		MU
A. veillonis Hallé	E		FU
Agrostophyllum			
A. sp.	P	e	FA
Anoectochilus			
A. imitans Schlechter	E		FI
Appendicula			
A. reflexa Blume	P	e	FI
Bulbophyllum			
B. aphanopetalum Schlechter	E	e	FI
B. atrorubens Schlechter	E	e	FI
B. baladeanum J. J. Smith	E	e	FI
B. betchei F. Mueller	P	e	FI
B. comptonii Rendle	E	e	FI
B. gracillimum (Rolfe) Rolfe	E	e	FI
B. hexarhopalos Schlechter	E	e	FI
B. keekee Hallé	E	e	FI
B. lingulatum Rendle	E	e	FI
B. longiflorum DuPetit Thouars	P	e	FI
B. lophoglottis (Guillaumin) Hallé	E	e	FI
B. neocaledonicum Schlechter	E	e	FI
B. ngoyense Schlechter	E	e	FI
B. pachyanthum Schlechter	P	e	FI
B. pallidiflorum Schlechter	E	e	FI
B. polypodioides Schlechter	P	e	FI
Calanthe			
C. balansae Finet	E		FI
C. langei F. Mueller	E		FI
C. neocaledonica Rendle	E		FI
C. oreadum Rendle	E		FI
C. triplicata (Willemet) Ames	P		FI
Ceratostylis			
C. micrantha Schlechter	P	e	FI
Chamaeanthus			
C. aymardii Hallé	E	e	FI
C. begaudii Hallé	E	e	FI
C. neocaledonicus (Rendle) Hallé	E	e	FI
Cheirostylis			
C. montana Blume	P		MI
Chrysoglossum			
C. neocaledonicum Schlechter	E		FA
Cleisostoma			
C. montanum (J. J. Smith) Garay	P	e	FI
Clematepistephium			
C. smilacifolium (H. Reichenbach) Hallé	E		FI
Coelogyne			
C. lycastoides F. Mueller & Kraenzlin	P	e	FA
Coilochilus			
C. neocaledonicus Schlechter	E		MU
Corybas			
C. neocaledonicus (Schlechter) Schlechter	E		FI
Cryptostylis			
C. arachnites (Blume) Hasskarl	P		FI
Dendrobium			
D. arthrobulbum Kraenzlin	E	e	FI
D. austrocaledonicum Schlechter	P	e	FI
D. camaridiorum H. Reichenbach	E	e	FI
D. camptocentrum Schlechter	P	e	FI
D. cleistogamum Schlechter	E	e	FI
D. crassicaule Schlechter	E	e	FI
D. crassifolium Schlechter	E	e	FI
D. deplanchei H. Reichenbach	E		MU
D. finetianum Schlechter	E	e	FI
D. fractiflexum Finet	E		FU
D. gracilicaule F. Mueller	P	e	FI
D. linguiforme Swartz	P	e	FI
D. macrophyllum A. Richard	P	e	FI
D. munificum (Finet) Hallé	E	e	FI
D. muricatum Finet	E	e	FI
D. ngoyense Schlechter	E	e	FU
D. oppositifolium (Kraenzlin) Hallé	E	e	FU
D. pectinatum Finet	E	e	FI
D. poissonianum Schlechter	E	e	FI
D. sylvanum H. Reichenbach	E	e	FI
D. virotii Guillaumin	E	e	FU
Didymoplexis			
D. minor J. J. Smith	P		FA

Diplocaulobium			
D. ouhinnae (Schlechter) Kraenzlin	P	e	FI
Earina			
E. deplanchei H. Reichenbach	E		MU
E. valida H. Reichenbach	P	e	FI
Epipogium			
E. roseum (D. Don) Lindley	P		FU
Eria			
E. aeridostachya H. Reichenbach ex Lindley	P	e	FU
E. karicouyensis Schlechter	E	e	FU
E. vieillardii H. Reichenbach	P	e	FI
Erythrodes			
E. oxyglossa Schlechter	P		FU
Eulophia			
E. moratii Hallé	E		FA
E. pulchra (DuPetit Thouars) Lindley	P		FA
Glomera			
G. macdonaldii (Schlechter) J. J. Smith	P	e	FI
Gonatostylis			
G. bougainvillei Hallé	E		FU
G. vieillardii (H. Reichenbach) Schlechter	E		MI
Goodyera			
G. rubicunda (Blume) Lindley	P		FA
G. scripta (H. Reichenbach) Schlechter	E		FA
G. subregularis (H. Reichenbach) Schlechter	E		FA
G. viridiflora (Blume) Blume	P		FA
Habenaria			
H. insularis Schlechter	E		FA
Hetaeria			
H. discoidea (H. Reichenbach) Schlechter	P		FA
Liparis			
L. caespitosa Lindley	P	e	FI
L. chalandei Finet	E		MU
L. condylobulbon H. Reichenbach	P	e	FI
L. elliptica Wight	P	e	FI
L. gibbosa Finet	P	e	FI
L. laxa Schlechter	E		MI
L. layardii F. Mueller	P		FA
L. phalacrocorax Hallé	E		MI
L. sula Hallé	E		FU
L. zosterops Hallé	E		FI
Malaxis			
M. taurina (H. Reichenbach) Kuntze	E		MI
Megastylis			
M. glandulosa (Schlechter) Schlechter	E		FI
M. latissima (Schlechter) Schlechter	E		FU
M. montana (Schlechter) Schlechter	E		FU
Microtatorchis			
M. oreophila Schlechter	E	e	FI
M. schlechteri Garay	E	e	FI
Moerenhoutia			
M. grandiflora (Schlechter) Schlechter	E		FI
Nervilia			
N. aragoana Gaudichaud	P		FA
N. platychila Schlechter	P		FA
Oberonia			
O. ensiformis (J. E. Smith) Lindley	P	e	MI
O. equitans (G. Forster) Mutel	P	e	FI
O. fissiglossa Hallé	E	e	FI
O. neocaledonica Schlechter	E	e	FI
Octarrhena			
O. oberonioides (Schlechter) Schlechter	E	e	FI
O. saccolabioides (Schlechter) Schlechter	E	e	FI
Pachyplectron			
P. arifolium Schlechter	E		FI
P. neocaledonicum Schlechter	E		MU
Peristylus			
P. minimiflorus (Kraenzlin) Hallé	E		FA
P. ngoyensis (Schlechter) Hallé	E		FI
Phajus			
P. daenikeri Kraenzlin	E		FU
P. neocaledonicus Rendle	E		FA
P. robertsii F. Mueller	E		FI
Pholidota			
P. pallida Lindley	P	e	FI
Phreatia			
P. hypsorhynchos Schlechter	E	e	FI
P. neocaledonica Schlechter	E	e	FI
P. oubatchensis Schlechter	E	e	FI
P. pachyphylla Schlechter	P	e	FI
P. paleata H. Reichenbach	E	e	FI
P. sublata Hallé	E	e	FI

Pristiglottis			
P. montana (Schlechter) Cretziou & J. J. Smith	P		FI
Pterostylis			
P. bureaviana Schlechter	E		FU
P. curta R. Brown	P		FA
Rhynchophreatia			
R. micrantha (A. Richard) Hallé	P	e	FI
Sarcochilus			
S. hillii (F. Mueller) F. Mueller	P	e	FI
S. koghiensis Schlechter	E	e	FI
Schoenorchis			
S. micrantha Reinwardt ex Blume	P	e	FI
Spathoglottis			
S. petri H. Reichenbach	P		FA
Spiranthes			
S. sinensis (Persoon) Ames	P		MI
Taeniophyllum			
T. graptolitum Hallé	E	e	FI
T. trachypus Schlechter	E	e	FI
Thrixspermum			
T. sp.	P	e	FI
Trachoma			
T. subluteum (Rupp) Garay	P	e	FI
Tropidia			
T. viridifusca Kraenzlin	E		FA
Zeuxine			
Z. francii Schlechter	E		FA
Z. vieillardii (H. Reichenbach) Schlechter	E		FA

Palmae

Actinokentia		
A. divaricata Dammer	E	FU
A. huerlimannii H. E. Moore	E	FU
Alloschmidia		
A. glabrata (Beccari) H. E. Moore	E	FA
Basselinia		
B. deplanchei Vieillard	E	FA
B. eriostachys Beccari	E	FI
B. gracilis Vieillard	E	FI
B. pancheri Vieillard	E	FU
B. surculosa Beccari	E	FU
B. tomentosa Beccari	E	FA
B. velutina Beccari	E	FI
Brongniartikentia		
B. lanuginosa H. E. Moore	E	FA
B. vaginata (Brongniart) Beccari	E	FU
Burretiokentia		
B. hapala H. E. Moore	E	FA
B. vieillardii H. E. Moore	E	FI
Campecarpus		
C. fulcitus Wendland	E	FU
Chambeyronia		
C. lepidota H. E. Moore	E	FI
C. macrocarpa Vieillard	E	FI
Clinosperma		
C. bractealis Beccari	E	FU
Cyphokentia		
C. macrostachya Brongniart	E	FI
Cyphophoenix		
C. elegans Wendland	E	FA
C. nucele H. E. Moore	E	FA
Cyphosperma		
C. balansae Wendland ex Bentham & J. D. Hooker	E	FA
Kentiopsis		
K. olivaeformis Brongniart	E	FA
Lavoixia		
L. macrocarpa H. E. Moore	E	FA
Mackeea		
M. magnifica H. E. Moore	E	FA
Moratia		
M. cerifera H. E. Moore	E	FA
Pritchardiopsis		
P. jeanneneyi Beccari	E	FU
Veillonia		
V. alba H. E. Moore	E	FA

Pandanaceae

Freycinetia		
F. brevifolia Martelli	E	FA
F. comptonii Rendle	E	FA
F. coriacea Warburg	E	FA
F. cylindracea Solms	E	FI
F. erythrostigma Solms ex Martelli	E	FU
F. graminifolia Solms	E	FI
F. hydra B. C. Stone	E	FI
F. longispica Martelli	P	FA
F. lorifolia Martelli	E	FU
F. microdonta Martelli	P	FA
F. monticola Rendle	P	FA
F. novocaledonica Warburg	E	FA
F. schlechteri Warburg	P	FI
F. spectabilis Solms	E	FI
F. sulcata Warburg	P	FI
F. verrucosa Warburg	E	FA
F. vieillardii Martelli	E	FA
Pandanus		
P. altissimus Solms	E	FI
P. aragoensis Solms	E	FA
P. balansae Solms	E	FU
P. bernardii St. John	E	FI
P. clandestinus B. C. Stone	E	FA
P. lacuum St. John	E	FU

P. mackeei St. John	E	FU
P. macrocarpus (Brongniart) Solms	E	FA
P. neocaledonicus Martelli	E	FA
P. oblongus (Brongniart) Solms	E	FI
P. pancheri (Brongniart) Solms	E	FU
P. sphaerocephalus (Brongniart) Solms	E	FI
P. verecundus B. C. Stone	E	FU
P. vieillardii Martelli	E	FA
Papilionaceae		
Arthroclianthus		
A. andersonii (Seemann) Schindler	E	
A. angustifolius Hochreutiner	E	FA
A. balansae Schindler	E	MI
A. caudatus Schindler	E	
A. comptonii E. G. Baker	E	FA
A. cuneatus Schindler	E	FU
A. grandifolius E. G. Baker	E	FA
A. ischnopodus Guillaumin	E	FA
A. leratii Schindler	E	
A. macrobothryosus Hochreutiner	E	FA
A. macrophyllus Schindler	E	
A. maximus Schindler	E	
A. microbotrys Hochreutiner	E	FA
A. ovalifolius Schindler	E	
A. tenuifolius Schindler	E	
A. vieillardii Schindler	E	
Mucuna		
M. gigantea (Willdenow) A. DC.	P	
M. neocaledonica E. G. Baker	E	FA
Paracryphiaceae		
Paracryphia		
P. alticola (Schlechter) Steenis	E	FI
Passifloraceae		
Passiflora		
P. aurantia J. R. Forster	P	MI
Phellinaceae		
Phelline		
P. balansae Baillon	E	
P. billardieri Pancher ex Loesener	E	FU
P. brachyphylla Baillon	E	FA
P. comosa Labillardière	E	FI
P. confertifolia Baillon	E	FA
P. dumbeensis Guillaumin	E	FU
P. erubescens Baillon	E	MI
P. floribunda Baillon	E	
P. indivisa (Baillon) Harms & Loesener	E	
P. lucida Vieillard ex Baillon	E	FU
P. macrophylla Baillon	E	FI
P. microcarpa Baillon	E	FU
P. robusta Baillon	E	FU
P. wagapensis Baillon	E	FA
Philesiaceae		
Eustrephus		
E. latifolius R. Brown ex Sims	P	FI
Geitonoplesium		
G. cymosum (Cunningham) R. Brown	P	FI
Piperaceae		
Peperomia		
P. baueriana Miquel	P	FA
P. caledonica C. DC.	E	FI
P. endlicheri Miquel	P	FA
P. insularum Miquel	P	
P. kanalensis C. DC.	E	
P. leptostachya W. J. Hooker & Arnott	P	FI
P. lifuana C. DC.	E	FA
P. reflexa A. Dietrich	P	FU
P. sarasinii C. DC.	E	FU
P. subpallescens C. DC.	E	
Piper		
P. austrocaledonicum C. DC.	P	FI
P. comptonii S. Moore	E	FA
P. paitense Schlechter	E	FU
P. staminodiferum C. DC.	E	FA
Pittosporaceae		
Pittosporum		
P. cheesmanniae Guillaumin	E	FA
P. coccineum (Montrouzier) Beauvisage	E	FU
P. dognyense Guillaumin	E	FA
P. dzumacense Guillaumin	E	MU
P. echinatum Brongniart & Gris	E	MI
P. heckeli Dubard	E	FA
P. leratii Guillaumin	E	FU
P. lifuense Guillaumin	E	FA
P. morierei (Vieillard) Guillaumin	E	FA
P. mouanum Guillaumin	E	FU
P. obovatum Guillaumin	E	FA
P. oreophilum Guillaumin	E	FA
P. oubatchense Schlechter	E	FA
P. paitense Guillaumin	E	FU

P. pancheri Brongniart & Gris	E	MA
P. paniculatum Brongniart & Gris	E	FA
P. paniense Guillaumin	E	FA
P. poueboense Guillaumin	E	FA
P. pronyense Guillaumin	E	MU
P. rhytidocarpum A. Gray	P	MA
P. sylvaticum Guillaumin	E	FA
P. thyense Guillaumin	E	FA
P. vieillardi Brongniart & Gris	E	FA
Podocarpaceae		
Acmopyle		
A. pancheri (Brongniart & Gris) Pilger	E	FI
Dacrycarpus		
D. vieillardii (Parlatore) de Laubenfels	E	MU
Dacrydium		
D. balansae Brongniart & Gris	E	MU
D. lycopodioides Brongniart &Gris	E	FU
Decussocarpus		
D. comptonii (Buchholz) de Laubenfels	E	FI
Falcatifolium		
F. taxoides (Brongniart & Gris) de Laubenfels	E	FI
Parasitaxus		
P. ustus (Vieillard) de Laubenfels	E	FI
Podocarpus		
P. decumbens N. Gray	E	MU
P. longefoliolatus Pilger	E	FU
P. lucienii de Laubenfels	E	FI
P. polyspermus de Laubenfels	E	FI
P. sylvestris Buchholz	E	FI
Prumnopitys		
P. ferruginoides (Compton) de Laubenfels	E	FI
Proteaceae		
Beauprea		
B. asplenioides Schlechter	E	FU
B. comptonii S. Moore	E	FA
B. crassifolia Virot	E	MA
B. filipes Schlechter	E	FI
B. gracilis Brongniart & Gris	E	MU
B. montana (Brongniart & Gris) Virot	E	MU
B. montisfontium Guillaumin	E	MU
B. neglecta Virot	E	MU
B. pancheri Brongniart & Gris	E	MU
B. spathulaefolia Brongniart & Gris	E	MU
Kermadecia		
K. elliptica Brongniart & Gris	E	FA
K. pronyensis (Guillaumin) Guillaumin	E	FU
K. rotundifolia Brongniart & Gris	E	FA
K. sinuata Brongniart & Gris	E	FA
Knightia		
K. deplanchei Vieillard ex Brongniart & Gris	E	MU
K. strobilina (Labillardière) R. Brown	E	FI
Macadamia		
M. francii (Guillaumin) Sleumer	E	FI
M. neurophylla (Guillaumin) Virot	E	FU
M. rousselii (Vieillard) Sleumer	E	FA
M. vieillardii (Brongniart & Gris) Sleumer	E	FA
Sleumerodendron		
S. austrocaledonicum (Brongniart & Gris) Virot	E	FI
Stenocarpus		
S. intermedius Brongniart & Gris	E	MU
S. rubiginosus Brongniart & Gris	E	FI
S. trinervis (Montr.) Guillaumin	E	MI
S. umbelliferus (Forster) Druce	E	MI
Virotia		
V. leptophylla (Guillaumin) L. Johnson & Briggs	E	FI
Rhamnaceae		
Alphitonia		
A. neocaledonica Guillaumin	E	MI
A. xerocarpa Baillon	E	FU
Ventilago		
V. buxoides Baillon	E	FI
Rhizophoraceae		
Crossostylis		
C. biflora J. R. & G. Forster	P	FI
C. grandiflora Pancher ex Brongniart & Gris	E	FA

C. multiflora Brongniart & Gris	E	FA
C. seberti Brongniart & Gris	E	FU
Rubiaceae		
Atractocarpus		
A. aragoensis Guillaumin	E	FA
A. bracteatus Schlechter & Krause	E	
A. cucumicarpus S. Moore	E	MU
A. heterophyllus Guillaumin & Beauvisage	E	FI
A. longistipitatus Guillaumin	E	FA
A. oblongus S. Moore	E	MU
A. platyxylon (Vieillard ex Pancher & Sébert) Guillaumin	E	FA
A. sessilifolius Guillaumin	E	FA
A. simulans Guillaumin	E	FU
Coelospermum		
C. balansaeanum Baillon	E	FA
C. monticolum Baillon ex Guillaumin	E	FA
C. nigrescens (Krause) Guillaumin	E	FA
Cyclophyllum		
C. baladense Guillaumin	E	FA
C. calyculatum Guillaumin	E	FA
C. cymosum S. Moore	E	FA
C. henriettae (Baillon) Guillaumin	E	FA
C. sagittatum (Baillon) Guillaumin	E	MU
C. vieillardi (Baillon) Guillaumin	E	FA
Gardenia		
G. mollis Schlechter	E	FA
Guettarda		
G. balansaeana Baillon	E	FU
G. eximia Baillon	E	FU
G. heterosepala Guillaumin	E	FU
G. hypolasia Baillon	E	FU
G. rhamnoides Baillon	E	FI
G. splendens Baillon	E	FU
G. vieillardii Guillaumin	E	FA
Ixora		
I. cauliflora Montrouzier	E	FA
I. collina (Montrouzier) Beauvisage	P	FI
I. comptonii S. Moore	E	FA
I. kuakuensis S. Moore	E	FU
I. lecardii Guillaumin	E	FA
I. longiloba Guillaumin	E	FU
I. montana Schlechter	E	FI
I. oligantha Schlechter & Krause	E	FI
I. vieillardii Guillaumin	E	FA
I. yahouensis Schlechter	E	FA
Morierina		
M. montana Vieillard	E	FU
M. propinqua Brongniart & Gris	E	MU
Morinda		
M. artensis Montrouzier	E	MU
M. billardieri Baillon	E	FI
M. choriophylla Baillon	E	FI
M. decipiens Schlechter	E	FA
M. elongata Schlechter	E	FU
M. fallax Schlechter	E	FA
M. forsteri Seemann	P	FA
M. kanalensis Baillon	E	FU
M. ligustrina S. Moore	E	FA
M. phylliraeoides Labillardière	E	FA
M. velutina Guillaumin	E	FI
M. vieillardi Guillaumin	E	FA
Mussaenda		
M. cylindrocarpa Burck	P	FA
Psychotria		
P. amieuensis Guillaumin	E	FA
P. arbutifolia (Baillon) Schlechter	E	FI
P. baillonii Schlechter	E	FA
P. baladensis (Baillon) Guillaumin	E	FA
P. bourailensis Guillaumin	E	
P. brachylaena (Baillon) Guillaumin	E	FA
P. calliantha (Baillon) Guillaumin	E	FI
P. calothyrsus (Baillon) Guillaumin	E	FI
P. canalensis (Baillon) Guillaumin	E	FA
P. collina Labillardière	P	FI
P. colnettiana Guillaumin	E	FA
P. damnatorum Guillaumin	E	FA
P. deplanchei (Beauvisage) Guillaumin	E	FU
P. douarrei (Beauvisage) Daeniker	E	FU
P. faguetii (Baillon) Schlechter	E	FI
P. frondosa S. Moore	E	FA
P. fuscopilosa Schlechter	E	FU
P. gabriellae (Baillon) Guillaumin	E	MU
P. gneissica S. Moore	E	FA

P. goniocarpa (Baillon) Guillaumin	E	FI
P. laxissima S. Moore	E	FA
P. lenormandi Schlechter	E	FU
P. lepidocalyx S. Moore	E	FA
P. leratii Guillaumin	E	FI
P. ligustrina (Baillon) Guillaumin	E	FA
P. lyciiflora (Baillon) Schlechter	E	FU
P. lycioides (Baillon) Guillaumin	E	FU
P. macroglossa (Baillon) Guillaumin	E	FI
P. microglossa (Baillon) Guillaumin	E	FI
P. monanthos (Baillon) Schlechter	E	FU
P. nathaliae (Baillon) Guillaumin	E	MU
P. nekouana (Baillon) Guillaumin	E	
P. nummularioides Guillaumin	E	FA
P. oleoides (Baillon) Schlechter	E	MU
P. oreophila Guillaumin	E	FA
P. oubatchensis Schlechter	E	FA
P. pancheri (Baillon) Schlechter	E	FA
P. papillosa Guillaumin	E	FA
P. paramaracarpa (Baillon) Schlechter	E	MU
P. patula S. Moore	E	FA
P. phyllanthoides Schlechter ex Guillaumin	E	MU
P. poissoniana (Baillon) Guillaumin	E	FI
P. pseudomicrodaphne (Baillon) Guillaumin	E	MU
P. pubituba S. Moore	E	FA
P. pulchrebracteata Guillaumin	E	FA
P. rarifolia S. Moore	E	FA
P. roseotincta S. Moore	E	FA
P. rosmarinifolia (Baillon) Schlechter	E	FA
P. rubefacta (S. Moore) Guillaumin	E	MU
P. rupicola (Baillon) Schlechter	E	MI
P. sagittalis (Baillon) Guillaumin	E	FA
P. saltiensis (S. Moore) Guillaumin	E	FU
P. schlechteriana Krause	E	FA
P. schumanniana Schlechter	E	FA
P. semperflorens (Baillon) Pancher ex Beauvisage	E	FI
P. speciosa (Montrouzier) S. Moore	E	MU
P. stenophylla Guillaumin	E	
P. stricta (Baillon) Guillaumin	E	FA
P. suaveolens S. Moore	E	FA
P. subpallens S. Moore	E	MU
P. subuniflora (Baillon) Guillaumin	E	FU
P. toninensis S. Moore	E	FA
P. trichopodantha (Baillon) Guillaumin	E	FA
P. trisulcata (Baillon) Guillaumin	E	FA
P. unioensis Guillaumin	E	FI
P. vieillardii (Baillon) Baillon	E	FA
P. wagapensis Guillaumin	E	FA
Randia		
R. artensis (Montrouzier) Guillaumin	E	FA
R. sezitat Guillaumin	E	FA
R. vieillardii Baillon	E	FA
Rhopalobrachium		
R. congestum Schlechter	E	FA
R. fragrans Schlechter & Krause	E	FA
Tarenna		
T. ignambiensis (Guillaumin) Jérémie	E	FA
T. lifouana (Daeniker) Jérémie	E	FA
T. truncatocalyx (Guillaumin) Bremekamp	E	MI
T. unioensis (Guillaumin) Jérémie	E	FA
Rutaceae		
Dutaillyea		
D. comptonii E. G. Baker	E	FA
D. oreophila Guillaumin	E	FI
D. sessilifoliola Guillaumin	E	MI
D. trifoliolata Baillon	E	FA
Geijera		
G. balansae (Baillon) Schinz & Guillaumin	E	FI
Melicope		
M. diversifolia Guillaumin	E	FA
M. glaberrima Guillaumin	E	FI
M. lasioneura Baillon	E	FU
M. leptophylla Guillaumin	E	
M. leratii Guillaumin	E	FU
M. montana E. G. Baker	E	FA

M. platystemon E. G. Baker	E	FA
M. triphylla Merrill	E	FI
Micromelum		
M. minutum (J. R. & G. Forster) Wight & Arnott	E	FA
Sarcomelicope		
S. argyrophylla Guillaumin	E	FI
S. sarcococca Engler	E	FI
Zanthoxylum		
Z. neocaledonicum E. G. Baker	E	FU
Z. pancheri P. S. Green	E	MI
Z. sarasinii Guillaumin	E	MA
Z. schlechteri Guillaumin	E	FI
Zieridium		
Z. melicopaefolium Guillaumin	E	FI
Z. pseudobtusifolium Guillaumin	E	FI
Santalaceae		
Amphorogyne		
A. celastroides Stauffer & Huerlimann	E	FI
A. spicata Stauffer & Huerlimann	E	FI
Daenikera		
D. corallina Huerlimann & Stauffer	E	FU
Exocarpos		
E. phyllanthoides Endlicher	E	MU
E. pseudocasuarina Guillaumin	E	MU
Santalum		
S. austrocaledonicum Vieillard	E	MI
Sapindaceae		
Alectryon		
A. carinatum Radlkofer	E	FA
Arytera		
A. arcuata Radlkofer	E	MA
A. gracilipes Radlkofer	E	MU
A. lepidota Radlkofer	E	MU
Cupaniopsis		
C. apiocarpa Radlkofer	E	MI
C. azantha Radlkofer	E	FA
C. chytradenia Radlkofer	E	MU
C. dictyophora Radlkofer	E	FA
C. ganophloea Radlkofer	E	FA
C. godefroyi Guillaumin	E	
C. guioides Guillaumin	E	
C. hypodermatica Radlkofer	E	
C. macrocarpa Radlkofer	E	MI
C. mareensis Guillaumin	E	FA
C. mouana Guillaumin	E	FU
C. oedipoda Radlkofer	E	FI
C. pennelii Guillaumin	E	FA
C. petiolulata Radlkofer	E	FA
C. psilocarpa Radlkofer	E	FA
C. subcuneata Radlkofer	E	MI
C. sylvatica Guillaumin	E	FU
C. trigonocarpa Radlkofer	E	MA
Elattostachys		
E. apetala (Labillardière) Radlkofer	E	FI
E. falcata (A. Gray) Radlkofer	P	
E. incisa Radlkofer	E	MA
Gongrodiscus		
G. sp.	E	FA
G. sufferrugineus Radlkofer	E	FI
Guioa		
G. crenata Radlkofer	E	MI
G. crenulata Radlkofer	E	MI
G. fusca Radlkofer	E	MU
G. glauca (Labillardière) Radlkofer	E	MI
G. gracilis (Pancher & Sebert) Radlkofer	E	MI
G. microsepala Radlkofer	E	FI
G. pectinata Radlkofer	E	MI
G. villosa Radlkofer	E	MI
Harpullia		
H. austrocaledonica Baillon	E	FA
Podonephelium		
P. balansae Guillaumin	E	FA
P. concolor Radlkofer	E	MU
P. homei (Seemann) Radlkofer	E	FI
Storthocalyx		
S. chryseus Radlkofer	E	FI
S. leioneurus Radlkofer	E	FU
S. sordidus Radlkofer	E	MU
Sapotaceae		
Beccariella		
B. balanseana (Pierre) Aubréville	E	MI
B. rubicunda (Pierre ex Baillon) Pierre	E	FA
Bureavella		
B. endlicheri (Montrouzier) Aubréville	E	FI
B. wakere (Pancher & Sébert) Aubréville	E	FI
Leptostylis		
L. grandifolia Vink	E	FU
Manilkara		
M. dissecta (L.) Dubard	E	FI
Mimusops		
M. elengi L.	P	FA

Niemeyera		
N. balansae (Baillon) Aubréville	E	FI
Ochrothallus		
O. gordoniaefolius (S. Moore) Aubréville	E	MI
O. multipetalus (Vink) Aubréville	E	FU
O. sarlinii Aubréville	E	FI
O. schmidii Aubréville	E	FU
O. sessilifolius (Pancher & Sébert) Pierre ex Guillaumin	E	FU
O. wagapensis (Guillaumin) Aubréville	E	FA
Pichonia		
P. balansana Pierre	E	FA
Planchonella		
P. amieuana (Guillaumin) Aubréville	E	FA
P. kuebiniensis Aubréville	E	FU
P. laetevirens (Baillon) Pierre ex Dubard	E	FU
P. lifuana (Baillon) Pierre ex Dubard	E	FA
P. linggensis (Burck) Pierre	P	FA
P. microphylla Pierre ex Dubard	E	MI
P. pronyensis Guillaumin	E	FU
P. saligna S. Moore	E	FA
P. skottsbergii Guillaumin	E	FU
P. thiensis Aubréville	E	FU
Pycnandra		
P. benthamii Baillon	E	FA
P. comptonii (S. Moore) Vink	E	FA
P. controversa (Guillaumin) Vink	E	FA
P. fastuosa (Baillon) Vink	E	FI
P. griseosepala Vink	E	FU
P. neocaledonica (S. Moore) Vink	E	FA
P. paniensis Aubréville	E	FA
P. vieillardii (Baillon) Vink	E	FA
Pyriluma		
P. sphaerocarpum (Baillon) Aubréville	E	FA
Sebertia		
S. acuminata Pierre ex Baillon	E	FU
Saxifragaceae		
Polyosma		
P. brachystachys Schlechter	E	FA
P. comptonii E. G. Baker	E	MU
P. discolor Baillon	E	FA
P. leratii Guillaumin	E	FA
P. pancheri Baillon	E	FU
P. podophylla Schlechter	E	FA
P. spicata Baillon	E	FA
Quintinia		
Q. major (Baillon) Schlechter	E	FI
Q. media (Baillon) Guillaumin	E	MU
Q. minor (Baillon) Schlechter	E	MU
Q. oreophila (Schlechter) Schlechter	E	FU
Q. parviflora (Schlechter) Schlechter	E	FA
Simaroubaceae		
Soulamea		
S. fraxinifolia Brongniart & Gris	E	FI
S. muelleri Brongniart & Gris	E	MU
Smilacaceae		
Smilax		
S. neocaledonica Schlechter	E	MI
S. orbiculata Labillardière	E	
S. plurifurcata A. DC.	E	MI
S. purpurata J. R. & G. Forster	E	MI
Solanaceae		
Duboisia		
D. myoporoides R. Brown	P	MI
Sphenostemonaceae		
Sphenostemon		
S. balansae Baillon	E	MU
S. comptonii E. G. Baker	E	MI
S. oppositifolius Huerlimann	E	FU
S. pachycladum Baillon	E	FI
Sterculiaceae		
Sterculia		
S. austrocaledonica J. D. Hooker	E	FI
S. bullata Pancher & Sébert	E	MA
S. comptonii E. G. Baker	E	FI
S. dzumacensis Guillaumin	E	FU
S. fatsioides (Schlechter) Guillaumin	E	FA
S. francii Guillaumin	E	MU
S. megaphylla Bureau & Poisson ex Guillaumin	E	FA
S. sp.	E	FU
S. sageniifolia (Schlechter) Guillaumin	E	FA

S. scheffleraefolia Guillaumin	E	FI
S. schumanniana (Schlechter) Guillaumin	E	FA
Strasburgeriaceae		
Strasburgeria		
S. robusta (Vieillard ex Pancher & Sébert) Guillaumin	E	FU
Symplocaceae		
Symplocos		
S. arborea (Vieillard) Brongniart & Gris	E	FA
S. baptica Brongniart & Gris	E	FI
S. caerulescens Brongniart & Gris	E	FA
S. flavescens Brand	E	FI
S. gracilis Brongniart & Gris	E	FA
S. montana Brongniart & Gris	E	MI
S. munda S. Moore	E	MI
S. pseudonitida Guillaumin	E	FA
S. tortuosa Vieillard	E	FI
Taxaceae		
Austrotaxus		
A. spicata Compton	E	FA
Thymeleaceae		
Lethedon		
L. balansae (Baillon) Kostermans	E	MI
L. calleana (Guillaumin) Kostermans	E	FU
L. cernua (Baillon) Kostermans	E	FU
L. ciliaris (Baillon) Kostermans	E	FU
L. comptonii (E. G. Baker) Kostermans	E	FA
L. cordatoretusa Aymonin	E	MI
L. oblonga (Schlechter) Kostermans	E	MU
L. ovata (Guillaumin) Kostermans	E	MU
L. tannensis Sprengel	E	MI
Microsemma		
M. calophylla Guillaumin & MacKee	E	FU
M. rhizophoraefolia Guillaumin	E	FA
Wickstroemia		
W. indica (L.) Mey	P	MI
Tiliaceae		
Trichospermum		
T. sp.	P	FA
Trimeniaceae		
Trimenia		
T. neocaledonica E. G. Baker	E	FA
Triuridaceae		
Sciaphila		
S. dolichostyla Schlechter	E	FA
S. neocaledonica Schlechter	E	FU
Ulmaceae		
Celtis		
C. balansae Planchon	E	FA
C. paniculata (Endlicher) Planchon	P	FA
Urticaceae		
Dendrocnide		
D. latifolia (Gaudichaud) Chew	P	FA
D. vitiensis (Seemann) Chew	P	FA
Procris		
P. pedunculata (J. R. Forster) Weddell	P	FI
Verbenaceae		
Gmelina		
G. lignumvitreum Guillaumin	E	FA
Oxera		
O. arborea Schlechter	E	FA
O. baladica Vieillard	E	FA
O. candelabrum Beauvisage	E	FI
O. comptonii S. Moore	E	FA
O. cordifolia Dubard	E	
O. coriacea Dubard	E	
O. crassiflora S. Moore	E	FA
O. crassifolia Virot	E	MU
O. floribunda Schlechter	E	FA
O. longifolia Vieillard	E	
O. merytaefolia Guillaumin	E	FA
O. microcalyx Guillaumin	E	FA
O. morierei Vieillard	E	FA
O. nuda Virot	E	MI
O. oblongifolia Vieillard	E	
O. oreophila Guillaumin	E	FU
O. ovata Vieillard	E	
O. palmatinervia Dubard	E	FU
O. pancheri Dubard	E	
O. robusta Vieillard	E	FI
O. rugosa Guillaumin	E	FU
O. suaveolens Guillaumin	E	MU
O. subverticillata Vieillard	E	
O. sulfurea Dubard	E	FA

Vitex		
V. neocaledonica Gandoger	E	
V. rapini Beauvisage	E	FU
Violaceae		
Agatea		
A. lenormandi Melchior	E	
A. schlechteri Melchior	E	FU
Hybanthus		
H. austrocaledonicus (Vieillard) Melchior	E	FU
H. caledonicus (Turcz) Cretz	P	MI
H. ilicifolius (Vieillard) Schinz & Guillaumin	E	FI
H. micranthus Guillaumin	E	FA
Winteraceae		
Belliolum		
B. crassifolium (Baillon) Tieghem	E	FA
B. rivulare Tieghem	E	FA
B. vieillardii Tieghem	E	FA
Bubbia		
B. auriculata Tieghem	E	FA
B. balansae (Baillon) Tieghem	E	FU
B. comptonii (E. G. Baker) Dandy	E	FA
B. deplanchei Tieghem	E	FA
B. heteroneura Tieghem	E	FA
B. isoneura Tieghem	E	FA
B. odorata (E. G. Baker) Dandy	E	FA
B. pauciflora (E. G. Baker) Dandy	E	FA
Exospermum		
E. stipitatum (Baillon) Tieghem	E	FA
Zygogynum		
Z. acsmithii Vink	E	FU
Z. baillonii Tieghem	E	FU
Z. bicolor Tieghem	E	FA
Z. mackeei Vink	E	FA
Z. pomiferum Baillon	E	FU

APPENDIX 2
RAIN FOREST GENERA IN NEW CALEDONIA AND THEIR PHYTOGEOGRAPHIC DISTRIBUTION.

For family assignments, see Appendix 1.

(Symbols: 1–present in area; 2–species concentration in area; 3, 4, 5 (column headed "Africa")–Madagascar, Mascarenes, Seychelles, respectively. Letters (column headed "Distribution type"): A–Pantropical; B–New Caledonian; C–Palaeotropical; D–Indo-Australian; E–Endemic; F–Indo-Malayan; G–Malesian-Papuan; H–Australian; I–Australian-Papuan: J–Pacific subantarctic; K–Subantarctic; L–Pacific. An asterisk before the name of a genus indicates taxon of infrageneric rank.)

	Africa	Asia	Malesia	New Guinea	Australia	Solomons	New Hebrides	Lord Howe	Norfolk	New Zealand	Fiji	North Pacific	Polynesia	Samoa-Tonga	America	Distribution type	New Caledonia
Graptophyllum	1			1	1	1	1				1			1		I	
Hemigraphis		1	1	1	1	1	1					1		1		F	
Justicia	1	1	1	1	1	1									1	A	
Pseuderanthemum	1	1	1	1	1	1	1	1			1	1			1	A	
Cordyline	3	1	1	1	1				1	1				1	1	K	
Alangium	1	1	1	1	1	1	1				1					C	
Periomphale				1												B	2
Campynema					1											L	
Campynemanthe																E	
Amborella																E	
Cayratia	1	1	1	1	1	1		1			1	1				C	
Cissus	1	1	1	1	1	1		1				1			1	A	
Euroschinus				1	1											I	2
Semecarpus		1	1	1	1	1	1				1	1		1		G	
Desmos			1	1	1						1					D	
Polyalthia	3	1	2	1	1	1	1				1	1		1		C	
Richella			1								1					G	
Uvaria	1	1	1	1		1	1									C	
Xylopia	1	1	1	1		1					1				1	A	
Alstonia	1		1	1	1	1	1				1		1	1		C	
Alyxia	1	1	1	1	1	1	1	1	1		1	1	1	1		C	
Artia		1	1													B	2
Cerbera	3	1	1	1	1	1	1				1	1	1	1		D	
Cerberiopsis																E	
Melodinus		1	1	1	1	1	1		1		1			1		D	
Neisosperma	5	1	1	1	1	1	1					1		1		D	2
Ochrosia	4	1	1	1	1	1	1	1			1	1	1	1		D	
Pagiantha		1	1			1					1					F	
Parsonsia		1	1	1	1	1	1	1		1	1					D	
Rauvolfia	1	1	1	1	1	1						1			1	A	
Ilex	1	1	1	1	1		1	1				1	1		1	A	
Epipremnum		1	2	1	1	1	1				1	1	1	1		D	
Apiopetalum																E	
Arthrophyllum		1	1													F	2
Botryomeryta																E	

	Africa	Asia	Malesia	New Guinea	Australia	Solomons	New Hebrides	Lord Howe	Norfolk	New Zealand	Fiji	North Pacific	Polynesia	Samoa-Tonga	America	Distribution type	New Caledonia
Delarbrea			1	1	1	1	1									B	2
Dizygotheca							1									B	2
Meryta				1	1	1	1		1	1		1	1	1		B	2
Myodocarpus																E	
Polyscias (Tieghemopanax)	1	1	1	1	1	1	1	1			1	1		1		C	
Schefflera	1	1	1	1	1	1	1			1	1	1		1	1	A	
Strobilopanax																E	
Agathis			1	1	1		1			1	1					I	2
Araucaria				1	1				1						1	J	2
Hoya		1	1	1	1	1	1				1	1		1		D	
Marsdenia	1	1	1	1	1			1				1			1	A	
Secamone	1	1	1	1	1		1									C	
Tylophora	1	1	1	1	1	1	1	1	1		1	1		1		C	
Nemuaron																E	
Balanophora	1	1	1	1	1		1				1	1	1			C	
Hachettea																E	
Balanops					1		1				1					B	2
Deplanchea			1	1	1											I	
Pandorea			1	1	1	1		1								I	
Bischofia		1	1	1	1	1	1				1		1	1		D	
**Canarium* (Sect. *Canariellum*)					1											H	2
Gymnostoma (Casuarina)			1	1	1	1	1				1					G	2
Cassine (Elaeodendron)	1	1	1		1		1	1	1		1				1	A	
Maytenus	1	1	1	1	1						1	1	1	1	1	A	
Menepetalum																E	
Salaciopsis																E	
Caesalpinia	1	1	1	1	1	1	1	1	1		1	1	1	1	1	A	
Intsia	1	1	1	1	1	1	1				1	1		1		C	
Mezoneurum	1	1	1	1	1	1						1				C	
Storckiella					1						1					L	
Ascarina	3		1	1		1	1			1	1		1	1		L	
Hunga				1												B	2
Aneilema	1	1	1	1	1	1					1		1	1	1	A	
Santaloides (Rourea)		1	1	1	1	1	1				1			1		D	
Corynocarpus				1	1		1			1						I	
Diplocyclos	1	1	1	1	1	1	1		1							C	
Acsmithia			1	1	1						1					I	2
Codia																E	
Cunonia	1															K	2
Geissois					1	1	1				1					H	2
Pancheria																E	
Weinmannia	1		1	1		1	1			1	1		1	1	2	K	
Libocedrus				1						1					1	J	
Baumea	4	1	1	1	1								1			C	
Carex	1	1	1	1	1	1	1	1	1	1	1	1	1	1	1	A	
Costularia	1		1	1	1											C	
Scleria	1	1	1	1	1	1	1				1	1		1	1	A	
Uncinia			1	1	1			1		1		1			1	K	
Hibbertia	1		1	1	1						1					H	
Tetracera	1	1	1	1	1										1	A	

	Africa	Asia	Malesia	New Guinea	Australia	Solomons	New Hebrides	Lord Howe	Norfolk	New Zealand	Fiji	North Pacific	Polynesia	Samoa-Tonga	America	Distribution type	New Caledonia
Dioscorea	1	1	1	1	1		1				1	1	1	1		C	
Diospyros	1	1	1	1	1	1	1				1	1		1	1	A	
Dubouzetia			1	1												G	
Elaeocarpus	1	1	1	1	1	1	1	1		1	1	1	1	1		C	
**Sloanea (s.g. Antholoma)*				1												G	
Dracophyllum					1			1		2						L	
Styphelia		1	1	1	2	1	1	1		1	1	1	1			H	
Agapetes		2	1	1	1	1					1					D	
Argophyllum					1											H	2
Antidesma	1	1	2	1	1	1	1				1	1		1		C	
Austrobuxus			1	1	1						1					H	2
Baloghia					1				1							B	2
Bocquillonia																E	
Claoxylon	1	1	1	1	1	1	1				1	1	1	1		C	
Cleidion	1	1	1	1	1	1	1				1				1	A	
Cleistanthus	1	1	2	1	1	1					1	1				C	
Cocconerion																E	
Codiaeum			1	1	1	1	1				1					I	
Croton	1	1	1	1	1	1	1				1	1		1	1	A	
Dendrophyllanthus																E	
Drypetes	2	1	1	1	1	1	1	1			1	1			1	A	
Glochidion	1	1	2	1	1	1	1				1	1	1	1	1	A	
Homalanthus		1	1	1	1	1	1	1	1	1	1	1	1	1		D	
Macaranga	1	1	1	1	1	1	1				1	1	1	1		C	
Mallotus	1	1	1	1	1	1	1				1	1		1		C	
Neoguillauminia																E	
Phyllanthus	1	1	1	1	1	1	1				1	1	1	1	1	A	
Nothofagus				1	1					1					1	J	
Casearia	2	1	1	1	1	1					1	1			1	A	
Homalium	1	1	1	1	1	1	1				1		1	1	1	A	
Lasiochlamys																E	
Xylosma	1	1	1	1	1	1	1				1	1	1	1	1	A	
Flagellaria	1	1	1	1	1	1		1			1	1		1		C	
Flindersia			1	1	1											I	
Coronanthera					1	1										B	2
Depanthus																E	
**Scaevola (Sect. Scaevola)*			1	1		1					1	1	1			L	
Centosteca	1	1	1	1	1	1	1				1	1	1	1		C	
Leptaspis	1	1	1	1	1	1					1					C	
Oplismenus	1	1	1	1	1	1	1	1	1	1	1	1	1	1	1	A	
Calophyllum	1	1	1	1	1	1	1	1	1		1	1	1	1	1	A	
Garcinia	1	1	2	1	1	1	1				1	1		1		C	
Montrouziera																E	
Ochrocarpus (Mammea)	1	1	1	1	1	1	1				1	1		1	1	A	
Hernandia	1	1	1	1	1	1	1				1	1	1	1	1	A	
Dicarpellum (Salacia)																E	
Apodytes	1	1	1		1											K	
Citronella			1	1	1	1	1				1			1	1	A	
Gastrolepis																E	
Adenodaphne																E	

	Africa	Asia	Malesia	New Guinea	Australia	Solomons	New Hebrides	Lord Howe	Norfolk	New Zealand	Fiji	North Pacific	Polynesia	Samoa-Tonga	America	Distribution type	New Caledonia
Beilschmiedia	1	1	2	1	1	1				1					1	A	
Cryptocarya	1	1	1	1	1	1	1	1			1	1		1	1	A	
Endiandra		1	1	2	2	1	1				1			1		D	
Litsea		1	2	1	1	1	1			1	1			1	1	A	
Barringtonia	1	1	1	2	1	1	1				1	1	1	1		C	
Arthropodium	3			1	1					1						H	
Astelia	4			1						1		1	1		1	L	
Dianella	3	1	1	1	1	1	1	1	1	1	1	1	1	1	1	K	2
Hugonia	1		1	1	1	1	1				1					I	2
Fagraea		1	2	1	1	1	1				1	1	1	1		D	
Geniostoma	4	1	1	1	1	1	1	1		1	1	1	1	1		C	2
Neuburgia			1	1		1	1				1	1				G	
Amyema			1	1	1	1	1					1		1		I	
Amylotheca				1	1									1		I	
Korthalsella	1	1	1	1	1			1	1	1	1	1	1	1		C	
Aglaia		1	2	1	1	1	1				1	1		1		D	
Anthocarapa (Amoora)				1	1											I	
Dysoxylum		1	2	1	1	1	1	1	1	1	1	1		1		D	
Hypserpa		1	1	1	1	1	1					1				D	
Pachygone		1	1	1	1						1	1				D	
Stephania	1		1	1	1	1	1						1	1		C	
Adenanthera		1	1	1	1	1	1				1	1	1	1		C	
Albizia	1	1	1	1	1	1	1					1			1	A	
Serianthes			1	1		1	1				1	1	1	1		L	
Hedycarya					1		1			1	1			1		B	2
Kibaropsis																E	
Ficus	1	1	1	1	1	1	1	1			1	1	1	1	1	A	
Sparattosyce																E	
Streblus	1	1	1	1	1	1	1			1	1		1	1		C	
Canacomyrica																E	
Maesa	1	1	1	1	1	1	1				1	1		1		C	
Rapanea	1	1	1	1	1	1	1	1	1	1	1	1	1	1	1	A	
Tapeinosperma			1	1	1	1	1				1					L	
Arillastrum																E	
Austromyrtus					1		1				1					L	
Carpolepis																E	
Cleistocalyx		1	1	1	1			1			1					D	
Cupheanthus																E	
Eugenia	1		1	1	1	1	1			1	1	1		1	1	A	
Mearnsia	1		1	1	1	1				1						G	
Metrosideros						1		1		1	1	1	1	1		L	
Piliocalyx							1				1					B	2
Pleurocalyptus																E	
Rhodomyrtus (Archirhodomyrtus)		1	1	1	1	1										I	
Syzygium (Caryophyllus, Jambosa)																C	
Tristaniopsis		1	1	1	1											D	
Xanthostemon			1	1	1	1										I	2
Nepenthes	3	1	2	1	1							1				C	
Pisonia (Calpidia)	1	1	1	1	1	1	1	1	1	1	1	1	1	1	1	A	
Jasminum	1	1	1	1	1	1	1	1	1		1	1	1	1		C	

	Africa	Asia	Malesia	New Guinea	Australia	Solomons	New Hebrides	Lord Howe	Norfolk	New Zealand	Fiji	North Pacific	Polynesia	Samoa-Tonga	America	Distribution type	New Caledonia
Linociera	1	1	1	1	1	1	1	1			1	1		1	1	A	
Olea	2	1	1		1			1								C	
**Osmanthus (*Sect. *Nothosmanthus)*																E	
Oncotheca																E	
Acanthephippium		1	1	1							1			1		E	
Acianthus				1	1	1				1						H	2
Agrostophyllum	5	1	1	1		1	1				1	1		1		F	
Anoectochilus		1	1	1	1		1				1					F	
Appendicula		1	1	1		1	1				1	1		1		F	
Bulbophyllum	1	1	1	1	1	1	1	1	1	1	1	1	1	1	1	A	
Calanthe	1	1	2	1	1	1	1	1			1	1	1	1	1	A	
Ceratostylis		1	1	1		1	1					1	1			E	
Chamaeanthus		1	1				1									G	
Cheirostylis	1	1	1	1	1						1	1		1		C	
Chrysoglossum		1	1	1		1	1				1					E	
Cleisostoma		1	1	1	1											G	
Clematepistephium																E	
Coelogyne		1	1	1		1	1				1	1		1		F	
Coilochilus																E	
Corybas		1	1	1	1	1	1			1	1	1	1			I	
Cryptostylis		1	1	1	1	1					1			1		D	
Dendrobium		1	1	1	1	1	1	1	1	1	1	1	1	1		D	
Didymoplexis		1	2	1	1						1	1		1		D	
Diplocaulobium			1	1	1	1					1	1		1		G	
Drymoanthus					1					1						L	
Earina							1			1	1		1	1		L	
Ephemerantha		1	1	1	1	1						1	1	1		F	
Epipogium	1	1	1	1	1		1									C	
Eria		1	2	1	1	1	1				1		1	1		D	
Erythrodes		1	1	1		1	1				1			1	1	A	
Eulophia	1	1	1	1	1						1	1			1	A	
Glomera			1	2		1	1				1	1		1		G	
Gonatostylis																E	
Goodyera	1	1	1	1	1	1	1				1	1		1	1	A	
Habenaria	1	1	1	1	1	1	1				1		1	1	1	A	
Hetaeria	1	1	1	1	1	1	1				1	1	1	1		C	
Liparis	1	1	1	1	1	1	1				1	1	1	1	1	A	
Malaxis	1	1	1	1	1	1	1				1	1	1	1	1	A	
Megastylis																E	
Microtatorchis		1	1	1							1	1		1		G	
Moerenhoutia		1	1	1	1		1				1	1	1	1		G	
Nervilia	1	1	1	1	1		1				1	1	1	1		C	
Oberonia	1	1	1	2	1		1		1		1	1	1	1		C	
Octarrhena		1	1	1		1	1					1	1			G	
Pachyplectron																E	
Peristylus		1	1	1	1							1		1		F	
Phajus	1	1	1	1	1	1	1				1	1	1	1		C	
Pholidota		1	2	1	1	1	1				1		1			F	
Phreatia		1	1	1	1	1	1		1	1		1	1	1		D	
Pristiglottis		1	1	1	1	1	1				1			1		F	

	Africa	Asia	Malesia	New Guinea	Australia	Solomons	New Hebrides	Lord Howe	Norfolk	New Zealand	Fiji	North Pacific	Polynesia	Samoa-Tonga	America	Distribution type	New Caledonia
Pterostylis			1	1	1			1		1						H	
Rhynchophreatia			1	1	1							1				G	
Sarcochilus		1	1	1	1	1					1		1			D	
Schoenorchis		1	1	1	1	1					1					D	
Spathoglottis		1	1	1	1	1	1				1	1	1	1		D	
Spiranthes	1	1	1	1	1	1	1			1				1	1	A	
Taeniophyllum	1	1	1	1	1	1	1		1		1	1	1	1		C	
Thrixspermum		1	1	1	1	1	1				1	1		1		D	
Trachoma		1	1		1		1									D	
Tropidia		1	1	1		1	1		1		1			1	1	A	
Zeuxine	1	1	1	1	1	1	1				1	1		1		C	
Actinokentia																E	
Alloschmidia																E	
Basselinia																E	
Brongniartikentia																E	
Burretiokentia																E	
Campecarpus																E	
Chambeyronia																E	
Clinosperma																E	
Cyphokentia																E	
Cyphophoenix																E	
Cyphosperma											1					L	
Kentiopsis																E	
Lavoixia																E	
Mackeea																E	
Moratia																E	
Pritchardiopsis																E	
Veillonia																E	
Freycinetia		1	1	1	1	1	1		1	1	1	1	1	1		D	
Pandanus	1	1	1	1	1	1	1	1			1	1	1	1		C	
Arthroclianthus																E	
Mucuna	1	1	1	1	1	1	1	1			1	1	1	1	1	A	
Paracryphia																E	
Passiflora		1	1	1	1	1	1	1	1	1	1			1	2	A	
Phelline																E	
Eustrephus				1	1											I	
Geitonoplesium			1	1	1	1	1	1	1		1					I	
Peperomia	1	1	1	1	1	1	1	1	1	1	1	1	1	1	1	A	
Piper	1	1	1	1	1	1	1	1	1	1	1	1	1	1	1	A	
Pittosporum	1	1	1	1	1	1	1	1	1	1	1	1	1	1		C	2
Acmopyle											1					L	
Dacrycarpus		1	1	1		1	1			1	1					F	
Dacrydium		1	1	1		1	1			1	1				1	J	
Decussocarpus			1	1		1					1				1	J	
Falcatifolium			1	1												G	
Parasitaxus																E	
Podocarpus	1		1	1	1		1			1	1			1	1	K	
Prumnopitys			1	1	1					1					1	J	
Beauprea																E	
Eucarpha (Knightia)																E	

	Africa	Asia	Malesia	New Guinea	Australia	Solomons	New Hebrides	Lord Howe	Norfolk	New Zealand	Fiji	North Pacific	Polynesia	Samoa-Tonga	America	Distribution type	New Caledonia
Kermadecia																E	
Sleumerodendron																E	
Stenocarpus			1	1	1											I	2
Virotia (Macadamia)					1											B	2
Alphitonia		1	1	1	1	1	1				1	1	1	1		D	
Ventilago	1	1	1	1	1		1				1	1			1	A	
Crossostylis						1	1				1		1	1		L	
Atractocarpus																E	
Coelospermum		1	1	1	1											D	
Cyclophyllum							1									B	2
Gardenia	1	1	1	1	1	1	1				1	1	1	1		C	
Guettarda	1	1	1	1	1	1	1				1	1	1	1	1	A	
Ixora	1	1	1	1	1	1	1				1	1	1	1	1	A	
Morierina																E	
Morinda	1	1	1	1	1	1	1				1	1	1	1	1	A	
Mussaenda	1	1	1	1		1	1				1	1	1	1		F	
Psychotria	2	1	1	1	1	1	1	1			1	1	1	1	1	A	
Randia	2	1	1	1	1	1	1	1			1	1		1	1	A	
Rhopalobrachium																E	
Tarenna	1	1	1	1	1	1	1				1	1	1	1		C	
Dutaillyea																E	
Geijera				1	1											I	
Medicosma					1											I	
Melicope		1	1	1	1	1	1	1		1	1	1		1		D	
Micromelum		1	1	1	1	1	1				1	1		1		D	
Sarcomelicope																E	
Zanthoxylum	1	1	1	1	1	1		1	1		1	1	1	1	1	A	
Zieridium																E	
Amphorogyne																E	
Daenikera																E	
Exocarpos		1	1	1	1			1		1	1	1	1			D	
**Santalum* (Sect. *Santalum)*			1	1	1		1				1	1		1		I	
Alectryon		1	1	1	1		1			1	1	1		1		D	
Arytera		1	1	1	1	1					1			1		D	
Cupaniopsis			1	1	1	1	1				1			1		I	
Elattostachys			1	1	1	1	1				1	1		1		I	
Gongrodiscus																E	
Guioa		1	1	1	1	1		1			1			1		D	
Harpullia		1	1	1	1	1	1				1			1		D	
Podonephelium																E	
Storthocalyx																E	
Beccariella			1	1	1											D	
Bureavella			1	1	1	1										G	
Leptostylis																E	
Manilkara	1	1	1	1	1		1				1	1		1	1	A	
Mimusops	1	1	1	1	1							1				C	
Niemeyera					1											H	
Ochrothallus																E	
Pichonia																E	
Planchonella		1	1	1	1	1	1	1	1	1	1	1	1	1		D	

	Africa	Asia	Malesia	New Guinea	Australia	Solomons	New Hebrides	Lord Howe	Norfolk	New Zealand	Fiji	North Pacific	Polynesia	Samoa-Tonga	America	Distribution Type	New Caledonia
Pycnandra																E	
Pyriluma																E	
Sebertia																E	
Polyosma		1	1	1	1	1										D	
Quintinia			1	1	1		1			1						D	
Soulamea	5		1	1		1	1				1	1				K	2
Smilax	1	1	1	1	1	1	1	1			1	1		1	1	A	
Duboisia					1											H	
Sphenostemon				1	1											I	
Sterculia	1	1	1	1	1	1	1				1	1		1	2	A	
Strasburgeria																E	
Symplocos		1	1	1	1	1	1	1			1	1			1	A	
Austrotaxus																E	
Lethedon (Microsemma)					1		1									B	2
Wickstroemia		1	1	1	1	1	1		1		1	1	1	1		D	
Trichospermum		1	1	1	1	1					1	1		1	1	A	
Trimenia			1	1		1					1		1	1		G	
Sciaphila	1	1	1	2		1	1					1			1	A	
Celtis	1	1	1	1	1	1	1	1	1		1	1	1	1	1	A	
Dendrocnide		1	1	1	1	1	1				1	1	1	1		D	
Procris	1	1	2	1		1	1				1	1	1	1		F	
Gmelina	1	1	1	1	1	1					1	1				C	
Oxera																E	
Vitex	2	1	1	1	1	1	1		1	1	1	1	1	1	1	A	
Agatea				1		1					1			1		B	2
Hybanthus	1	1	1	1	1		1								1	A	
Belliolum						1										I	
Bubbia				1	1			1								I	
Exospermum																E	
Zygogynum																E	

LEPIDOPTERA AND THE MELANESIAN ARCS

J. D. Holloway
Commonwealth Institute of Entomology
London S.W.7, UK

INTRODUCTION

Three hypotheses of the geographical evolution of the Melanesian tropics must be considered: (1) the fragmentation of a previously more considerable extent of land, a vicariance model; (2) stability of present complexity with little major change through geological time, involving considerable biotic dispersal; (3) a dynamic complexity as suggested by recent findings in plate tectonics, involving both the separation of old land masses and marginal island arc processes, i.e. both vicariance and dispersal in biological terms.

Biogeographical methodology is at present undergoing a welcome phase of critical reappraisal, though with a tendency to polarize workers into 'vicariance' or 'dispersal' schools. The vicariance approach emphasises a search for a deductive method that confronts the data without ad hoc interpretation, thus divorcing pattern from process. The aim is to explore the relationships of areas in terms of the phylogenetic relationships of endemic organisms, generating a hypothesis that can be set against evidence from other sources such as geology. The dispersal approach seeks to explore the relationship between process and pattern by modelling processes, attempting to simulate patterns that can then be tested in reference to real data. The processes inevitably involve concepts of biotic exchange between areas and therefore dispersal.

The biogeography of Melanesia provides an ideal testing ground for both approaches, and this may in time lead to some sort of synthesis between the two. Distribution patterns in Melanesia are of the sort that would suggest even in vicariance methodology that dispersal has taken place: there is much sympatry and considerable overlap of diverse general patterns. Some of these general patterns may prove to have resulted through vicariance, others through dispersal. An integral part of either approach must be to assess these general patterns: as generalized tracks; as specific or generic faunal elements; or, ideally, in congruent cladistic structure with respect to areas. This last ideal will be hard, perhaps in many cases impossible, to attain in practice. Given general patterns, general explanations can be sought, either through reference to data from other sources, e.g. geological, or through attempts to simulate the processes that gave rise to them.

HYPOTHESES OF THE GEOLOGICAL EVOLUTION OF MELANESIA AND THEIR BIOGEOGRAPHIC IMPLICATIONS

Fixed or Constant Geography

A hypothesis of fixed geography is perhaps the null hypothesis considering the wealth of contrary geological evidence. It supposes that the extent and juxtaposition of land areas have remained unchanged through geological time barring minor changes through erosion, orogeny, or sea-level fluctuations, the last only significant in continental shelf areas such as the Sunda or Sahul shelves.

In an archipelagic area such as Melanesia and Malesia, a hypothesis of constant geography must, in the absence of high to total endemism in each land area, imply considerable dispersal and exchange of biotic elements between areas; the degree of dispersal might be assessed crudely from group to group in terms of the number of endemic species relative to the number of widespread ones. MacArthur and Wilson (1967) modelled several processes for exchange between islands, and their exponential model for passive dispersal has been tested in an analysis of the relationships of the biota of Norfolk Island to the three main source areas (Australia, New Zealand, and New Caledonia) (Holloway, 1977). The analysis showed that dispersal processes were predictive, rather than random, and that powers of dispersal were correlated with the proportions of widespread and endemic taxa for various taxonomic groups.

Thus one must ask whether the congruence of area relationships in the Indo-Australian tropics for species of birds and butterflies, revealed by Holloway and Jardine (1968) and re-examined in detail by Nelson and Platnick (1981), reflect in common a geological fragmentation of the area or a general pattern of barriers to exchange of similarly flighted organisms. The measure of comparison used by Holloway and Jardine was based on the proportion suggested by the Norfolk Island analysis to be related to dispersal and therefore to biotic exchange between areas. It is therefore perhaps not surprising that two-dimensional non-metric multidimensional scaling analyses (ordination) of the two sets of "biotic distance" measures gave arrays of points that approximated closely to the geography of the areas represented. Holloway and Jardine suggested that any significant departures from present geography might be interpreted as indications of past geography in such analyses.

These analyses were based on consideration of numerous non-endemic taxa, a factor ignored by Nelson and Platnick (1981) in their later discussion of the relevance of non-endemic taxa to the vicariance method. In generating area cladograms later in this paper I shall employ the constraints of their Assumption 1 (p. 421) because a species common to two areas such as islands, with a sister-species on a third island, must indicate, in terms of an area cladogram, closer relationship between the first two areas than of either to the third given that the species common to the two areas is defined by an autapomorphy. In the absence of this constraint (Assumption 2 of Nelson and Platnick) the whole vicariance method fails, as it must then logically be possible to disregard apomorphies defining taxa anywhere to generate congruence between area cladograms of different groups.

With fixed geography one might predict that the evolution of patterns of distribution by processes of dispersal and exchange would be repetitive. It might be possible to model this process for a few simple systems, enabling general predictions to be made on the development of endemism and phylogenetic patterns that could be tested in comparison with actual situations.

Vicariant Geography

A school of phytogeographers, exemplified by Steenis (1962) but also followed with regard to Melanesia by Corner (1975), proposed that the Melanesian Archipelago seen today, indeed all the Indo-Australian island complexes, are the remnant fragments of a much more extensive, even continuous, tongue of mountainous land that extended from Southeast Asia to Australia and, as the Melanesian Foreland, out into the Pacific. This hypothesis is based purely on the assumption that dispersal across large water gaps of the sort seen today in the area is an impossibility for the majority of plant taxa, particularly those of the rain forest. The hypothesis never had a firm foundation in terms of geological evidence.

Nevertheless, it is the closest approach to a vicariance explanation for the development of endemism in Melanesian biotas and will be considered in that light here. The hypothesis may even enjoy some current support from members of the vicariance school, such as Cracraft (1980), who would seek to express it in terms of fragmentation due to earth expansion.

The breakup of this tongue of land, whether by subsidence and marine transgression, fragmentation through minor drift and dispersal (perhaps through earth expansion), or mainly through gradual erosion, should be apparent in the distribution of organisms today. There should be a degree of biological harmony throughout the area, such as dipterocarps in the lowland forests of eastern Melanesia, *Nothofagus* and *Araucaria* in the lower montane forests of Southeast Asia, and Ericaceae mingling with *Metrosideros* on mountains throughout the area. In fact there is considerable disharmony in plant and animal groups (Holloway, 1970) and in ecological components such as the understory insect fauna (Holloway, 1983). Palynological evidence is also contrary, particularly with reference to the Bornean montane flora (Flenley, 1979).

In higher taxa that do show a degree of uniformity of distribution throughout the area, such as the plant families Apocynaceae, Araliaceae, Euphorbiaceae, Lauraceae, Leguminosae, Moraceae, Myrtaceae, Palmae, Rubiaceae, Sapotaceae, and Urticaceae, there should, if the vicariance hypothesis is correct, be a high degree of congruence of area relationships. Any such general pattern would not be repetitive in phylogeny, but might be expected to contrast with patterns of modern derivation in organisms known to be dispersive such as insects. At present no cladistic analyses of these plant groups are available, so the hypothesis has yet to be tested rigorously.

Plate Tectonic Geography

The hypothesis of the geological evolution of Melanesia that is currently meeting with most support is that of the independent development of two Melanesian island arcs, subsequently convergent and, in New Guinea, fused (Fig. 1). It is a hypothesis that has attracted biogeographers for several decades, three of the earliest being Lieftinck (1949), Toxopeus (1950), and Ross (1956:170). It is strongly supported by recent geological evidence [reviewed by Holloway (1979) and Coleman (1980)].

The more northerly Outer Arc is thought to have extended from the northern Moluccas (Halmahera to Obi) through the Vogelkop and northern mountain ranges of New Guinea to the Bismarcks, thence via the Solomons and Vanuatu (New Hebrides) to Fiji and Tonga. The early Outer Arc may well have been a string of small islets that has only over the past ten or so million years developed to its current extent. Parts, such as Vanuatu, may have only presented significant land area over the past one to two million years.

The Inner Arc consists of the central ranges of New Guinea and its southeastern peninsula, the Rennell group, New Caledonia, the Norfolk Ridge, and Lord Howe Rise, extending south to New Zealand. Islands such as Ceram, Buru (southern Moluccas), Tenimber, Kei, and Timor may, at least in part, represent a westerly extension of the Inner Arc as may eastern Sulawesi. New Caledonia, New Zealand, and possibly southeastern New Guinea and parts of the Lord Howe Rise are known to have had a long geological history with extensive sedimentary succession, and they may have had a subaerial history extending back into the Mesozoic. Prior to the opening of the Tasman and Coral seas and the Southern Ocean (between New Zealand and Antarctica), dated roughly from 80 to 50 m.y.a. (million years ago), all these older areas were adjacent to (as offshore arc systems) or contiguous with the eastern coast of Australia (Taylor and Falvey, 1977; Holloway, 1979).

Initial separation of the two Arcs may have been from 50 to 30 m.y.a. (Holloway, 1979) or even to as late as 10 m.y.a. (Coleman, 1980) by the formation of sea floor between them. At some time between the major change in plate motions, thought to have occurred globally at around 30 m.y.a., and the 10 m.y.a. date favoured by Coleman, the inter-Arc zone became quiescent and thereafter was reduced by subduction along the southern edge of the Outer Arc to as far east as the southern islands of Vanuatu. A zone of shear between southern Vanuatu and northern Fiji has led to Vanuatu becoming distanced from Fiji whilst converging with New Caledonia. Collision of the Arcs in New Guinea progressed from west to east from a date given by Jaques and Robinson (1977) as early Miocene (about 20 m.y.a.), though actual fusion of land areas could well have postdated this. The independent structure of the two Arcs in New Guinea is still indicated by the two parallel mountain systems. The picture in New Guinea is complicated by the possibility that the central and southeast portions of the Inner Arc component may have been separated by a water gap, the Aure Trough, at some time in the past (Duffels, 1983). The northern margin of the Outer Arc has interacted with the margin of the westering Pacific Plate, resulting in disturbance, uplift, and left lateral shear.

If this geological picture is broadly accurate one might expect to find three classes of biogeographic patterns. Firstly there will be ancient patterns involving the older Inner Arc areas, New Zealand and Australia. These patterns may be vicariant and therefore have congruent area cladograms. A number of groups restricted to some or all of these areas have been examined in this light (Cracraft, 1980; Thornton, 1980; Winterbourn, 1980; Humphries, 1981) but no striking or really informative congruence has emerged. At most, one can say that the restriction of numerous diverse groups to these and other areas of Gondwanaland is suggestive of an ancient, southern history.

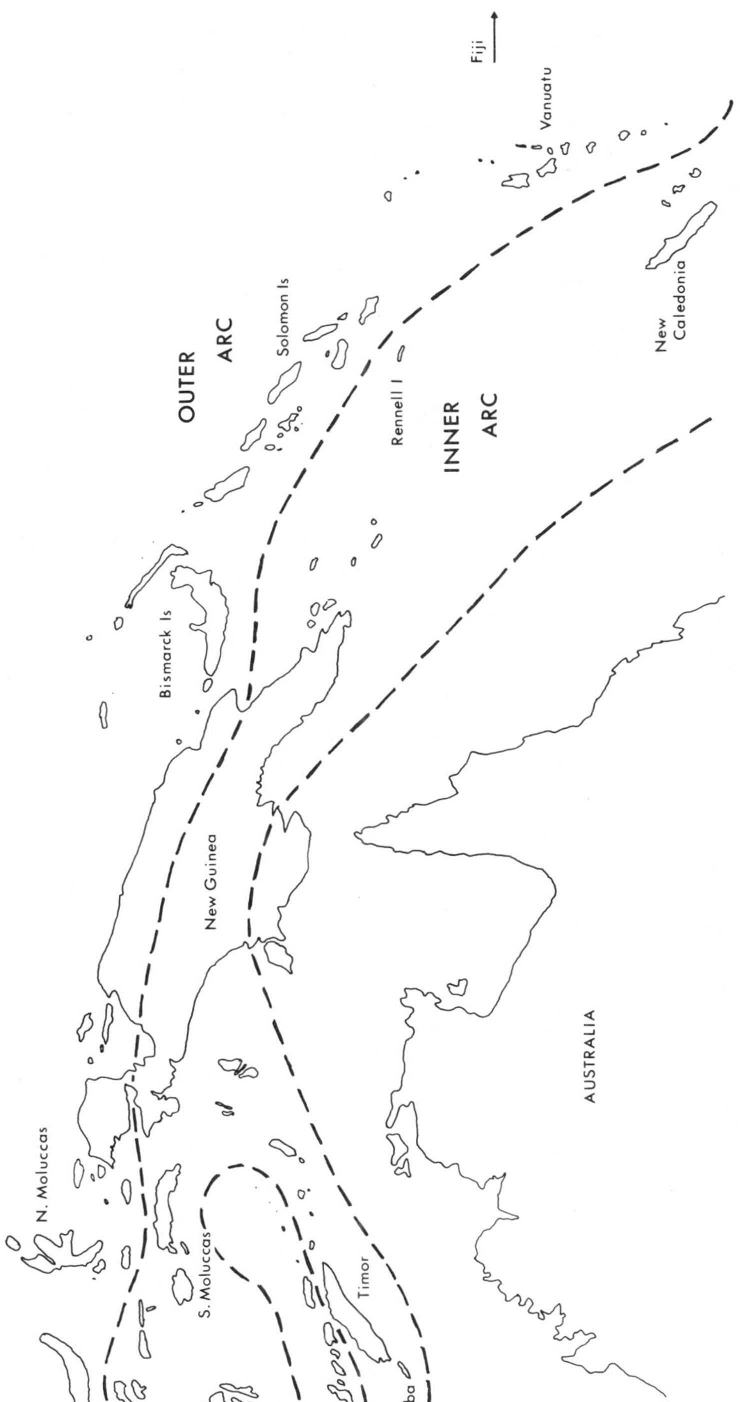

Fig. 1. Present day geography of Melanesia showing the Inner and Outer Arcs. Inclusion of parts of Sulawesi and the Lesser Sundas in the Inner Arc is tentative and speculative.

The second class of patterns will be of intermediate age involving Inner and Outer Arcs in independence, reflecting their isolation from each other. These patterns may nevertheless involve the same ancestral taxa in some cases, resulting in sister groups of which one reflects the configuration of the Outer Arc and the other that of the Inner Arc. Such sister groups would provide the strongest biological corroboration of the geological hypothesis of two arcs. Both this and the first class of pattern may become modified or obscured by subsequent dispersal.

A segregation of polytypic Melanesian genera into Inner or Outer Arc categories as suggested by Holloway (1979) is far weaker evidence because, though consistent with the hypothesis, it could equally well have an ecological basis. Indeed, modern Melanesian species distributions in the macrolepidoptera can be seen to grade from an 'Inner Arc' type (Coral Sea element) to an 'Outer Arc' type (Melanesian elements) in the numerical analysis by Holloway (1979). These presumably modern species distributions conceivably fragment through time, and their components diverge and redisperse to give polytypic groups with Inner or Outer Arc patterns.

Such segregation would begin to be more convincing as evidence for two arcs if segregation were manifest within New Guinea, Outer Arc groups showing their greatest diversity in the northern ranges and Inner Arc groups in the central ranges and southeast. Examples of the former may be found in the pandan genus *Sararanga* (Stone, 1982) and in the platymantine frogs (Zweifel and Tyler, 1982).

A classic example of sister pairs of genera related to each arc, together with a degree of segregation in New Guinea, has been provided by the studies by Duffels (1977, 1983) on the cicada subtribe Cosmopsaltriaria. The sister group to the arc groups is in Sulawesi.

The third class of pattern will provide a reflection of modern geography derived from recent dispersals. Indeed, both second and third classes of dispersal pattern, ranging over moderately isolated archipelagos, would require a considerable degree of dispersal to establish. The patterns or tracks resulting would reflect past *juxtaposition* of archipelagos rather than past *connections.*

Instead of the fragmentation and dispersal of land areas envisaged for most vicariance scenarios, one is faced with a complex blend of emergence, increment, convergence, and fusion of land areas. The construction of a geological area cladogram in such instances would, if not impossible, involve such oversimplification as to cast doubt on its value. In such an instance one is perhaps forced to accept that, in general, geological evidence for the distribution and extent of past land areas is much 'harder' than any biological evidence. One should perhaps attempt to model the development of distribution patterns in archipelagos where the geological history is well understood. Then the understanding so gained might be extrapolated to arbitrate in situations where there are conflicting geological interpretations.

In Melanesia the best hope is to try to identify groups in sequence that, having dispersed through the whole area, have diverged and speciated more in the initial stages of colonization than recently, the latter perhaps a reflection of the influence of competition from later colonists or, on islands, a general loss of vagility. The groups would tend to reflect the geography at the time of their dispersal. In effect, a series of snapshots of past geography preserved in the current distributions of organisms is sought (Holloway, 1983).

The sequencing of dispersals into Melanesia presents a problem in that an individual, isolated pattern cannot be dated except by reference to geological information, a circular argument in the context of the present discussion. Some evidence for sequencing can be provided when sister groups overlap through the area concerned, one consisting of very similar, allopatric species and another, morphologically more diverse, with more sympatry and diversity in any one area. The second group is likely to be the more ancient. In very few instances does such an ideal situation present itself; one is left with recourse only to the rather unsatisfactory general equating of antiquity with species diversity plus morphological diversity.

In the survey of Lepidoptera geography that follows, I shall treat ancient and modern distribution patterns before dealing with putative two-arc ones.

INNER ARC GROUPS, POSSIBLY ANCIENT

Groups of Oriental Affinity

Most of the major diversifications of Lepidoptera within Melanesia appear to be restricted to New Guinea; my survey revealed very few indications of extensive speciation within the more easterly Outer Arc archipelagos, and these will be discussed later. The New Guinea groups sometimes extend weakly to the Bismarcks, even to the Solomons, but it is unusual in such circumstances to determine the presence of a distinct Outer Arc component.

A major problem arises through the lack of any detailed anatomical survey of these large endemic groups. There is thus no information available on their general affinities, let alone any precise sister-group relationships. Many have become morphologically very divergent, adding to the problem. I will illustrate this point by discussing the preliminary findings of a study of genitalic morphology of the colorful, presumably day-flying ennomine geometrids that are so diverse in New Guinea; they often extend weakly westwards into the Oriental Region. The biological reasons behind the evolution of the striking patterns and complicated mimicry rings, of a degree of development that rivals any in South America, are not the concern of this paper, but they appear even more striking given the relationships my survey revealed.

The type species of each of these ennomine geometrid genera was dissected and, in many cases, also some of the species more divergent from the type within a genus.

The genus *Automolodes* Warren contains three species, two in New Guinea plus the Moluccas and one known only from Guadalcanal in the Solomons. They have striking deep orange banding on a dark brown background, perhaps involved in a mimicry complex with *Milionia* species. Dissection revealed them to have male genitalia typical of the Oriental representatives of the pantropical genus *Semiothisa* Hubner.

The monobasic Sulawesi genus *Bandobena* Walker resembles *Automolodes* in facies, but in male genitalia characters it is close to *Pseudeuchromia* Schultze (Sulawesi and Philippines) and the large New Guinea genus *Ctimene* Boisduval;

both of these latter genera contain numerous small species, most of which have two to three orange or yellow irregular patches on dark brown forewings and one on the hindwing. *Ctimene* has extended as far as the Solomons (one species) and has several species in N. Australia and three in New Caledonia. It is therefore predominantly an Inner Arc group. Its affinities are with the large Oriental complex of genera that includes *Sabaria* Walker and *Hypochrosis* Guenée, having similar distinctive characteristics of uncus, juxta, vinculum, and valves, the most definitive character being the extension ventrally of the slender vinculum in a double loop below the saccus [Oriental representatives are illustrated by Holloway (1976)].

The genera *Burtina* Walker (Sulawesi and N. Moluccas) and *Bursadopsis* Warren (several New Guinea species) have similar facies to *Ctimene* but are genitalically distinct; chalcosiine zygaenids in the genus *Caprima* Walker and allies also have this facies type.

The majority of genera belong to the *Milionia* Walker complex, most having a bifid uncus, a setose projection from the valve sacculus linked by a weakly sclerotized band across the valve to another subapical one, and often a swelling or lobe from the center of the valve costa invested with fine, hairlike setae. The general affinities of these genera are probably with genera centered in the Oriental Region, such as *Dilophodes* Warren, *Pogonopygia* Warren, *Elphos* Guenée, *Amblychia* Guenée and *Medasina* Moore, where similar characteristics of the male genitalia occur. Typical *Milionia* have transverse orange bands on midnight blue wings; the subapical and saccular projections to the valves are ornamented with irregular elongate patches of stout setae and the aedeagus vesica contains a single, short, stout cornutus. The last character is shared only with *Eucharidema* Rothschild, possibly therefore the sister genus, which is also restricted to New Guinea. Other more distantly associated genera are *Lobocraspeda* Warren, *Bordeta* Walker, *Pseudeusemia* Weymer, *Eubordeta* Rothschild, and *Praesos* Walker, all more or less restricted to New Guinea. *Praesos* is part of a mimicry ring centered on the geometrine genus *Dysphania* Hubner, also very diverse in New Guinea.

Dysphania is likewise well represented in the Oriental Region, where species are also involved in mimicry complexes, one, a "blue with dark spots" complex that involves an agaristid, *Longicella mollis* Walker, and several species of ennomines currently attributed to the genus *Arycanda* Walker. These last are mentioned here as they highlight the problems and complexities associated with the *Milionia* complex in New Guinea. The three Bornean species I examined (Holloway, 1976) fall into the two remaining groups of the *Milionia* complex to be discussed: *arycandata* Walker into the *Craspedosis* Butler group, defined in the male genitalia by a single uncus, elongate parallel-sided valves with a large, spine-like, apically directed subcostal projection and only weak saccular ornamentation, and an aedeagus vesica that bears a row of moderate, slender cornuti apically; *omissa* Warren and *georgiata* Guenée into the *Cosmethis* Hubner group, defined in the male genitalia by a prominent lobe to the valve costa and an elaboration of the central sclerotized band of the valve into a setose lobe dorsally, and also characterised by a more regular, confined spining to the subapical projection and the absence of ornamentation on the aedeagus vesica. The genus *Arycanda* thus appears to be polyphyletic. Most of the larger rusty streaked species in Sulawesi and New Guinea should be transferred to the *Cosmethis* complex and the smaller, darker blue, more irregularly marked ones may prove on dissection to be referable to *Craspedosis.*

The *Cosmethis* and *Craspedosis* groups both appeared on first inspection to have rather aberrant representatives in the Solomons, which were examined in case they offered evidence for Inner/Outer Arc sister groupings. In fact they both proved on dissection to belong to the *Cosmethis* group.

The species *disrupta* Warren, with complicated bright yellow markings on a black background, had been described in the genus *Xanthomima* Warren, a genus of numerous New Guinea species where the forewings are usually basally broadly pale yellow. Typical *Xanthomima* belong to the *Craspedosis* group. The Solomons species is best placed in the genus *Bracca* Hubner, which contains several very similar species from N. Australia, the S. Moluccas, New Guinea, the Bismarcks, and Sula Is., where only the hindwing has yellow markings. The Solomons species may be sister to the rest, but the pairing is far from being strong evidence for Inner versus Outer Arc development. The genus *Bracca* is defined within the *Cosmethis* complex by reduction of the uncus, development of the subapical valve projection into a rounded knob evenly distributed with stout setae, and, principally, by a long, curved blade arising from the sclerotized band where it joins the sacculus.

The second species, *woodfordi* Butler, with white spots on a black background, is referable to *Cosmethis* itself, defined by the presence of one or two long, blade-like setae on the sacculus, but its precise relationships are unclear.

There is an undescribed, cryptically patterned species in Vanuatu attributable to *Cosmethis. Milionia* species with red and white bands and lacking bluish iridescence, such as *tricolor* Felder and allies, also proved on dissection to belong to the *Cosmethis* group. Several new combinations in the group are presented in Appendix 1 (Note 1).

Patterns of association between these diverse New Guinea groups and taxa in the Oriental Region are very varied. The relatives of the genus *Ctimene* are found mainly in the N. Moluccas, Sulawesi and the Philippines as are the westerly taxa of the *Cosmethis* complex, though with somewhat greater diversity in Sulawesi. The lymantriid genus *Dura* Moore is very diverse in New Guinea and Sulawesi, has a few Oriental species, but is poorly represented in the Philippines. In the amathusiine butterflies the diverse New Guinea *Taenaris* Hubner complex is probably a sister group to *Faunis* Hubner and allies in Southeast Asia and Sundaland but there is no representation in the Philippines and Sulawesi.

The role of Sulawesi and the Philippines in connecting the Melanesian Arcs with Sundaland and Southeast Asia therefore needs to be investigated in depth. The sister group to the Melanesian Arc groups in the cicada subtribe Cosmopsaltriaria is found in Sulawesi (Duffels, 1977, 1983). There are several apparently monophyletic groups restricted to Sulawesi, the lesser Sundas, and Philippines, such as the cicada genus *Lembeja* (Jong and Duffels, 1981), a complex of rat genera (Musser, 1981), the bird genera *Leucotreron* and *Tanygnathus,* and two subgroups of the butterfly genus *Parantica* (Ackery and Vane-Wright, 1984; see also Fig. 10). Several show some association with New Guinea.

Penetration by Australian Genera

Thornton (1980), in studies of the Psocoptera, drew attention to groups that are diverse within the older land areas of the Inner Arc, including New Zealand, yet are absent from the archipelagic areas of the Outer Arc. The hypocystine complex of genera within the satyrine butterflies (Holloway, 1974) and the micropterigid moth genus *Sabatinca* Walker provide examples for the Lepidoptera. Such patterns may be vicariant, dating from the separation of Gondwanaland, but as stated earlier, no detailed congruence of area cladograms for such groups has yet been observed. Other Australian-centered Lepidoptera genera extend into Melanesia in a more complex way, suggestive of a degree of dispersal.

The noctuid genus *Calathusa* (Holloway, 1979) has five species in New Caledonia, two of which are shared with Australia and three with sister species there. One of the three occurs also on Vanuatu and another is part of a complex found also on Vanuatu, the Bismarcks, New Guinea, and Sulawesi. Further, rather divergent species are found in Fiji and Tonga, the Solomons, and New Guinea.

The ennomine geometrid genus *Casbia* Walker, associated with the plant family Rhamnaceae, has numerous Australian species and some extension into New Guinea (Holloway, 1979). It is much more restricted in distribution elsewhere in Melanesia. There are three species in New Caledonia, one shared with Australia and New Guinea, one with distinct races in Australia and Vanuatu, and one endemic, its sister species probably Australian. In Fiji there flies a sister pair of species whose relationships are obscure, but not evidently with any of the New Caledonian taxa.

The ennomine geometrid genus *Idiodes* also has numerous species in Australia, though I have been able to examine only two of these. The genus is defined in the male genitalia by a bicornute juxta bearing spines on each side apically, and a slender apical projection of the valve costa (Holloway, 1979: fig. 42:5); the aedeagus vesica bears two moderate, separate cornuti. The type species, *apicata* Guenée, feeds on bracken (*Pteridium*) in Australia (McFarland, 1979) apparently exclusively. It differs from the only other Australian species examined, *tenuicorpus* Prout, in having a single central corema on the male first sternite. *I. apicata* flies also in New Guinea and very similar species occur in Sulawesi (*unilinea* Warren) and Luzon (*noctuodes* West); the sister species to this trio is probably the New Caledonian *serpentina* Holloway (wrongly attributed to *Nadagara* in the original description; see Appendix 1, Note 2). The New Guinea *I. robusta* Warren has male genitalia of a different pattern and may be misplaced in *Idiodes.*

In none of these cases can one detect patterns that accord with a vicariance hypothesis for the evolution of the groups based on the separation of the elements of Gondwanaland. The only feasible explanation is that of a complex series of dispersals, some very recent, between Australia and New Caledonia and other islands in Melanesia. The alternative of vicariance poses all sorts of problems concerning the degree of, or absence of, divergence between vicariant populations in Australia and New Caledonia following the opening of the Tasman Sea in the late Cretaceous.

Groups of Uncertain Affinity

There are several small groups that are distributed over Inner Arc areas: New Guinea, New Caledonia, the tropics of northern Australia, and sometimes Ceram and Buru. They are not found to any great extent in more temperate parts of Australia. Most were identified by Holloway (1979), but further investigation has tended to reveal situations of greater complexity than was suspected originally.

For example, there is a quartet of genera in the geometrid subfamily Oenochrominae: *Encryphia* Turner, *Physetostege* Warren, *Callipotnia* Warren, *Caledophia* Holloway. These were associated together by Holloway (1979) as a potentially natural group distributed over the Inner Arc tropics.

These genera belong to a larger complex of medium-sized, delicately built species, usually of gray or pale brown, found throughout the tropics and subtropics of the world. The group is defined by a distinctive ventral looping of the tegumen in the male genitalia, such that the uncus appears to be supported by a slender sclerotized column within a broad area of delicate membrane. The uncus is usually complex with lateral flanges. The valve usually has a spur subapically, an extension of the costa. In the majority of genera the male abdomen has a complex series of modifications, probably secondary sexual characters, especially on the third and fourth sternites; the eighth segment has a pair of curved, sausage-shaped zones that are finely corrugate, and is itself usually slightly expanded. No general diagnostic features were found in the female genitalia.

South American genera included are *Ophiogramma* Hubner, *Dolerophyle* Warren, *Dolichoneura* Warren, *Leptoctenopsis* Warren, *Entogonia* Warren, and *Pycnoneura* Warren. In Africa the group is represented by *Panagropsis* Warren, *Apatadelpha* Prout, *Brachytrite* Swinhoe, and *Conolophia* Warren, the last with two species also in the Oriental Region. In the Indo-Australian tropics the genera *Alex* Walker and *Noreia* Walker range from the Asian mainland to the Solomons. *Noreia venusta* Warren, from Borneo and Malaya, is distinct from its congeners in male and female abdominal and genitalia characters and probably merits generic status. Typical *Noreia* species share with New World *Ophiogramma* the possession of a prominent ridge on the male seventh abdominal sternite and a round, heavily spined bursa copulatrix in the female; there are also general similarities in wing pattern. The two are probably sister genera.

The four Australasian genera are all monophyletic. *Physetostege* (Queensland, New Guinea) and *Caledophia* (New Caledonia) are monobasic. *Callipotnia* (three New Guinea species) is defined by the presence, in the male genitalia, of a long rod-like structure on the juxta, a heavily sclerotized, produced, acute gnathus, weak coremata at the base of each valve, a sinuous convolute nature to the spur from the valve costa, and a secondary spur from the longitudinal flange in the center of the valve. The hindwings have the postmedial acutely angled. The female of only one species was dissected and the bursa copulatrix bore a central ring of sclerotization with numerous broad spines directed distad.

Encryphia has one Australian species, *frontisignata* Walker, probably restricted to Queensland but with a distinct subspecies in New Caledonia. The species *pulverosa* Warren, in New Guinea, was found misplaced in *Noreia* (see Appendix 1, Note 3); a further, undescribed species, known from females only, occurs in the New Hebrides. The genus is distinguished in the male genitalia by having spurs from both costa and sacculus on the valve, linked basally by a transverse sclerotized band that bears a smaller spur centrally; the anellus membrane is scobinate. The female genitalia are distinguished in having a small scobinate

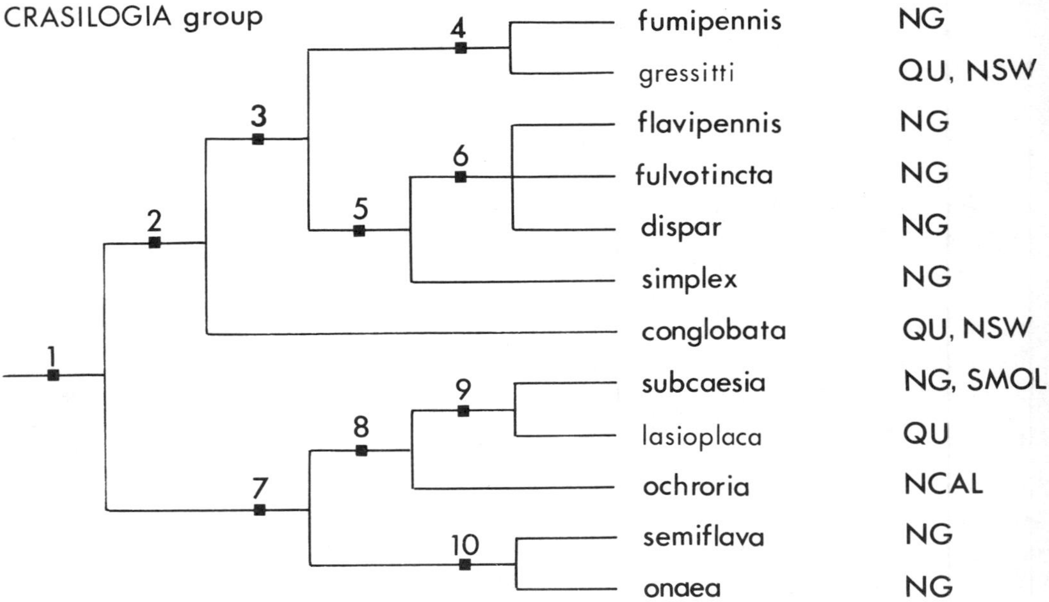

Fig. 2. Cladogram of relationships for species in the *Crasilogia* group of genera. The numbers refer to presumed apomorphies listed in Appendix 2. Abbreviations for distributions are as listed in legend to Fig. 10.

signum in the bursa and ductus bursae that is basally much sclerotised and broadly triangular.

No characters were found to indicate that these four Australasian genera form a monophyletic group. The only sister-group relationships evident were with genera outside the group. *Callipotnia* shares with *Conolophia* the characteristic coronate ring in the bursa copulatrix, a character that I have not seen even approached elsewhere in the Geometridae. *Conolophia* is a homogeneous genus, all species being very similar in facies, defined particularly by the presence of a round, dark brown subdorsal spot on the forewing just exterior to the postmedial; this is only weakly and diffusely present in *Callipotnia*. *Conolophia* species also share a unique paired series of lobes and hair pencils, four on each side, on the male third abdominal sternite.

Encryphia and *Physetostege* have no obvious sister groups. The latter has a highly developed subbasal fovea to the male forewing, overlain by much larger scales. This character is only found elsewhere, and then weakly, in *"Noreia" venusta,* but it occurs widely in the Ennominae and so may be plesiomorphous.

Caledophia is only known from males. In modification of the fourth abdominal sternite and of the posterior margin of the sixth, together with the structure of the genitalia, it resembles the African genus *Brachytrite.* In possession of a pair of eversible coremata between the seventh and eighth abdominal segments it resembles the South American genera *Leptoctenopsis, Dolichoneura,* and *Pychnoneura.* In the rather slender uncus and rhomboidal, peaked juxta it resembles *Conolophia.* The structure of the female genitalia, when known, may narrow the field somewhat.

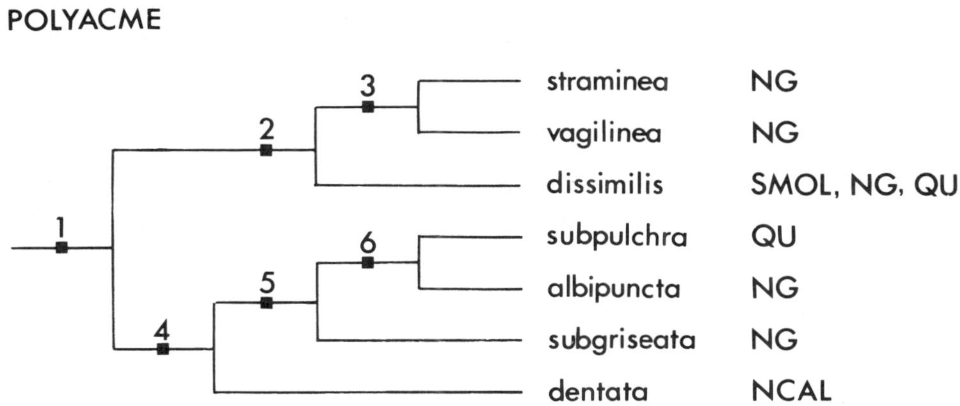

Fig. 3. Cladogram of relationships for species in the genus *Polyacme.* The numbers refer to presumed synapomorphies listed in Appendix 2. Abbreviations for distributions are as listed in legend to Fig. 10.

The distinctiveness of the four genera and the potentially African relationship of *Callipotnia* suggest they may have had a long history in Australasia that may date from the early stages of the dispersal of Gondwanaland.

Two further geometrid groups extend over Inner Arc areas, the *Crasilogia* Warren group of genera in the Larentiinae and the genus *Polyacme* Warren in the Ennominae. Both are most diverse in New Guinea. The presumed phylogeny of the *Crasilogia* group is shown in Fig. 2 with the characters on which it is based listed in Appendix 2. The species are listed in detail with one new combination in Note 4 of Appendix 1. The presumed phylogeny of *Polyacme* is shown in Fig. 3, the characters listed in Appendix 2, and the systematics set out, with a generic synonymy, in Note 5 of Appendix 1.

There is little information available yet on the genitalic morphology of most geometrid genera so it was not possible to pinpoint sister groups for either complex. No potential sister groups are apparent in the cross section of Oriental genera studied by Holloway (1976), which included most of the Sunda Shelf larentiine genera. The identification of derived characters in the cladistic analyses is based on experience gained by these studies of Oriental and Australian geometrids.

These cladistic analyses have revealed a common pattern where a group of New Guinea species is sister to a more widespread group extending over New Guinea, Queensland, and New Caledonia. The degree of sympatry is indicative of complex exchange between northern Australia and New Guinea. The widespread group species are endemic to the three areas concerned and give an area cladogram where New Guinea and Queensland associate to the exclusion of New Caledonia (Fig. 4). This is also seen in the *gelon-macleayanum* group of *Graphium* Scopoli (Fig. 9), a large genus of papilionid butterflies studied by Saigusa et al. (1977). In *Crasilogia* and the *Graphium* group the South Moluccas are also involved as sister area to New Guinea.

A different relationship of these areas is apparent within the noctuid genus *Leucogonia* Hampson. This genus has been placed until now in the subfamily Acronictinae but it should be transferred to the Agaristinae (Appendix 1, Note

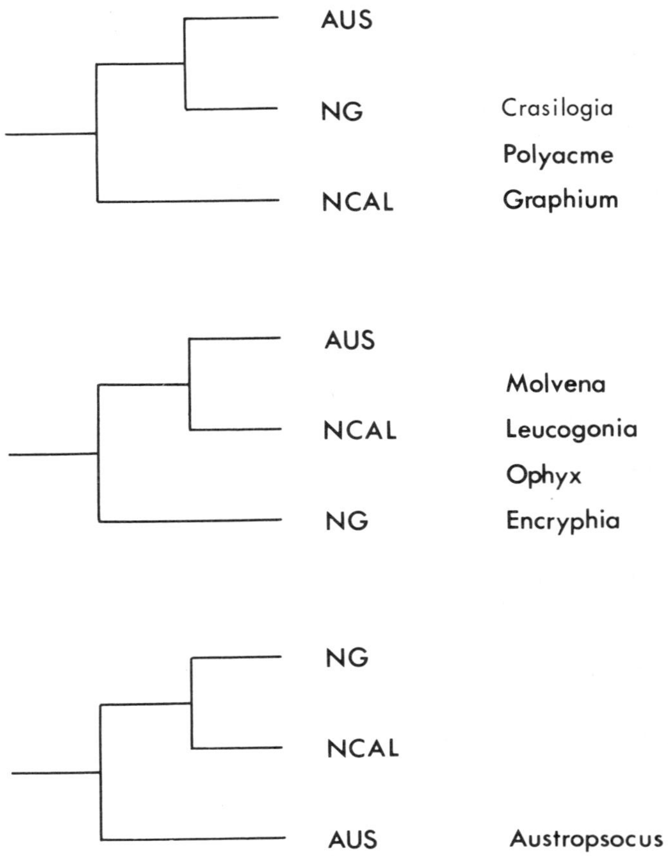

Fig. 4. The three possible area cladograms for Australia, New Guinea, and New Caledonia. Seven Lepidoptera genera and one Psocoptera genus with species groups showing these patterns are listed.

6). Its precise relationships are unclear but in facies of wings, thorax, and abdomen it has derived characteristics in common with agaristines such as *Tuerta* Walker in Africa and *Neotuerta* Kiriakoff in South America. The male genitalia are of a type that occurs in a number of agaristine genera (Kiriakoff, 1977) such as the Australasian *Ipanica* Hampson and *Idalima* Turner, the African *Syfanoidea* Bartel and *Massaga* Walker, the Malagasy *Musurgina* Jordan and *Arrothia,* Jordan, and the South American *Neotuerta, Copidryas* Grote, and *Eusscirrhopterus* Grote; however, the simple valves with a spinulose margin, the simple, curved, subbasal harpe, and simple, slender curved uncus are probably plesiomorphous within the Noctuidae, recurring in other subfamilies [see figures 30, 49, 87, and 213 in Holloway (1976)]. The genitalia of *Tuerta* species are modified and, with the partly green forewings, indicate the monophyly of this African group.

Leucogonia species (here taken to include the monobasic genus *Metaxanthiella;* Appendix 1, Note 6) are defined by the dull orange forewings with marginal and tornal borders and the pale fasciation at the tornus in the dark border of the yellow hindwings. *L. kebeensis* Bethune-Baker (New Guinea) and *L. ekeikei* Bethune-Baker (New Guinea, Queensland, Kei) are probably sister species as they have very similar facies, especially the even width to the forewing margins, and male genitalia. In the latter the valve is broader centrally and more curved over the apical two thirds than in the other species, and the harpe is set very basally. In the female genitalia they both lack sclerotization anterior to the ostium bursae or in the ductus bursae. Relationships of the other two species, *cosmopis* Lower (North Queensland) and *amarginata* Holloway (New Caledonia), are unclear but they are generally more similar to each other in facies and genitalia than to the other pair. So in *Leucogonia* there is weak indication of an area cladogram associating Queensland and New Caledonia to the exclusion of New Guinea (assuming that the presence of *ekeikei* in Queensland is due to dispersal.

The noctuid genera *Molvena* Walker and *Ophyx* Guené, also primarily Inner Arc groups and also with affinities outside Australasia as yet undetermined, share with *Leucogonia* the area cladogram with New Caledonia and Queensland associated to the exclusion of New Guinea; this also occurs in *Encryphia* (Fig. 4). The third possibility, association of New Guinea and New Caledonia to the exclusion of Queensland, has not been found in a Lepidoptera group but is seen in the psocopteran genus *Austropsocus* (Thornton, 1980) (Fig. 4).

Can the Australia-New Caledonia dichotomy in these area cladograms be related to the vicariant event of the opening of the Tasman Sea, dating it at about 80 m.y.a.? Then earlier branchings on the cladograms would predate this. A problem is presented by the fact that the Australian species of the *Graphium* complex has been recorded from Norfolk and Lord Howe islands, on the New Caledonian side of the Tasman Sea. The area cladograms do suggest that Inner Arc relationships between New Guinea and New Caledonia have been indirect via northern Australia. Isolation of Australia from New Caledonia may have been lessened at times in the Tertiary. Coleman (1980) has indicated that, in the middle of the Tertiary, the Queensland Plateau and Chesterfield Plateau (the northern extremity of the Lord Howe Rise) were large shoal areas with islands, a possible route for biological exchange between Queensland and New Caledonia.

APPARENT MODERN PATTERNS

Modern patterns of distribution in Melanesia resulting from colonization from outside might be expected to retain much of the character of the original dispersal: that of a single taxon found throughout the area. Divergence of populations throughout the range of this ancestral taxon, and therefore its fragmentation into daughter taxa, might be expected to reflect the geography of the area, i.e. physical barriers to continuation of genetic exchange between populations in areas of suitable habitat. Larger discontinuities would permit less genetic exchange. The increase in endemism with decrease of dispersal powers from group to group observed on Norfolk Island is a corollary of this.

This process of fragmentation might well lead to congruent area cladograms for the taxa concerned and, in a somewhat different way, be the basis for congruence of the "area cladograms" for butterflies and birds derived by Holloway and Jardine (1968).

Several common patterns of distribution have resulted from what might be considered recent dispersal and fragmentation in the Indo-Australian tropics (Hol-

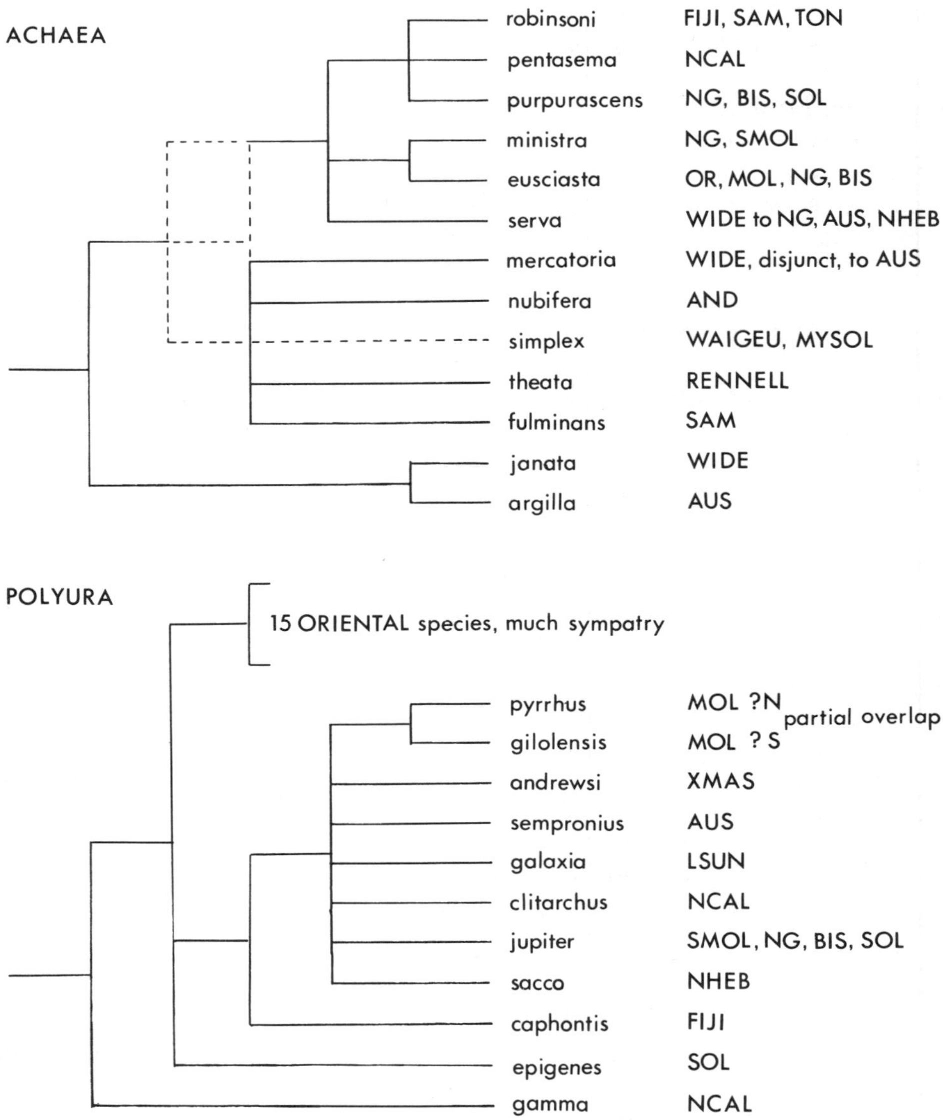

Fig. 5. Cladograms of relationships for species in the genera *Achaea* (Holloway, 1982, 1983) and *Polyura* (Smiles, 1982). Dotted lines indicate ambiguity or uncertainty. Abbreviations for distributions are as listed in legend to Fig. 10.

loway, 1983): Oriental/Melanesian (e.g. in Fig. 9; *empedovana/codrus*) and Widespread/Australian (e.g. in Fig. 5; *janata/argilla*) sister pairs are often encountered. Variations on an Asian/Sunda/Lesser Sunda/Philippines/Sulawesi/Melanesia theme are also frequent. More local allopatric distribution patterns in Melanesia often reflect the common distribution patterns within Melanesia identified numerically by Holloway (1979), namely New Guinea plus Bismarcks plus Solomons, Fiji plus Samoa plus Tonga, and New Caledonia plus Loyalties, the last two categories both including Vanuatu. Triplets of species or subspecies distrib-

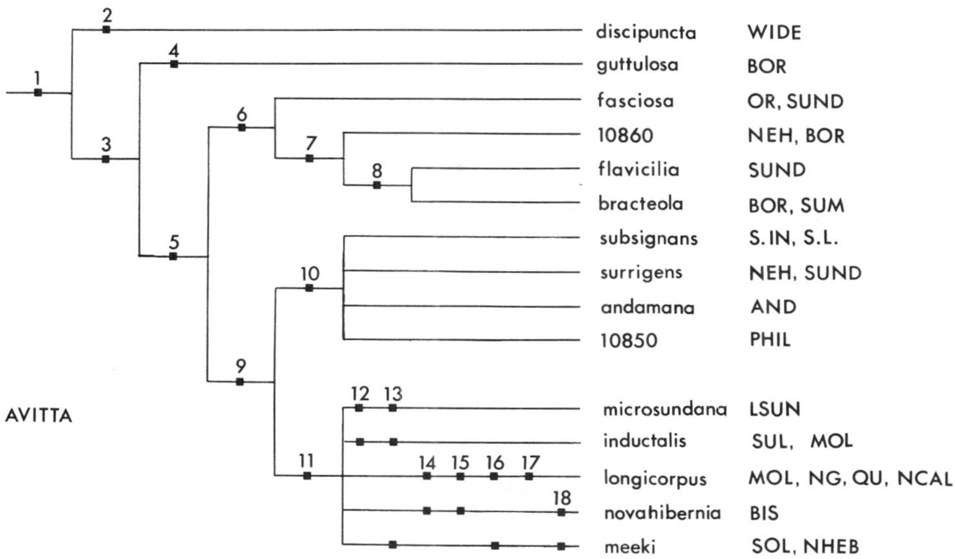

Fig. 6. Cladogram of relationships for species in the genus *Avitta.* The numbers refer to presumed synapomorphies listed in Appendix 2. Large numbers at branch ends refer to undescribed species known only from females, and are BM genitalia slide numbers. Abbreviations for distributions are as listed in legend to Fig. 10.

uted allopatrically among these three categories are seen in the *Achaea pentasema* Prout group (Fig. 5), the *Thalassodes chloropis* Prout group, and the races of the noctuid species *Thyas regia* Lucas and *Ophiusa disjungens* Walker.

A more fragmented allopatric array of species in Melanesia is exhibited by the *jupiter* group of the nymphalid genus *Polyura* Billberg (Smiles, 1982, and Fig. 5) and in the *subsignans* group of the noctuid genus *Avitta* Walker, analyzed for this paper. In both cases there is limited overlap of taxa in the Moluccas.

A cladistic analysis of *Avitta* s. str. is presented in Fig. 6 and the characters presumed to be apomorphic are presented in Appendix 2. The species are listed in Appendix 1, Note 7, with several described as new, and the characters of the valve of the male genitalia are shown in Fig. 7. An extensive survey of genitalic morphology in the Ophiderinae has yet to be undertaken; indeed it is not clear that other species generally placed in *Avitta* and referable to subgenera *Asta* and *Pantura* (Holloway, 1979) are the sister group to the species considered here. However, the characters of male and female genitalia listed are distinctive within the Noctuidae and there is thus a high probability of monophyly for the group.

The only overlap in the otherwise allopatric *subsignans* group occurs, as stated, in the Melanesian subgroup in the Moluccas, where *inductalis* Snellen (also Sulawesi) flies with the more easterly species *longicorpus* Prout. The Lesser Sunda and Sulawesi/Moluccas species may be sisters, defined by character 12, and a sister group to the other three species, defined by combinations of two or more of characters 14, 15, 16, and 18. However, the most interesting point about these three eastern taxa is that the presumed apomorphic characters defining them overlap each other (Fig. 8), a situation that is consistent with the concept of a

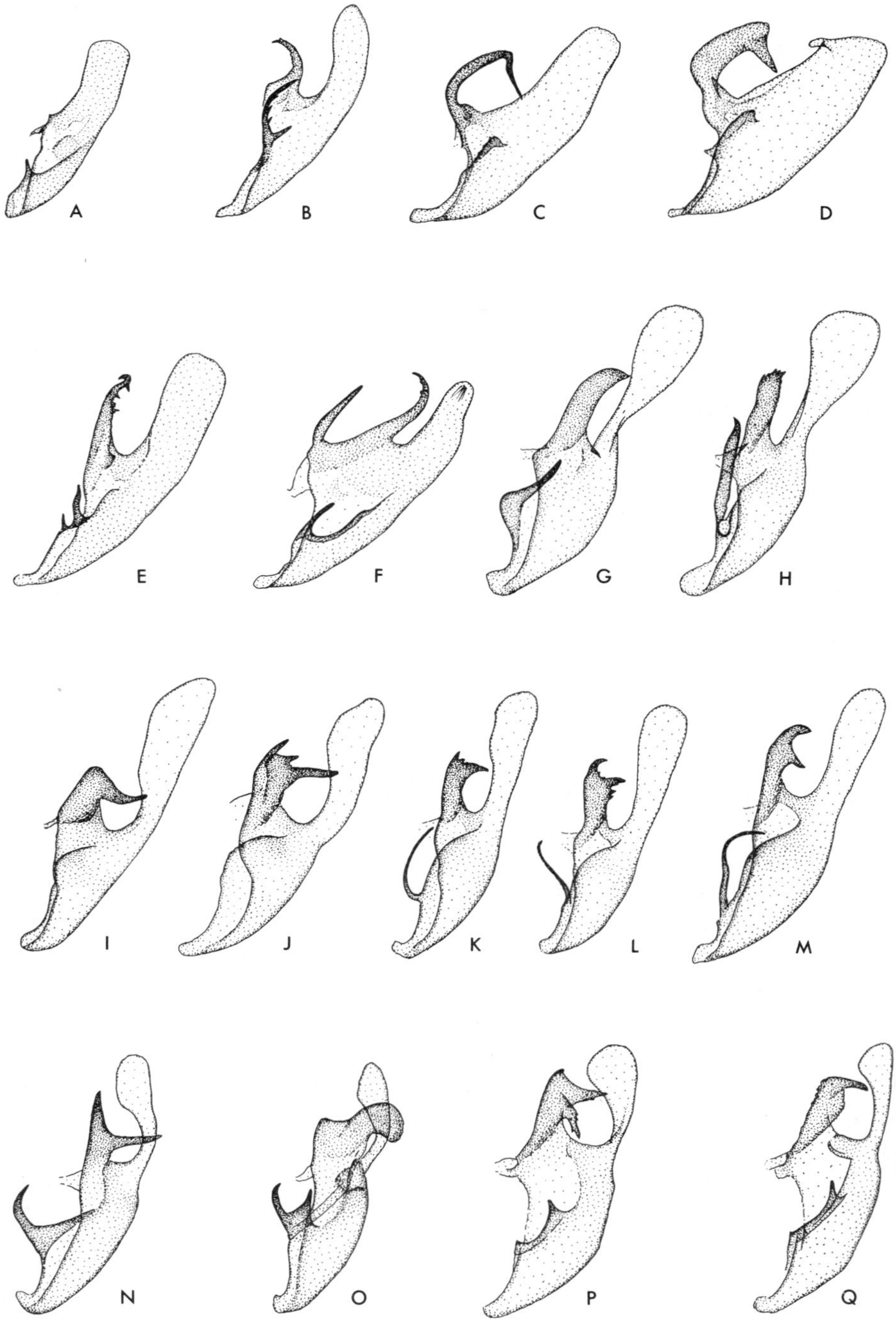

Fig. 7. Valves of the male genitalia of *Avitta* species as follows: A, *discipuncta* (Borneo); B, *guttulosa* (Borneo); C, *fasciosa* (Borneo-Sundaland form); D, *fasciosa* (N.E. Himalaya—typical mainland Asian form); E, *flavicilia* (Sumatra); F, *bracteola* (Borneo); G, *meeki meeki* (Solomons); H, *meeki vanuatua* (Vanuatu); I, *longicorpus* (Queensland); J, *novahibernia* (Bismarcks); K, *microsundana* (Flores); L, *microsundana* (Timor); M, *inductalis* (Batjan); N, *andamana* (Andamans); O, *subsignans* (Sri Lanka); P, *surrigens* (Sumatra—typical Sundaland form); Q, *surrigens* (N.E. Himalaya—mainland form).

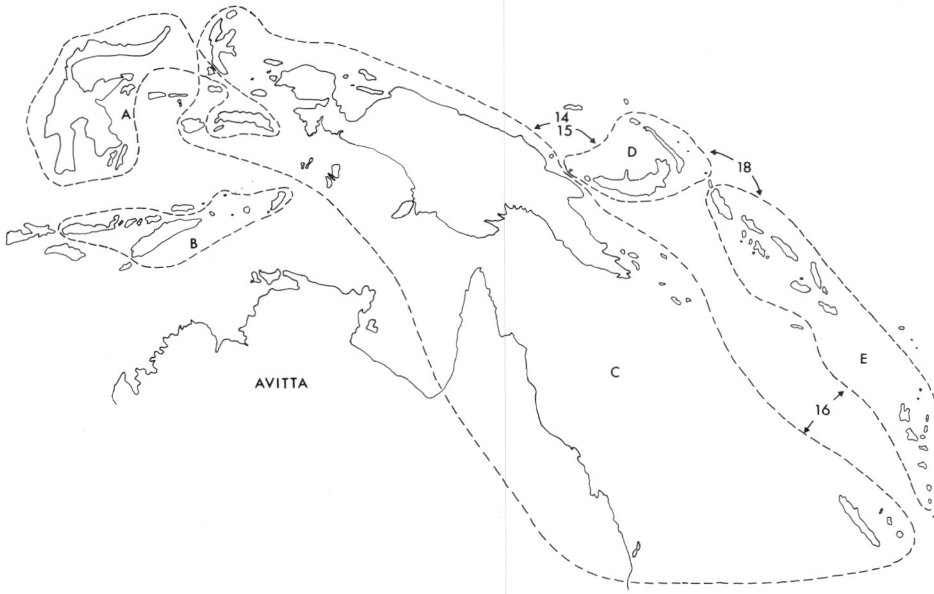

Fig. 8. Distribution of species of *Avitta* in Melanesia: A, *inductalis;* B, *microsundana;* C, *longicorpus;* D, *novahibernia;* E, *meeki.* The numbers indicate apomorhic characters as in Appendix 2 (reference Fig. 6), shared between the taxa concerned.

gradual fragmentation of a single widespread taxon through a decline of genetic exchange between neighbouring populations, but not with an abrupt termination of exchange through vicariance, unless some sort of clinal situation prior to vicariance is postulated.

Other groups appear to have widespread, allopatric arrays of closely related species throughout the Indo-Australian tropics, with a moderate degree of localised endemism within Melanesia. The noctuid genus *Eutamsia* Fletcher is one and the ennomine geometrid genus *Aplochlora* Warren is another, but in both cases material is at present too limited to permit any firm conclusions. Species of the ennomine *Petelia medardaria* Herrich-Schaffer complex are also probably distributed allopatrically through Melanesia, but investigation is hampered by the absence of good morphological characters.

PUTATIVE TWO-ARC PATTERNS

The strongest evidence for the two-arc hypothesis, as mentioned earlier, would be provided by the discovery of sister pairs of genera or species groups, one with each type of distribution, of roughly equal polytypy and complexity of endemisms. The classic example provided by the Cosmopsaltriaria cicadas has yet to be duplicated, though studies in progress on the cicada genus *Baeturia* are promising (Duffels, pers. commun.). No such sister groups have as yet appeared in the studies on Psocoptera by Thornton and Smithers (e.g. Thornton, 1980). The bird families Psittacidae and Columbidae, especially diverse in Australasia (Raven and Axelrod, 1974), may provide examples, as may the Gesneriaceae and some of the

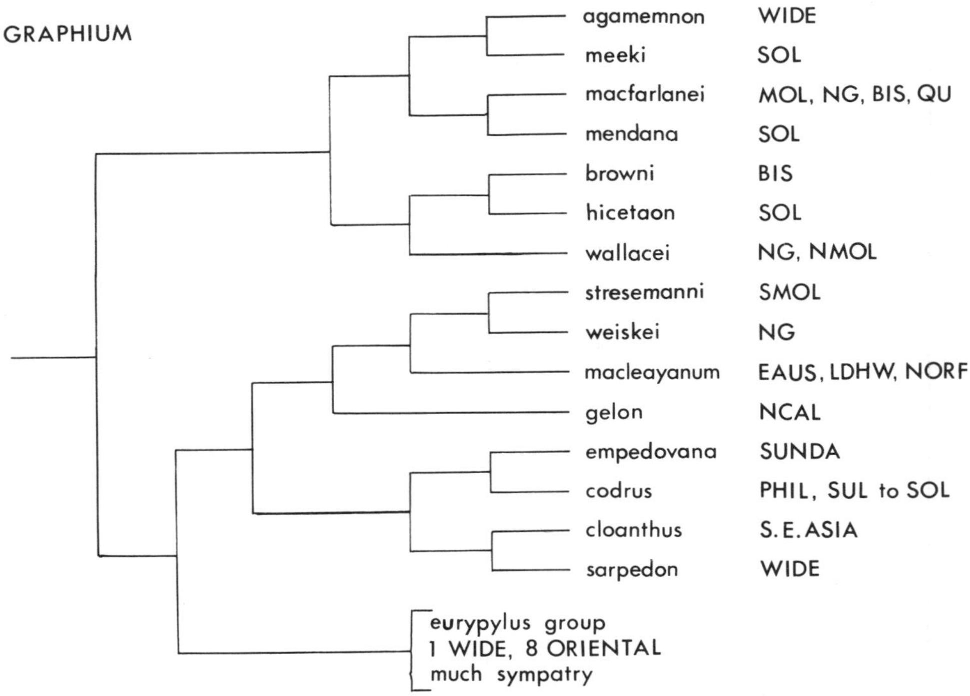

Fig. 9. Cladogram of relationships for species in the genus *Graphium* (Saigusa, et al., 1977). Abbreviations for distributions are as listed in legend to Fig. 10.

widespread plant families mentioned earlier, though no clear examples of sister pairs are evident in the works of Corner and Moore on figs and palms reviewed by Holloway (1979).

A survey of the macrolepidoptera has provided a few rather weak instances of sister groups. Very few cladistic analyses have been performed on such genera. The study of *Graphium* Scopoli by Saigusa et al. (1977) (Fig. 9) indicated subgroups in different lineages with Inner or Outer Arc distributions but not sister groups. A forthcoming study of the danaine genus *Parantica* (Ackery and Vane-Wright, 1984) reveals two groups of species extending through the Indo-Australian tropics (Fig. 10). The oldest group, determined as such by the diversity and sympatry criterion discussed earlier, has a small Melanesian group that cannot be attributed clearly to one or other Arc and contains one species, *pumila,* shared between New Caledonia and the New Hebrides, straddling the Arcs.

In a search for further groups attention was focused on those showing a degree of endemic development in the Solomons. Endemic speciations identified for Fiji (Robinson, 1975; Holloway, 1979, 1983) are either taxonomically very isolated or occur in extremely complex groups needing extensive revisional work.

In the Geometridae very few of the endemic Solomons taxa pinpointed potential Inner/Outer Arc sister groups. The monotypic or small endemic genera are either morphologically isolated with no obvious sister groups (e.g. *Heteresthes* Warren, monobasic; *Crypsimetalla* Warren, three species, one shared with Bismarcks; *Solomonophila* Prout, monobasic) or closely related to, and only dubiously sep-

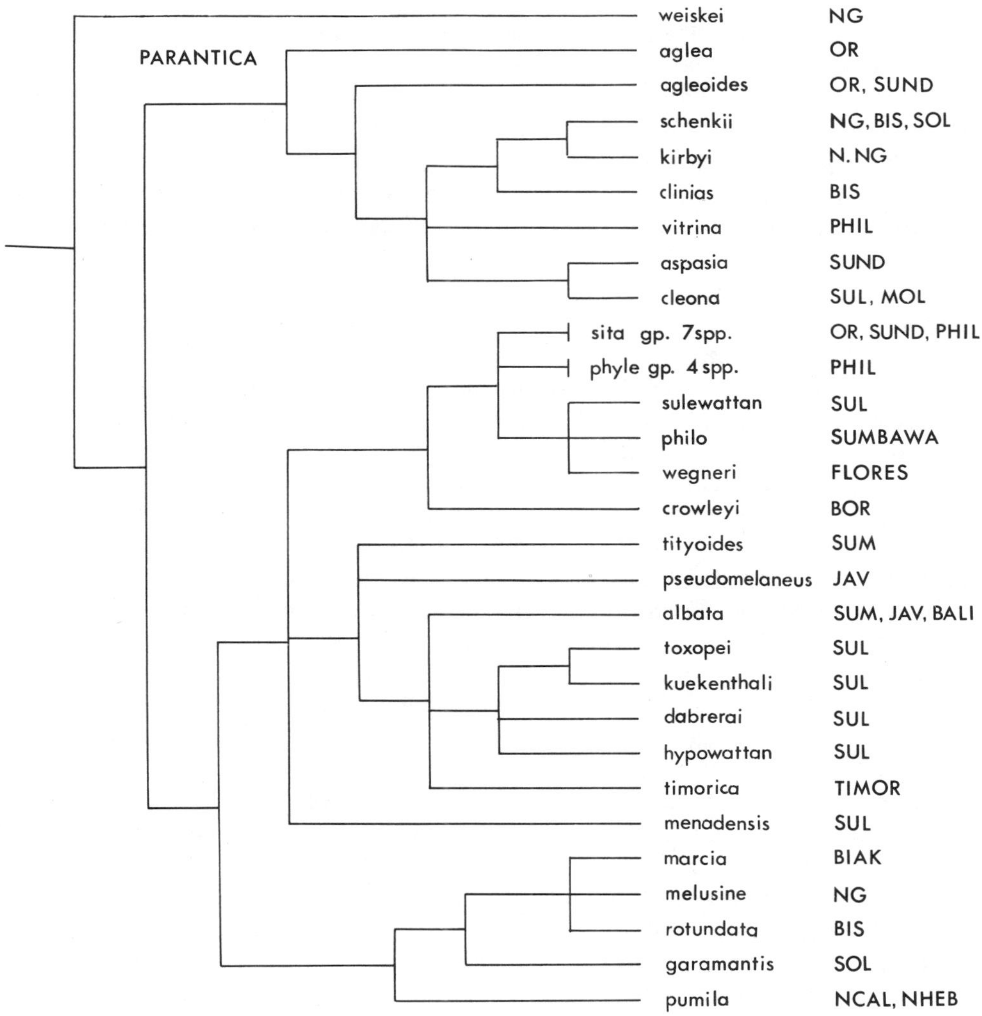

Fig. 10. Cladogram of relationships for species in the genus *Parantica* (Ackery and *Vane-Wright, 1984). Abbreviations for distributions are as follows: AND, Andamans; AUS, Australia; BIS, Bismarck Is., BOR, Borneo; EAUS, eastern Australia; JAV, Java; LDHW, Lord Howe I.; LSUN, Lesser Sunda Is. from Lombok to Timor, Tenimber, Kei; MAL, Malay Peninsula; MOL, N. + S. Moluccas; NCAL, New Caledonia; NEH, N.E. Himalaya; NG, New Guinea; NHEB, New Hebrides; NMOL, northern Moluccas (Halmahera, Batjan, Ternate); NORF, Norfolk I.; NSW, New South Wales; OR, Oriental tropics in mainland Asia; PHIL, Philippines; QU, Queensland; SAM, Samoa; S.IN, S. India; S.L., Sri Lanka; SMOL, South Moluccas (Ceram, Buru, Amboina); SOL, Solomon Is., SUL, Sulawesi; SUM, Sumatra; SUND, Sundaland, all areas on the Sunda Shelf; TON, Tonga; WIDE, Widespread in Indo-Australian tropics.

arated from, other more widespread genera [e.g. *Uranodoxa* Butler, monobasic and very close to *Scardamia* Guenée; *Prorocorys* Warren, monobasic, shared with the Bismarcks, and inseparable from *Ziridava* Walker in male genitalia characters, falling within the *xylinaria* Walker complex discussed by Holloway (1982) (see Appendix 1, Note 8)]. *Ctenistochlora* Warren (monobasic) is close to *Aplochlora*

Warren, a genus that may prove interesting biogeographically when, as suggested above, more material becomes available.

The genus *Metallochlora* Warren is well represented in the Solomons, and the *Cosmethis/Bracca* complex is represented by two species, but the former showed no definite groupings attributable to one or other Arc and the latter has yet to be examined in detail (see page 137); *Bracca* itself provides preliminary indications of independent development on the two Arcs, but this needs confirmation by more detailed morphological study.

Only the genus *Nadagarodes* Warren contains two subgroups that may be referable to one or other Melanesian Arc, though the situation is not very clear cut and several species, including two from the Solomons, are probably best placed outside the two groups. The genus belongs to the *Luxiaria* Walker group of genera, characterised by a deeply divided valve and rather 'shouldered' uncus in the male genitalia, as well as an overall similarity of wing marking. The sister genus to *Nadagarodes* is probably *Probithia* Warren, sharing with it the very slender, upcurved, tapering lower portion to the valve which bears a finely scobinate swelling subbasally (lacking in some *Nadagarodes,* perhaps secondarily lost). *Nadagarodes* itself is most clearly defined as monophyletic by characters on the male aedeagus, which bears apically two prominent spines, each with a number of small secondary spines adpressed over the apical part, and two slender but long cornuti, usually ornamented, occur in aedeagus vesica. The only exception to this is the Solomons species *flavipectus* Warren where the vesica has a single cornutus only, and the projections at the aedeagus apex are very small. The dorsal arms of the valve are asymmetrical, that of the right much broader. The insect itself is a steely gray color, contrasting with the reddish fawn, yellow, and brown of other species.

The remaining taxa, with the exception of the New Guinea species *regula* Prout, all have a subapical ventral spur to the round-ended dorsal process to the valve, and it is on modification of this that the species can be subdivided further. Five species have the valve process distal to the spur rather elongate, narrow, with the spur itself very slender. They are *subpulchrata* Warren from the Solomons, an undescribed species (genitalia slide 11477) from the Bismarcks, and a triplet with a distinctive uncus like a vulture head, consisting of *tentilinea* Prout from the Bismarcks and *mysolata* Walker and *duplicipuncta* Warren from New Guinea, both widely distributed and occurring on the southeast islands, the former also extending to Mysol and Amboina. This group of five more or less accords with the Outer Arc.

Three species have the apex of the dorsal process more rounded, more heavily setose, with the spur arising from the rounded part and being much more substantial. They are all found in New Guinea and may represent an Inner Arc sister to the previous group, though *sordida* Warren extends to the Bismarcks and Amboina. The other two species are *camura* Joicey and Talbot and *simplex* Warren; a fourth species, *tumida* Warren, is known only from two New Guinea females but may also belong in this group.

This picture is confused somewhat by a further Solomons species, robust in both appearance and male genitalia, *purpuraria* Warren. The spur to the dorsal process is manifest only as an angle subapically, and there are a few unique characters to the genitalia such as the anterior flexure to the long, even-width

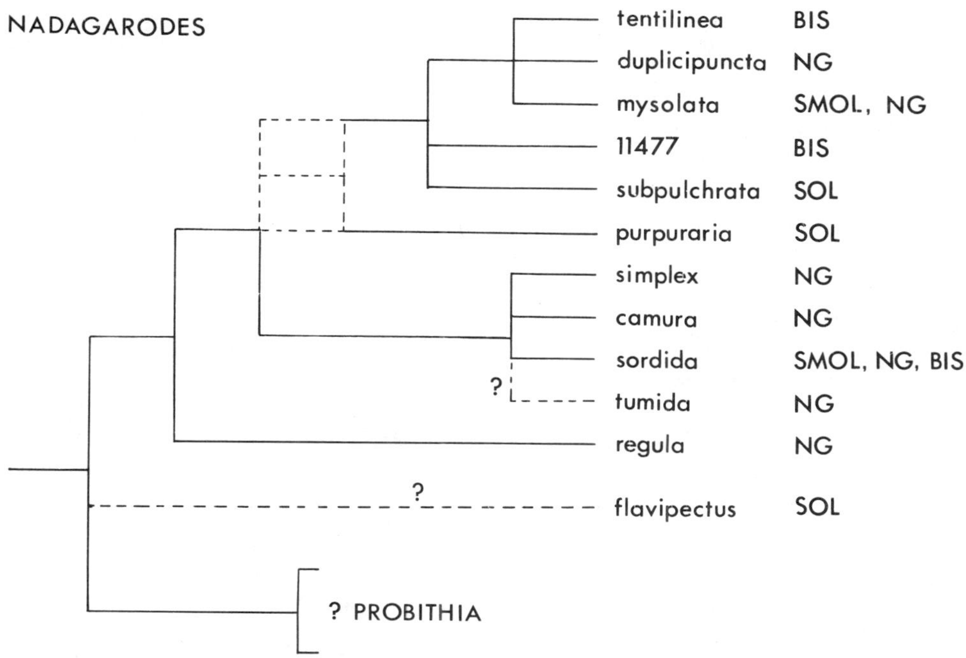

Fig. 11. Cladogram of relationships for species of *Nadagarodes* as discussed in the text. Dashed lines indicate ambiguity or uncertainty. Abbreviations for distributions are as listed in legend to Fig. 10.

uncus. The position of this species in relation to the others is unclear and further work on the morphology of the group and its relatives is needed. The Outer Arc group and *purpuraria* share a subapical white or yellow patch on the forewing underside lacking in other taxa. It is also lacking in *Probithia* yet occurs in some *Luxiaria* to a less marked extent, so it need not necessarily be an apomorphic trait within *Nadagarodes.* If it were, then *purpuraria* could be assigned to the Outer Arc group.

A tentative assessment of relationships in *Nadagarodes* is shown in Fig. 11. One species previously assigned to *Nadagarodes, turpis* Warren from Bali and Sumba, must be transferred to *Probithia* along with the Fijian species *sesquilinea* Prout, previously placed in *Luxiaria* (e.g. originally and by Robinson, 1975) (Appendix 1, Note 9). *Probithia* can again be defined by characters of the male genitalia such as the development of the scobinate subbasal swelling of the ventral process of the valve into more of a knob and the possession of a spined flange, somewhat akin to a broom head, ventrally at the center of the dorsal process of the valve. The genus is completed by the Oriental type species *exclusa* Walker and its Melanesian (Moluccas to Bismarcks) subspecies *perichla* Prout. Further examination may reveal it to have a typically allopatric yet contiguous distribution through the Indo-Australian tropics akin to those discussed by Holloway (1983).

If *Probithia* is the sister group of *Nadagarodes,* it may enable the attainment of the ancestral range of *Nadagarodes* in Melanesia to be assigned an earlier date than that of *Probithia.*

In the Arctiidae only the Lithosiinae provided groups of interest, though the genera *Euchromia* Hubner and *Hyalaethea* Butler in Ctenuchinae have yet to be examined and could be of interest.

The lithosiine genus *Zygaenosia* Hampson is very distinctive, containing numerous small blackish species, often marked with red or with partially transparent forewings. The genus probably has a mimetic association with the zygaenid genus *Apantocephala* Felder. It is diverse in the northern Moluccas and has two species in the Solomons but no definite monophyletic group distributed along the Outer Arc can be recognized; a group of transparent-winged species, probably monophyletic, is centered in southeast New Guinea.

In the *Damias* Boisduval complex of genera there are much clearer indications of Inner/Outer Arc sister groups. Four genera can be recognized, possibly all monophyletic but two, *Byrsia* Walker and *Scaptesyle* Walker, clearly so on characters of the male genitalia. The species are probably mainly day-flying though there is little biological information available. They are brightly colored, with either yellow (*Scaptesyle, Byrsia,* and *Neoscaptia* Hampson) or white (*Damias*) ground color. The bodies and broad wing margins are black, that of the forewing containing an area of orange or red.

The complex overall can be defined on several characters of male and female genitalia. In the male the valve apex is elastic, corematous, and there is a slender, curved process subbasally on the valve costa set dorsally to the valve, concave ventrad. A rounded setose flap, almost a calcar, arises from, and basally unites, the valve bases. Broad lateral lobes extend dorsally from the vinculum to overlap the tegumen. In the female there is a large pouch containing a pair of scale masses just anterior to the ostium bursae. The apophyses of the ninth segment are triangular. The ductus and bursa are short, barely extending beyond the scale pouch, the bursa usually having an area of moderate spines subbasally. The complex as recognized here is set out in Appendix 1, Note 10.

The complex consists of four genera, one mainly Oriental, one extending from Southeast Asia to the Solomons, and two mainly Melanesian with distributions reflecting the two Melanesian Arcs.

The Oriental genus is *Scaptesyle,* defined primarily by the possession of a fan of setae subbasally and ventrally on the uncus. In all except the single New Guinea species (where the hindwing is almost brown) the hindwing borders are narrow. The rather dull red apex to the forewing is simple, delineated by a transverse black bar interiorly that is indented centrally by yellow. The species are listed in Appendix 1, Note 10 and, apart from the single species in New Guinea (Mount Goliath), are all Oriental (Fig. 12).

The widespread genus is *Byrsia,* of all four groups the most clearly defined by derived characters. It lacks the lobes from the male vinculum mentioned above and may therefore be the sister group to the rest if this character is derived, though the lobes are also missing or reduced in subgroup 2 of *Damias.* The male valves have a pair of coremata each basally and are invested over the elastic portion with a dense mass of swollen, clublike scales. The aedeagus contains one very large cornutus. The yellow ground color is more orange than in *Scaptesyle* or *Neoscaptia.* The forewing black border tapers towards the posterior and encloses a triangular bright orange area. The taxa are allopatric except in New Guinea

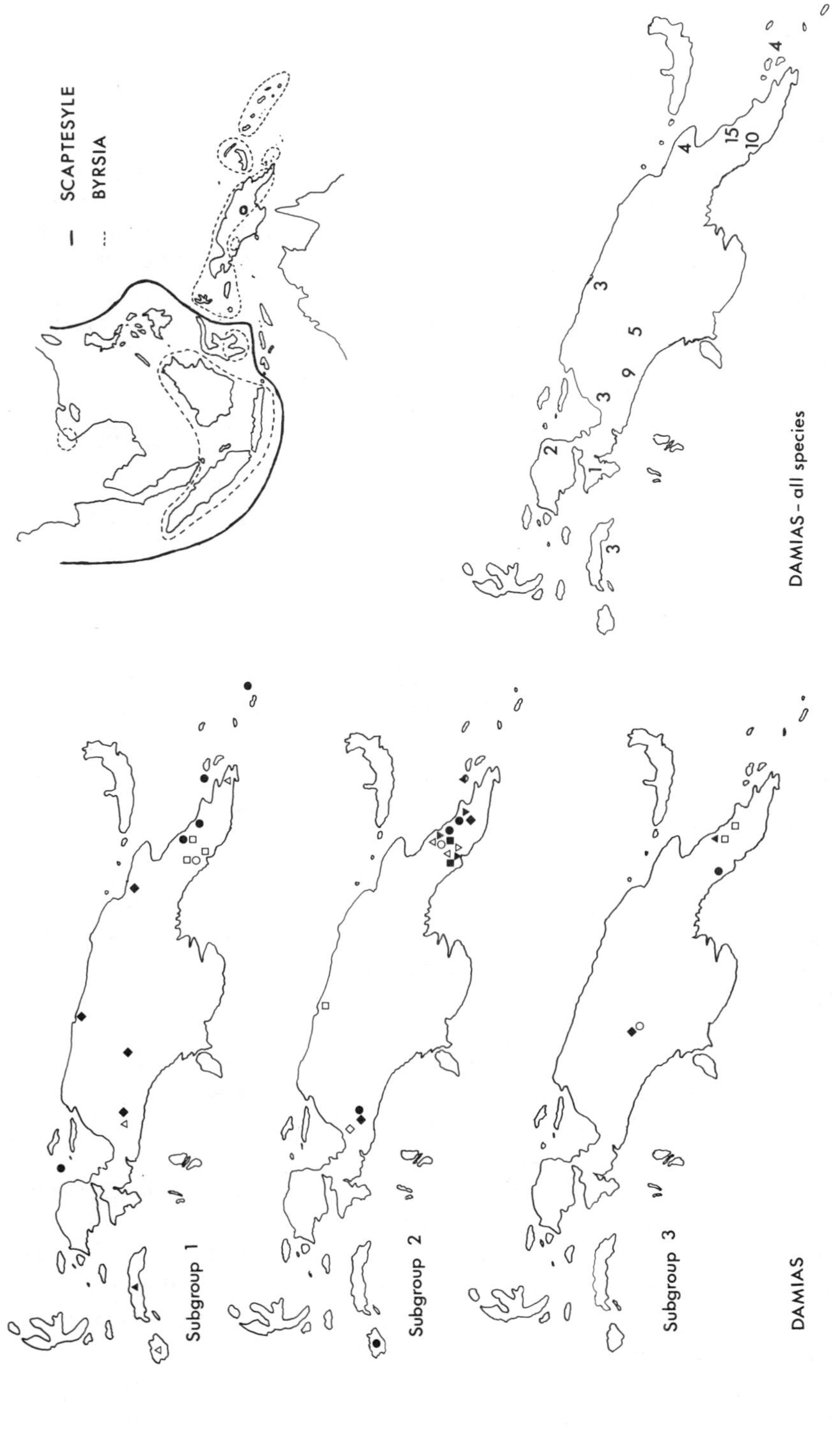

Fig. 12. Mapped distributions of species in the *Damias* group of genera. Top right, distributions of the genus *Scaptesyle* and species of the genus *Byrsia*. Bottom right, numbers of species of *Damias* in various localities. Left, distributions of species of the three subgroups recognized for *Damias*: solid circles, *bipuncta* (group 1), *flavicaput* + *torquata* (2), *poecila* (3); solid diamonds, *bimaculata* (1), *androconiata* (2), *albicollis* (3); open squares, *germana* (1), undescr. sp. (2), *affinis* (3); open circles, *aenea* (1), *apicipuncta* (2), *basinitens* (3); solid triangles, *quadripuncta* (1), *monoleuca* (2), *fasciconitens* (3); open triangles, *biguttata* (1), *albata* (2); solid squares, *leucodera* (2); solid inverse triangles, *unipunctata* (2); open inverse triangles, *aequalis* (2); open diamond, undescr. sp. (2).

where three or four taxa occur, overlapping each other only slightly if at all as far as can be told from the limited material available (Fig. 12). The taxa are listed in Appendix 1, Note 10.

Good, definitive characters for the two remaining genera are lacking and there are only rather tenuous indications that they are sister groups in that both show a tendency for the dark borders of the wings and enclosed orange or red of the forewing to be expanded, often obliquely or irregularly; in both there is a degree of sexual dimorphism, the male forewings tending to be produced apically; and in both there are species with a metallic iridescence though this is leaden purplish in *Damias* and blue to green in *Neoscaptia.* The male genitalia provide no good characters to define the groups but are of some use in distinguishing subgroups within *Damias.*

Neoscaptia species have a bright yellow ground color and the apical portions of the forewing are also usually bright red or deep orange. Taxa from the Bismarcks and Solomons and *esthla* from the islands southeast of New Guinea are larger than the rest and all show a degree of metallic iridescence on forewing bases, thorax, and abdomen. The New Guinea species are smaller and males in two of them (*mixta* and *postvitrea*) share the possession of a whitish hyaline area in the yellow of the hindwing. The New Guinea species are found throughout the island and the distribution of *calida* is primarily southern and extends to Queensland and the South Moluccas. Other single species occur in Sulawesi and Sundaland, as listed in Appendix 1. The distributions of the species are shown in Fig. 13.

The genus *Damias* is defined primarily by the development of a white rather than yellow ground color. The only exception to this is *flavicaput* where the ground is dull, rather pale yellow; this species shares the male secondary sexual characters of other species in one of the subgroups in *Damias* so it is possible this yellow color represents a reversal. The white ground color represents the only synapomorphy found so far for the whole group, though three subgroups within the genus are more clearly defined.

Subgroup 1 is defined by possession of a white spot in the forewing discal area and a double one subapically. The aedeagus usually contains two or three moderate, slender cornuti and the costal process of the valve is usually broadened or bifid subapically. This subgroup is the most widely distributed of the three, with one of the New Guinea species extending to the Bismarcks and distinct taxa on Ceram and in Sundaland (Fig. 12; Appendix 1).

Subgroup 2 has, with the exception of *monoleuca,* a single subapical white spot on the forewing and may therefore be the sister group of the first group. Many of the species have a metallic iridescence. The most definitive character is the possession by the male of scent scale patches and attendant modifications on both fore and hindwings. There is a fold with scent scales on the forewing subcostally at one third on the underside; the forewing is lobed posteriorly to overlap a modified area at the anterior of the hindwing. The male genitalia have very long, sharply curved processes to the valve costa, as long or longer than the valve. The lateral lobes of the tegumen are slight only. The majority of species are recorded from southeastern New Guinea (Fig. 12; Appendix 1).

Subgroup 3 contains a few species with slender, rather oblong forewings bearing a submarginal band of blue iridescence. In the male genitalia the processes of the

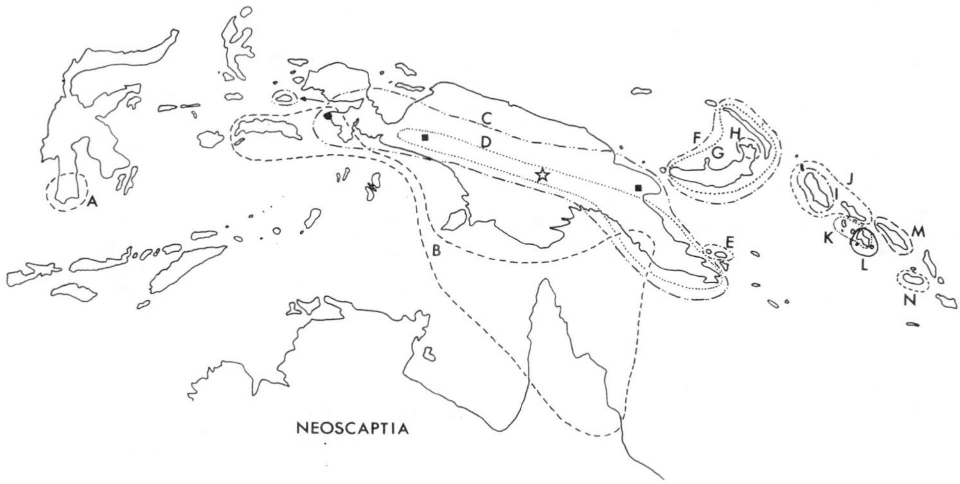

Fig. 13. Distribution of species of *Neoscaptia* in Melanesia: A, *flavicollis;* B, *calida;* C, *mixta;* D, *punctata*; E, *esthla;* F, *peculiaris;* G, undescr. sp. slide 3170; H, undescr. sp. slide 3207; I, *bougainvillei;* J, *choiseuli;* K, *insignis;* L, *occidentalis;* M, *isabella;* N, *caerulescens;* solid circle, *collateralis;* solid square, *simillima;* star, *postvitrea.*

valves are long, almost straight over the basal two thirds, curved over the apical third. Three of the species are restricted to southeastern New Guinea and the other two are known only from Mount Goliath (Fig. 12; Appendix 1).

The genus contains a miscellany of fourteen further species. It is most diverse as a whole in the southeast and in the central ranges of New Guinea and might therefore be considered characteristic of the Inner Arc (Fig. 12).

It is thus possible that *Neoscaptia* and *Damias* are, respectively, Outer and Inner Melanesian Arc sister groups. The pattern of distribution has broken down subsequently to some degree in New Guinea with species from both groups found throughout the island; *Neoscaptia* particularly shows little concentration in northern areas and both groups extend westwards to Sundaland where each is represented by a single species.

Both *Damias* and *Neoscaptia* are involved in mimicry with genera in the chalcosiine Zygaenidae. The genus *Caprima* Walker has a white group in New Guinea that mimics *Damias* and a yellow group in New Guinea and the Bismarcks that, with the monobasic *Hadrionella* Jordan (Bismarcks), mimics the yellow *Neoscaptia* group. Chalcosiinae are involved in numerous mimicry complexes and, as the distribution of *Caprima* spans only a small part of the range of the orange-tipped yellow lithosiine group, it is more probable that the chalcosiines have converged on the lithosiine pattern. If the reverse were true it would imply that the putative two-arc pattern would have only a secondary rather than primary geographical basis (the two groups being separated primarily on facies characters).

Cyana is a large and diverse Indo-Australian tropical arctiid genus that has a clearly defined subgroup in Melanesia that shows some segregation over the two Melanesian Arcs (Appendix 1, Note 11; Fig. 14). The segregation into two sections, one distributed over each of the Melanesian Arcs, needs confirmation by more detailed anatomical studies for, although the putative Inner Arc group is clearly defined, the definition of the Outer Arc group is more tenuous. These species have the patch of scent scales on the underside of the male forewing costa characteristic of the numerous Oriental species typical of *Cyana* and the male genitalia are also compatible. But there are several unique characteristics that define the group as a natural entity within the complex.

The black ground color with broad blocks of orange, red, or iridescent metallic patches is in striking contrast to the typical pattern of narrow orange, yellow, or red bands on white. The male genitalia are remarkably uniform within the group and are characterised by a triangular flap from the center of the dorsal half of the valve that overlaps the ventral half; the two halves of the valve are divided over its distal half as in more typical *Cyana.*

The Inner Arc group is defined by a longitudinal band of orange or red on the forewing that almost fills the wing. There are two species in Queensland and a group of five in New Guinea. The latter are defined by having this forewing red band slightly excavated over the outer half of its length posteriorly and orange or red on the thorax restricted to the patagia and tegulae, a completely orange thorax being taken to be plesiomorphic. The recorded distributions of these species are illustrated in Fig. 14.

The Outer Arc group generally has transverse bars or spots of color on the wings and prominent sexual dimorphism in wing pattern. Bismarcks and Solomons species usually have the forewing red areas much reduced in the male and in all species except *basiflava* the thorax is wholly black with a metallic iridescence. The Solomons species are all very similar in facies and sexual dimorphism and probably represent a natural group within the Outer Arc complex.

The New Guinea species attributed to the Outer Arc group have wholly orange thoraces and are represented by males only in the case of *bicolor* and *thoracica* and by females only in the case of *croceizona.* The banded pattern of *croceizona* is reminiscent of the numerous species in the geometrid genera *Milionia* and *Ctimene* that are found in New Guinea but not in the Outer Arc island groups. The two species represented by males only have much more elongate orange patches on their forewings than do other Outer Arc species except some females of Solomons species. Nevertheless the distribution of the three is generally more northerly in New Guinea than that of the putative Inner Arc taxa (Fig. 14).

The only host plant record for the group is of *fulvia* on *Dysoxylum* (Meliaceae), a plant genus that probably has occurred in the Melanesian area through most of the Tertiary (Holloway, 1979).

Only one example of a putative two-arc distribution has been located in the Noctuidae, in the genus *Ophyx* Guenée. It is possible that more will be discovered in the subfamily Ophiderinae when it has been better worked. Investigation of *Ophyx* has led to much generic synonymy (the diverse array of male secondary sexual modifications led to this situation) and description of several new species (Appendix 1, Note 12). The genus overall is defined on the basis of several good characters, probably derived (Appendix 2), but the immediate relatives of the group have still to be detected. The best characters are in the male genitalia, illustrated for the New Caledonian species by Holloway (1979), who at that time was unaware of the extent of the genus.

Fig. 14. Distributions of species in the Melanesian complex of the genus *Cyana* discussed in the text. Above, the Outer Arc group: A, *croceizona*; B, *thoracica;* C, *bicolor;* D, *nigromarginata;* E, *basiflava;* F, *scintillans;* G, *miles;* H, *securizonis;* I, *guizonis;* J, *omissa.* Below, the Inner Arc group: solid circles, *tricolor;* solid squares, *fulvia;* solid triangles, *postdivisa;* open circle, *metamelas;* open diamond, *dinawa;* solid stars, *meyricki;* open star, undescr. sp.

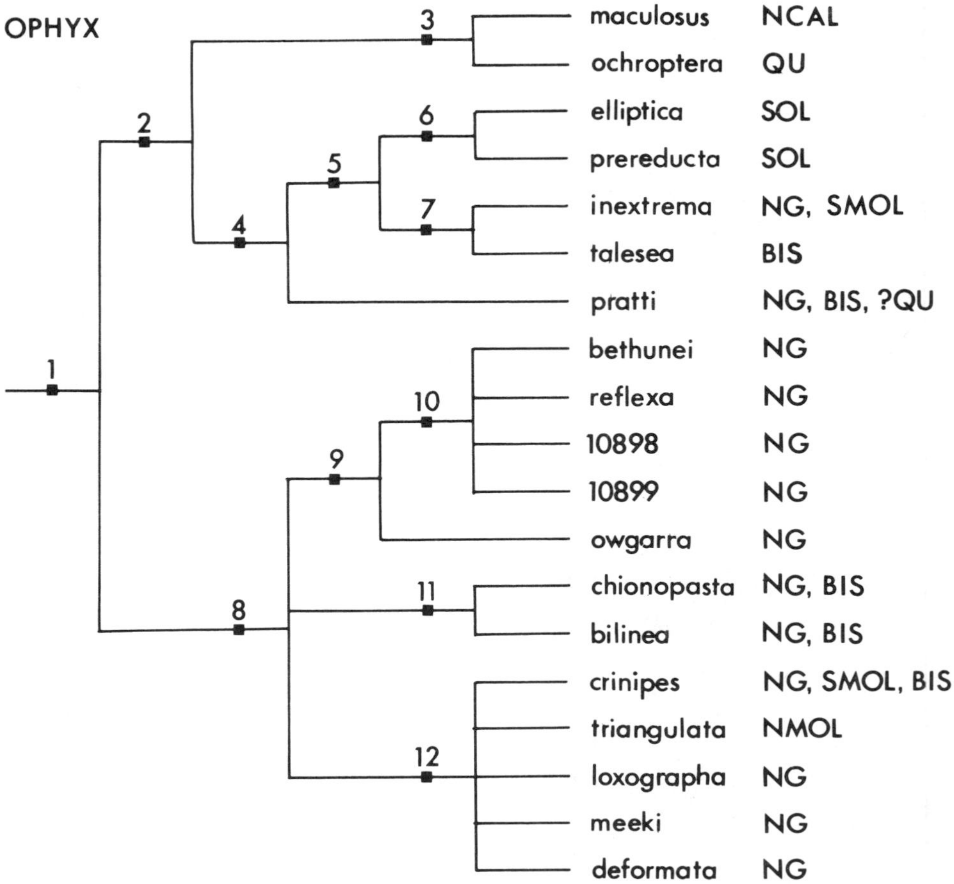

Fig. 15. Cladogram of relationships for species in the genus *Ophyx*. The numbers refer to presumed apomorphies listed in Appendix 2. Abbreviations for distributions are as in the legend to Fig. 10. Large numbers at branch ends refer to undescribed species as mentioned in Appendix 1.

Characters within the group are such that a preliminary cladistic analysis was attempted. The cladogram is shown in Fig. 15, and the character states presumed to be apomorphic are listed in Appendix 2.

Two subgroups are recognised, the largest concentrated in New Guinea with ten species recorded from the southeastern part, five only known from there; three species, however, extend to the Bismarcks. The smaller subgroup contains a New Caledonia-Australia pair, a species found widely through New Guinea and the Bismarcks (it may also occur in Queensland if the species *pseudoptera* Lower proves to be a synonym, as suggested by the original description), and a quartet distributed over Outer Arc archipelagos, though the New Guinea species occurs also in the South Moluccas.

The large species groups attributed to *Asura* and *Miltochrista* in the Lithosiinae and the genus *Euproctis* in the Lymantriidae may furnish further examples, but both present immense taxonomic problems in that they are very diverse yet little studied. Some zygaenid and callidulid genera are diverse in Melanesia.

The pierine butterflies include two genera that would repay study. The *Delias* group as revised by Talbot (1928–1937) was subdivided into several sections which, in Melanesia, can be divided readily into Inner Arc versus Outer Arc types. A full cladistic analysis would be an immense undertaking but potentially rewarding. The genus *Elodina* is represented in New Caledonia and widely elsewhere in Melanesia with endemicity, though not high diversity, in Outer Arc archipelagos.

The concentration of diversity of *Damias* and the New Guinea group of *Ophyx* in the southeast of New Guinea contrasts somewhat with the more even, perhaps westerly centered cicada genus *Cosmopsaltriaria,* and this may indicate a degree of independence in the biological development of central and southeastern New Guinea in the Inner Arc, and an isolating role for the Aure Trough mentioned on p. 132.

CONCLUSIONS

It is evident that within Melanesia there can be found a bewildering array of different distribution patterns. For the Lepidoptera at least there is no indication of a prevalent area cladogram for Melanesia that would support a hypothesis of vicariant geography such as the break up of the proposed Melanesian Foreland. It remains to be seen whether phytogeographic patterns provide a different picture.

The array of different patterns indicates that the geography of the area has probably been dynamic rather than fixed. Groups of organisms restricted to, or more diverse in, the older Inner Arc areas may well provide evidence of vicariance when sufficient have been studied, but the area cladograms generated for Australasia from various groups to date have been contradictory and uninformative.

There are some indications that the older radiations within Melanesia followed a different pattern from more recent ones. The majority of butterfly and moth genera endemic to Melanesia are restricted to or strongly centered in New Guinea, with perhaps only minor extensions to Outer Arc archipelagos. A very high proportion of the species found in faunas from the Solomons to Tonga are widespread either throughout the Indo-Australian tropics or throughout Melanesia (Robinson, 1975; Holloway, 1979). This might be expected, for the Outer Arc, both in age and extent if the geological history presented earlier is largely correct, may never have offered such opportunities for colonization and speciation by plants and animals as did Inner Arc New Guinea. It is perhaps not surprising therefore that putative two-arc sister patterns are rare. None of the examples presented for the Lepidoptera provide such strong concordance with the hypothesis as the Cosmopsaltriaria cicadas. Indeed one can perhaps only say that such patterns in the Lepidoptera could be explained by the geological hypothesis rather than support it. Nevertheless, I hope this rather preliminary and cursory study may stimulate investigations in other plant and animal groups that may be more productive.

ACKNOWLEDGMENTS

I am grateful to the Trustees of the British Museum (Natural History) for access to the collections there and to the Lepidoptera Staff for their support and helpful comments. My thanks go to Dr. R. de Jong of the Rijksmuseum van Natuurlijke Historie, Leiden, for arranging the loan of the type material of *Avitta [Epizeuxis] inductalis* Snellen. Dr. J. P. Duffels, Dr. C. J. Humphries, Dr. J. Rawlins, and Mr. R. I. Vane-Wright were kind enough to scrutinize and criticize a draft of this paper. My gratitude also goes to my wife who typed the final draft.

I would also like to pay tribute to the late Dr. Lin Gressitt whose works on Pacific and Melanesian biogeography were a stimulus and inspiration to me as a student and subsequently when, following a week of fieldwork together in New Caledonia, we became friends, colleagues, and frequent correspondents.

LITERATURE CITED

Ackery, P. R. & R. I. Vane-Wright. 1984. *Milkweed butterflies, their cladistics and biology.* Br. Mus. Nat. Hist. and Cornell Univ. Press.

Coleman, P. J. 1980. Plate tectonics background to biogeographic development in the Southwest Pacific over the last 100 million years. *Palaeogeogr. Palaeoclimat. Palaeoecol.* 31: 105–21.

Corner, E. J. H. 1975. *Ficus* in the New Hebrides. *Phil. Trans. R. Soc. Lond.* B, 272: 343–67.

Cracraft, J. 1980. Biogeographic patterns of terrestrial vertebrates in the Southwest Pacific. *Palaeogeogr. Palaeoclimat. Palaeocol.* 31: 353–69.

Duffels, J. P. 1977. A revision of the genus *Diceropyga* Stal, 1870 (Homoptera: Cicadidae). *Monogr. Ned. Entomol. Ver.* 8: 1–127.

———. 1983. Taxonomy, phylogeny and biogeography of the genus *Cosmopsaltria,* with remarks on the historic biogeography of the subtribe Cosmopsaltriaria (Homoptera: Cicadidae). *Pac. Insects Monogr.* 39: 1–27.

Flenley, J. R. 1979. *The equatorial rainforest: a geological history.* Butterworths, London and Boston.

Holloway, J. D. 1970. The biogeographical analysis of a transect sample of the moth fauna of Mount Kinabalu, Sabah, using numerical methods. *Biol. J. Linn. Soc.* 2: 259–86.

———. 1974. The endemic Satyridae (Lepidoptera: Rhopalocera) of New Caledonia. *J. Entomol. Ser.* B 43: 89–101.

———. 1976. *Moths of Borneo with special references to Mount Kinabalu.* Malayan Nature Society, Kuala Lumpur.

———. 1977. The Lepidoptera of Norfolk Island, their biogeography and ecology. *Ser. Entomol, The Hague.* 13.

———. 1979. A survey of the Lepidoptera, biogeography and ecology of New Caledonia. *Ser. Entomol., The Hague.* 15.

———. 1982. Taxonomic appendix. *In:* Barlow, H. S., *An introduction to the moths of South East Asia.* Kuala Lumpur.

———. 1983. Mobile organisms in a geologically complex area: Lepidoptera in the Indo-Australian tropics. *Zool. J. Linn. Soc.* 76: 353–373.

Holloway, J. D. & N. Jardine. 1968. Two approaches to zoogeography: a study based on the distributions of butterflies, birds and bats in the Indo-Australian area. *Proc. Linn. Soc. Lond.* 179: 153–88.

Humphries, C. J. 1981. Biogeographical methods and the southern beeches (Fagaceae: *Nothofagus*). *In:* Funk, V. A. & D. R. Brooks, eds, *Advances in cladistics.* New York Botanical Garden.

Jaques, A. L. & G. P. Robinson. 1977. The continent/island arc collision in northern Papua New Guinea. *Bur. Miner. Resour. J. Aust. Geol. Geophys.* 2: 289–303.

Jardine, N. & D. McKenzie. 1972. Continental drift and the dispersal and evolution of organisms. *Nature, Lond.* 234: 20–24.

Jong, M. R. de & J. P. Duffels. 1981. The identity, distribution and synonymy of *Lembeja papuensis* Distant, 1897 (Homoptera: Tibicinidae).

Kiriakoff, S. C. 1977. (1976). Lepidoptera Noctuiformes: Agaristidae (I–III). *Das Tierreich,* 97, 98, 99.

Lieftinck, M. A. 1949. Synopsis of the odonate fauna of the Bismarck Archipelago and the Solomon Islands. *Treubia* 20: 319–74.

MacArthur, R. H. & E. O. Wilson. 1967. *The theory of island biogeography.* Princeton University Press, Princeton.
McFarland, N. 1979. Annotated list of larval foodplant records for 280 species of Australian moths. *J. Lepid. Soc.* 33: Supplement.
Musser, G. G. 1981. The giant rat of Flores and its relatives east of Borneo and Bali. *Bull. Am. Mus. Nat. Hist.* 169: 67–176.
Nelson, G. & N. Platnick. 1981. *Systematics and biogeography, cladistics and vicariance.* Columbia University Press, New York.
Raven, P. H. & D. I. Axelrod. 1974. Angiosperm biogeography and past continental movements. *Ann. Mo. Bot. Gard.* 61: 539–673.
Robinson, G. S. 1975. *Macrolepidoptera of Fiji and Rotuma, a taxonomic and geographic study.* E. W. Classey, Faringdon.
Ross, H. H. 1956. *Evolution and classification of the mountain caddisflies.* University of Illinois Press, Urbana.
Saigusa, T., A. Nakanishi, H. Shima, & O. Yata. 1977. Phylogeny and biogeography of the subgenus *Graphium* Scopoli (Lepidoptera: Papilionidae, *Graphium). Acta Rhopalocerologica* 1: 1–32.
Smiles, R. L. 1982. The taxonomy and phylogeny of the genus *Polyura* Billberg (Lepidoptera: Nymphalidae). *Bull. Br. Mus. Nat. Hist. Entomol.* 44: 115–237.
Steenis, C. G. G. J. van. 1962. The land-bridge theory in botany. *Blumea* 11: 235–372.
Stone, B. C. 1982. New Guinea Pandanaceae: first approach to ecology and biogeography. p. 401–36 *In:* Gressitt, J. L., ed., *Biogeography and ecology of New Guinea.* Monographiae Biologicae 42. W. Junk, The Hague.
Talbot, G. 1928–37. *A Monograph of the pierine genus* Delias. John Bale, Danielsson and British Museum (Natural History), London.
Taylor, L. M. H. & D. A. Falvey. 1977. Queensland Plateau and Coral Sea Basin: stratigraphy, structure and tectonics. *APEA J.* 17: 13–29.
Thornton, I. W. B. 1980. Plate tectonics and the distribution of the insect family Philotarsidae (order Psocoptera) in the Southwest Pacific. *Palaeogeogr. Palaeoclimat. Palaeoecol.* 31: 251–66.
Toxopeus, L. J. 1950. The geological principles of species evolution in New Guinea. p. 508–22 *In: Proc. 8th Int. Congr. Entomol. Stockholm,* 1948.
Winterbourn, M. J. 1980. The freshwater insects of Australasia and their affinities. *Palaeogeogr. Palaeoclimat. Palaeoecol.* 31: 235–49.
Zweifel, R. G. & M. J. Tyler. 1982. Amphibia of New Guinea. p. 759–801 *In:* Gressitt, J. L., ed., *Biogeography and ecology of New Guinea.* Monographiae Biologicae 42. W. Junk, The Hague.

APPENDIX 1
TAXONOMIC NOTES

In the course of the work described in this paper the need for numerous taxonomic changes became apparent. In the genera *Avitta* Walker and *Ophyx* Guenée several new taxa are involved. Therefore the following notes are appended to the main text to establish a number of new combinations and synonymies and publish brief descriptions of the new taxa. Where new combinations are presented, the original genus of the species concerned is given in square brackets. Reasons for the changes are presented in the main text.

Note 1. New combinations in the *Cosmethis* Hubner (Geometridae: Ennominae) group of genera.

Bracca disrupta Warren, **comb. n.** [*Xanthomima*].

Craspedosis arycandata Walker, **comb. n.** [*Panaethia*].

Cosmethis tricolor Felder, **comb. n.** [*Automolodes*].

Xanthomima Warren should be associated with *Craspedosis* Butler, possibly as a subgenus.

Arycanda Walker and *Bracca* Hubner should be associated with *Cosmethis* Hubner, possibly as subgenera.

Note 2. A new combination in *Idiodes* Guenée (Geometridae: Ennominae).
Idiodes serpentina Holloway, **comb. n.** [*Nadagara*].
Note 3. A new combination in *Encryphia* Turner (Geometridae:Oenochrominae).
Encryphia pulverosa Warren, **comb. n.** [*Noreia*].
Note 4. New synonymies and combinations involved in the *Crasilogia* Warren group of genera (Geometridae: Larentiinae), a checklist of species and description of a new species.

Gonanticlea horia Prout is a junior subjective synonym of *Heterochasta lasioplaca* Lower, **syn. n.** The former name is currently in combination with *Papuanticlea* Holloway, therefore the latter should be combined with this genus as *Papuanticlea lasioplaca,* **comb. n.**

One syntype of *Polyclysta hypogrammata* Guenée (in BMNH) is a specimen of the sister species to *Crasilogia fumipennis* Warren, described below; the lectotype (in MNHN Paris) is conspecific with *Protaulaca scythropa* Meyrick, the latter being the junior synonym, **syn. n.** Therefore *Protaulaca* Meyrick is a junior subjective synonym of *Polyclysta* Guenée, **syn. n.**

Crasilogia gressitti, **sp. n.** ♂♂ 16–17mm, ♀♀ 16–17mm. The species resembles *C. fumipennis* Warren closely in forewing pattern, but the medial zone is narrower and stands out dark brown between paler zones in a generally more contrasted arrangement. The male hindwing is slightly reduced in size relative to that of the female and has modified fine scales distally as in *fumipennis,* but the margin is rounded rather than angled and the color uniform pale brown rather than dark brown basally and white distally. In the male genitalia the elongate flange on the valve transtilla is acutely angled rather than gently rounded, and the anterior process of the tegumen+vinculum is half the size of that of *fumipennis.* The female genitalia differ from those of *fumipennis* in having a smaller signum, a ductus bursae sclerotized over the distal third rather than two-thirds, and in lacking a central band of fine scobination in the bursa. *Holotype* ♂ : AUSTRALIA, Queensland, Kuranda (*Dodd*) 1910, BM geometrid slide 10864. *Paratypes:* 7 ♂♂, 5 ♀♀ as holotype.

The genera and species of the *Crasilogia* Warren group are as follows:

Crasilogia Warren: *simplex* Warren; *flavipennis* Warren; *fulvitincta* Joicey and Talbot; *dispar* Warren; *fumipennis* Warren; *gressitti,* **sp. n.**

Heterochasta Meyrick: *conglobata,* Walker.

Papauanticlea Holloway: *lasioplaca* Lower; *ochroria* Holloway; *subcaesia* Warren (+ ssp. *neutralis* Prout); *onaea* Prout; *semiflava* Warren.

Note 5. *Anisographe* Warren, a synonym of *Polyacme* Warren (Geometridae: Ennominae).
Polyacme Warren, 1896, type species *dentata* Warren.
Anisographe Warren, 1897, type species *dissimilis* Warren, **syn. n.**

The following species therefore become combined with *Polyacme: dissimilis* Warren, *albipuncta* Warren, *subpulchra* Warren, **combs. n.**

Polyacme punctilinea Joicey and Talbot, represented only by the holotype female from the Arfak Mts., is probably the female of *Polyacme vagilinea* Warren, represented only by males in the BMNH series; the underside markings on both wings are almost identical. Therefore *punctilinea* has been taken to represent the female of *vagilinea* in the analysis.

Note 6. *Metaxanthiella* Collins, a synonym of *Leucogonia* Hampson (Noctuidae: Agaristinae, **comb. n.**).
Leucogonia Hampson, 1908, type species *ekeikei* Bethune-Baker.
Metaxanthiella Collins, 1962, objective replacement name for *Metaxanthia* Hampson, 1908, type species *cosmopis* Lower, **syn. n.**

The facies of both wings of both these genera are very similar, and unique within the Noctuidae; the structure of the male genitalia is also similar. They are therefore brought into synonymy, leading to the following new combination: *Leucogonia cosmopis* Lower, **comb. n.** [*Britha*]. The reasons for transference to Agaristinae are given in the main text.

Note 7. New taxa in the genus *Avitta* Walker (Noctuidae: Ophiderinae), a change in status for an established taxon, and a new synonymy.

The new taxa described here all resemble *subsignans* Walker in wing markings. Diagnostic characters are presented in the cladistic analysis (Fig. 6; Appendix 2) and in the illustrations of the valves of the male genitalia in Fig. 7. The descriptions here are therefore limited to reference to the relevant genitalia illustration and designation of a holotype. Two further taxa assessed in the analysis are represented by females only; their description must await the discovery of males.

Other species included in the analysis are *surrigens* Walker; *guttulosa* Swinhoe; *fasciosa* Moore; *longicorpus* Prout; *flavicilia* Holloway; *bracteola* Holloway.

Avitta andamana, **sp. n.** Fig. 7:N. *Holotype* ♂, Andamans, BM noctuid slide 10832.

Avitta microsundana, **sp. n.** Fig. 7:K, L. *Holotype* ♂, [Timor] Oinainisa, Nov–Dec 1891 (*W. Doherty*), BM noctuid slide 10835. Also Alor, Dammer, Flores, Tenimber.

Avitta novahibernia, **sp. n.** Fig. 7:J. *Holotype* ♂, [Bismarcks] New Ireland, November 1923 (*A. F. Eichhorn*), BM noctuid slide 6170. Also New Britain, Rook I., Nissan I.

Avitta meeki, **sp. n.** Fig. 7:G. *Holotype* ♂, [Solomon I.] Guadalcanal iv.01 (*A. S. Meek*), BM noctuid slide 10846. Also Kulambangra, Choiseul, Isabel, Bougainville.

Avitta meeki vanuatua, **ssp. n.** Fig. 7:H. *Holotype* ♂, New Hebrides [Vanuatu]: Santo, Apouna R., Camp 3—1000 ft, 9–11.ix. 1971 (*G. S. Robinson*) Roy. Soc. Exped., BM noctuid slide 7093. Also Efate, Aneityum.

Avitta longicorpus australis Holloway, **stat. n.** The similarity of the male genitalia of *australis* to those of *longicorpus* Prout in relation to the diversity of structure in other taxa, together with the sharing of derived characters as mentioned in Appendix 2, indicates that *australis* (New Caledonia, Loyalties) should be placed as a subspecies of *longicorpus*. Described from the S. Moluccas, *longicorpus* is also found in the N. Moluccas, Kei, New Guinea, and Queensland.

Avitta discipuncta Felder, 1874, [*Sympis*]; *Avitta pastea* Hampson, 1902, **syn. n.**; *Ceromacra dinawa* Bethune-Baker, 1906, **syn. n.** The facies of the types of these species are all very similar and no difference could be detected in the male genitalia of specimens from extremes of the widespread Indo-Australian tropical range.

The following name may also be referable to this complex but the type was not readily available for study: *Zethes leitimorensis* Pagenstecher (Amboina, based on a female), possibly a senior synonym of *longicorpus*.

The species *inductalis* Snellen was originally described in *Epizeuxis* and was based on two syntypes, a male from the Celebes and a female from Java. The male from Celebes is hereby designated LECTOTYPE; the Javan specimen is referable to *surrigens* Walker. *A. inductalis* was, together with *surrigens* and *fasciosa*, placed as a synonym of *subsignans* by Swinhoe [1917, *Ann. Mag. Nat. Hist.* (8), 19:340].

Note 8. *Prorocorys* Warren, a synonym of *Ziridava* Walker (Geometridae:Larentiinae).

Ziridava Walker, 1863, type species *zylinaria* Walker.

Prorocorys Warren, 1899, type species *gemmata* Warren, **syn. n.**

Ziridava gemmata Warren, **comb. n.**

Note 9. Two new combinations in *Probithia* Warren (Geometridae:Ennominae).

Probithia turpis Warren, **comb. n.** [*Nadagarodes*].

Probithia sesquilinea Prout, **comb. n.** [*Luxiaria*].

Note 10. Genera and species of the *Damias* Boisduval complex (Arctiidae:Lithosiinae), including several new combinations. Species where the male genitalia have been examined have been marked with an asterisk (*); where a female has been examined, this is specified by a dagger (†).

Scaptesyle Walker (type species *tricolor** Walker). Species included are *tricolor** Walker; *plumosus** Rothschild; *dentata* Talbot; *thestias* Snellen; *ixias** Hampson; *mirabilis** Hampson; *integra* Swinhoe; *szetscwana* Draesecke; *bicolor* Walker; *incerta* Semper; *luzonica* Swinhoe; unassigned specimens from Luzon* and Sulawesi in BMNH.

*Scaptesyle postnigra** Rothschild, **comb. n.** [*Caprimima*].

Byrsia Walker (type species *dotata** Walker). Species included are *aurantiaca** Snellen; *celebensis* Rothschild and Jordan, **stat. n.**; *dotata** Walker, a name that may prove to embrace also the New Guinea taxa *reducta* Rothschild, *ornata* Rothschild and Jordan, *pallidior* Rothschild, and *papuana* Rothschild and Jordan; *latiplaga** Rothschild; *amoena** Rothschild and Jordan; undescribed taxa from Tonkin, Goodenough I., and the Bismarcks.

Neoscaptia Hampson (type species *collateralis* Hampson, ♀†). Syn. *Atopomima* Meyrick (type species *philocosma* Meyrick, a junior synonym of *caerulescens* Butler). Apart from *collateralis* Hampson, all the species listed under this genus are in **new combination:** *tenuis** Rothschild and Jordan, **stat. n.** [*Caprimima*]; *flavicollis** Snellen, **stat. n.** [*Hypocrita*]; *calida** Walker [*Scapesyle*]; *punctata* Rothschild [*Caprimima*] (ab. *milnensis* Strand of *calida* is probably referable to this species); *mixta* Hampson [*Caprimima*] (the type from Mysol was not examined but a long series from localities in New Guinea is attributed to this species in the BMNH); *postvitrea** Rothschild [*Damias*] (possibly the male of the next species); *simillima* Rothschild [*Caprimima*]; *esthla** Prout [*Caprimima*]; *peculiariş** Rothschild [*Caprimima*]; *caerulescens* Butler [*Scapesyle*]; *occidentalis** Rothschild [*Caprimima*]; *bougainvillei** Rothschild [*Caprimima*]; *insignis** Rothschild [*Caprimima*]; *isabella** Rothschild and Jordan [*Caprimima*]; *choiseuli** Jordan [*Caprimima*]. There are also two undescribed taxa from the Bismarcks (ref. BM arctiid slides 3170, 3207).

Damias Boisduval (type species *elegans** Boisduval). Syn.*Caprimima* Hampson (type species *albicollis* Pagenstecher = *elegans*):

Subgroup 1: *occidentalis** Rothschild and Jordan; *quadripuncta* Rothschild; *bipuncta* Rothschild; *germana* Rothschild; *bimaculata** Rothschild, **comb. n.** [*Neoscaptia*]; *biguttata* Rothschild; *aenea* Jordan.

Subgroup 2: *flavicaput** Rothschild, **comb. n.** [*Neoscaptia*]; *torquata* van Eecke, **comb. n.** [*Neoscaptia*]; *monoleuca** Hampson; *albata** Jordan, **comb. n.** [*Neoscaptia*]; *apicipuncta* Rothschild, **comb. n.** [*Neoscaptia*]; *aequalis** Jordan, **comb. n.** [*Neoscaptia*] *unipunctata* Rothschild; *leucodera* Jordan **comb. n.** [*Neoscaptia*]; *androconiata* Rothschild, **comb. n.** [*Neoscaptia*]; also two undescribed species.

Subgroup 3 (all the following are in **comb. n.,** transferred from *Neoscaptia*): *poecila* Jordan; *fasciconitens** Rothschild; *affinis* Rothschild; *affinis* Rothschild; *albicollis** Rothschild; *basinitens* Rothschild.

Miscellaneous species: *fuliginosa* Rothschild; *marginipuncta* Rothschild; *elegans** Boisduval; *rufobasalis* Rothschild; *peramitatrix* Rothschild; *peraffinis** Rothschild; *pseudaffinis* Rothschild; *postexpansa* Rothschild; *metallica* Jordan; *shawmeyeri* Rothschild; *obliqua* Rothschild and Jordan; *cupreonitens* Rothschild; *pseudogelida* Rothschild.

At present a motley collection of additional species, mainly from Australia, are attributed to the four genera above. They do not appear to be related. The genus *Chiriphe* Walker (type species *monogrammaria* Walker) is available for them until such a time as they can be examined in detail and classified correctly.

Note 11. Species of *Cyana* Walker (Arctiidae: Lithosiinae) discussed in the text.

Inner Arc Group

New Guinea subgroup: *metamelas* Hampson; *fulvia* Linnaeus; *dinawa* Bethune-Baker; *postdivisa* Rothschild; *tricolor* Druce.

Australian subgroup: *meyricki* Rothschild and Jordan and an undescribed species from Kuranda.

Outer Arc Group

Solomons subgroup: *omissa* Rothschild and Jordan; *guizonis* Jordan; *securizonis* Butler; *miles* Butler.

Miscellaneous species: *scintillans* Rothschild; *basiflava* Rothschild; *nigromarginata* Rothschild; *bicolor* Rothschild; *thoracica* Rothschild and Jordan; *croceizona* Hampson.

Note 12. The genus *Ophyx* Guenée (Noctuidae: Ophiderinae): generic synonymy, new combinations and the descriptions of several new taxa.

Ophyx Guenée, 1852, type species *ochroptera* Guenée.

Sinariola Bethune-Baker, 1906, type species *owgarra* Bethune-Baker, **syn. n.**

Pseudophyx Bethune-Baker, 1906, type species *pratti* Bethune-Baker, **syn. n.**

Hirsutipes Bethune-Baker, 1906, type species *trifasciata* Swinhoe (*crinipes* Felder), **syn. n.**

Lasiopoderes Bethune-Baker, 1906, type species *pratti* Bethune-Baker, **syn. n.**

Cyttarolopha Bethune-Baker, 1908, type species *loxographa* Bethune-Baker, **syn. n.**

Temnoptera Bethune-Baker, 1908, type species *meeki* Bethune-Baker, **syn. n.**

The type species of these genera resemble each other in the derived characteristics listed in the text, although they show great variety of secondary sexual structures in the males. New combinations are therefore erected and one replacement name is assigned as follows:

Ophyx crinipes, Felder, **comb. n.** [*Remigia*].

Ophyx pratti Bethune-Baker, **comb. n.** [*Pseudophyx*].

Ophyx inextrema Bethune-Baker, **comb. n.** [*Pseudophyx*].

Ophyx owgarra Bethune-Baker, **comb. n.** [*Sinariola*].

Ophyx loxographa Bethune-Baker, **comb. n.** [*Cyttarolopha*].

Ophyx chionopasta Hampson, **comb. n.** [*Mecodina*].

Ophyx striata Hampson, **comb. n.** [*Mecodina*].

Ophyx bethunei, **nom. n.** for *Lasiopoderes pratti* Bethune-Baker, here brought into secondary homonymy with *Pseudophyx pratti* Bethune-Baker.

The taxa *ochroptera* Guenée and *maculosus* Holloway are already in combination with *Ophyx*. It has not been possible to examine material of the taxa *pseudoptera* Lower or *dochmotoma* Turner (both Australia), but the description of the former suggests it may be a subspecies of *pratti* Bethune-Baker. The following new taxa were included in the analysis:

Ophyx inextrema simplicior, **ssp. n.** Male resembles typical *inextrema* (Buru) in facies but lacks the hair scales subdorsally on the underside of the forewing; some specimens have a trapezoid cream area based on the interior of the dark medial line of the forewing. *Holotype* ♂: Goodenough I., 2500–4000 ft, April 1913 (*A. S. Meek*), BM noctuid slide 10887. Also New Guinea: Arfak Mts., Snow Mts., Kapaur.

Ophyx talesea, **sp. n.** Male coloration and markings as for *inextrema* but forewings significantly deeper, medial less oblique, with a distinctive sagittate dark brown mark subdorsally on the postmedial. The forewing below has a straw-colored tuft of hair scales arising subbasally between Cu1b and anal vein. The hindwing has a straw tuft in the humeral area. *Holotype* ♂: Talesea, New Britain, January 1925 (*A. F. Eichhorn*), BM noctuid slide 10893.

Ophyx elliptica, **sp. n.** General coloration of male as in *talesea*, including sagittate mark on forewing postmedial, but has area between postmedial and medial grading darker brown dorsad and an area of cream suffusion basal to the medial. The forewing below has a scent pencil of long, straw-colored scales overlapping an elongate brand in area between Cu1b and anal vein, and a broad, diffuse tuft of scales submarginally over Cu1b; there is a slight fringe of scales anteriorly from Sc over the cell. The hindwing has a massive elliptical brand of very small scales between Sc and the cell in an otherwise scaleless anterior area. *Holotype* ♂: North side of Choiseul I. [Solomon Is.] Dec. 1903 (*A. S. Meek*), BM noctuid slide 10889. Also Bougainville, Isabel, Guadalcanal.

Ophyx prereducta, **sp. n.** Male forewings reduced and hindwings enlarged such that the former are roughly half the area of the latter, distorted somewhat, with rounded apices. Forewings with violet gray ground and dark brown diffusing away distad from sharp, straight, transverse medial boundary; two faint, diffuse brown fasciae traverse basal half. Forewing with large, long scent pencil of straw-colored scales based on an elliptical area straddling the anal vein subbasally below. Hindwings with large (half area of wing) semicircular area of small, glossy, transparent scales anteriorly, above. *Holotype* ♂: North side of Choiseul I. [Solomon Is.], Dec. 1903 (*A. S. Meek*), BM noctuid slide 10890.

Ophyx triangulata, **sp. n.** Male forewings angled marginally, a brownish gray, tinged violet basal to the postmedial; a diffuse, straight, transverse dark brown medial passes basal to reniform which, as in *crinipes*, is a cream dot edged with dark brown; the postmedial is faint except for a dark brown spot 5 mm from the costa. The hindwing is the same color as the forewing, unmarked except for a large, central, dark brown triangle associated with the postmedial. Forewing with extensive area on underside posterior to cell devoid of scales, surrounded anteriorly by fringe of long scales. Hindwing with dorsum prominently lobed and invested with long scales; a large straw-colored hair pencil arises between Sc and cell above, set in a concavity seen as a slight 'blister' from below, and making the anterior margin slightly lobed also. *Holotype* ♂: Batchian, March 1892 (*W. Doherty*), BM noctuid slide 10888.

Ophyx deformata, **sp. n.** Male with forewings reduced relative to hindwings as in *prereducta*, but the forewings are more elongate, less triangular, and both are generally a mottled, rich, dark brown, irrorated with violet, and with sinuous, broken, thick, dark brown postmedials over the posterior half of both wings; the forewing has a dark ochreous fawn spot, 2 mm deep, at the reniform. The hindwing upperside bears a massive, untidy, straw-colored hair pencil in space between Sc and cell, extending over a large semicircular area of small, satiny scales. *Holotype* ♂: [Papua New Guinea] Biagi, Mambare R., 5000 ft, B.N.G., Feb. '06 (*A. S. Meek*), BM noctuid slide 10891.

Ophyx bilinea, **sp. n.** Male with similarly oblique, straight, dark brown medial and postmedial on forewing and postmedial on hindwing, all invested with violet scales on the basal side that diffuse away basad; there is further violet suffusion submarginally. Forewing below with brush of long scales from Sc that interacts with hindwing structures. Hindwing with hair pencil in concavity as in *triangulata*. *Holotype* ♂: New Ireland, January 1924 (*A. F. Eichhorn*), BM noctuid slide 10896. Also New Guinea: Aroa R., Snow Mts.

Ophyx reflexa, **sp. n.** The male is very similar in reduction of the hindwing to *bethunei*, but with hindwing margins diagnostically scrolled upwards. Forewing postmedial diagnostically invested with a dark brown triangle posterior to the costal rectangular pale patch, and a dark brown crescent subdorsally. Forewing below has more of a fringe of scales from Sc than *bethunei* and also, unlike *bethunei* and the two undescribed taxa below, retains a straw hair pencil at the anterior of the hindwing. *Holotype* ♂: [Papua New Guinea] Hydrographer Mts., Brit. N.G., 2500 ft, Feb. 1918 (*Eichhorn Bros.*), BM noctuid slide 10897. Also from Aroa R.

Two further series of specimens close to *bethunei* and *reflexa* may represent good species but their distinctiveness from *bethunei* is not so clear cut as that of *reflexa*. Therefore they are not described but indicated in the cladogram (Fig. 15) by their genitalia slide numbers (10898, 10899).

The secondary sexual characteristics of males of previously described species are as follows:

O. maculosus: Absent.

O. ochroptera: Forewing below with slight brand behind cell basal to origin of Cu1b, with a dark brown patch posterior to it, and a tuft of longer scales distally but still posterior to Cu1b. Hindwing with anterior straw-colored hair pencil as in other taxa.

O. inextrema: See description of ssp. *simplicior* above.

O. pratti: Forewing with extensive fringe of long scales from Sc below, and a massive elliptical patch of white hair scales in space between Cu1b and anal vein, the posterior margin being expanded and curved to accommodate it; Cu1b bears a small tuft of straw-colored hair scales. Hindwing above has a broad, scaleless semicircular area anteriorly with a fringe of hairs from the anterior margin overlapping the forewing; there is an elliptical area of brown sclerotization in the scaleless area at anterior angle of cell but outside it.

O. chionopasta: Both wings of male uniform dull brown with basal and marginal thirds suffused with pale blue scales, the marginal zone bounded basad by a straight or slightly sinuous, thin, pale blue postmedial. The forewing below has a scaleless area as in *triangulata*, but with a small, brown, elliptical brand of very small scales posterior to cell just basal to origin of Cu1b. Hindwing with scent pencil in concavity as in *triangulata*.

O. meeki: Forewing with no modification. Hindwing underside with posteriorly directed tuft of longer, straw-colored scales over elongate area between Cu1a and Cu1b.

O. loxographa: Forewing below with fringe of long scales from Rs in cell; triple groove full of small, rounded, shiny, dark scales in oval patch at bases of M1–3 in cell. Hindwing anterior margin scrolled downwards, fringed; above with anterior hair pencil present, the anterior half with an extent of white, satiny scales; below with posteriorly protrusive discoidal patch in cell.

O. crinipes: Forewing below with weak fringe on Sc. Hindwing without modifications.

O. owgarra: Forewing very long fringe from Sc below that interacts with hindwing structures. Hindwing reduced anteriorly and posteriorly so central portion of margin appears produced: anterior third below covered with a dense mat of scales creating a semicircular pocket; above, scales over anterior third sparse, downy, more frequent and longer at anterior margin where they are directed forward.

O. bethunei: Hindwings reduced to half size; anterior half below invested with rough scales with straw-colored hair pencil reduced. Both *bethunei* and *reflexa* have a massive tuft of hair scales on the hind-tibia and some also on the mid-tibia. Tibial ornamentation of this nature, but less prominent, is also found in males of *crinipes, deformata,* and *chionopasta*.

O. striata is known only from the female holotype (W. Irian: Fak Fak). The presence of hindwing markings places it in the large New Guinea subgroup and in general forewing facies it is akin to *meeki* or *loxographa*. It is not included in the cladistic analysis.

Postscript (1984): *Ophyx dochmotoma* Turner has since been examined (type in Australian National Insect Collection, Canberra) and is a subspecies or synonym of *Hypospila bolinoides* Guinee.

APPENDIX 2
PRESUMED SYNAPOMORPHIES USED IN SOME CLADOGRAMS

Reference Fig. 2. Presumed synapomorphies used in the cladogram for the *Crasilogia* group of genera.

1. A prominent anterior projection from the ventral part of tegumen fused with dorsal part of vinculum in male genitalia. General similarities of facies and male genitalia.
2. The anterior projections from tegumen and vinculum are hooked downwards, the vinculum more obviously involved than in group 7. The ductus seminalis joins the bursa copulatrix basally.
3. Corematous hairs from Y-shaped base on membrane between male genitalia and eighth segment. Male eighth sternite produced into slender projections on each side anteriorly. Much of male hindwing invested with slightly raised scales larger than elsewhere, giving a slightly furry appearance to the wing (not in *simplex).* Process present on valve extending dorsally from transtilla.
4. Process on valve transtilla an elongate flange. Line of scobination present at apex of aedeagus but vesica immaculate. Bursa with scobinate ring of sclerotisation basally, surrounding aperture to ductus, and a heavily scobinate signum in distal half.
5. Distinctive forewing facies, e.g. very large discal spot. Aedeagus vesica always with cluster of slender cornuti. Process on valve from transtilla slender, outcurved.
6. Male with dark patch of modified scales at anal angle of hindwing. Process on valve from transtilla curved over basal third, more or less straight over distal two thirds.
7. Projections from fused tegumen and vinculum straight, though small in *onea* and lacking in *semiflava.* Aedeagus with one very large cornutus. Center of valve just inside margin with patch or lobe bearing long fine setae (absent in *ochroria*). Ductus seminalis joins bursa copulatrix centrally.
8. A pair of sclerotized teeth on posterior margin of male eighth sternite (present but not sclerotized in *onea*). Short, rod-like structure with rounded end running out over central part of exterior surface of valve. Spur directed inwardly from center of interior surface of valve near base.
9. Leaden grayish tint to hindwing. Interior spur at center of valve straight rather than upturned. Aedeagus vesica cornutus elongate, irregular.
10. Sharp anterior angle to forewing antemedial. Subcostal indentation to forewing postmedial. Submarginal setose lobe on valve (see 7) well developed such that the valve tends to fold at that point.

Reference Fig. 3. Presumed synapomorphies used in the cladogram for *Polyacme.*

1. Posterior margin of male eighth sternite broadly bilobed. Signum of bursa copulatrix a single sinuous spine, usually large but lost in *subpulchra.* General facies and wing shape similarities.
2. Process on ventral margin of valve very distal. Subcostal, subbasal flange on valve present.
3. Lobes on eighth sternite themselves bilobed. Distal ventral process on valve small, single rather than bifid. Subcostal flange to valve serrate to evenly setose, rounded, rather than sparsely and irregularly setose, produced into slender process distally. Uncus slender versus broadly triangular.
4. Projection on ventral margin of valve set centrally. No ornamentation in costal area of valve (probable plesiomorphy). Aedeagus vesica with at least one lateral lobe. Ostium bursae with a prominent pair of projecting plates, one on each side, directed posteriorly.
5. Anellus sclerotized, sparsely scobinate. Uncus with very stout setae. Projection from ventral margin of valve a round-ended spur directed distally instead of triangular, acute, directed dorsally.
6. Anterior vaginal lamella sclerotized and crinkled in an elongate transverse zone where it overlies the ostium bursae. Hindwing submarginal straight rather than angled.

Reference Fig. 6. Presumed synapomorphies used in the cladogram for *Avitta.*

1. Several parallel fasciae on forewings, diffuse, irregular. Reniform bilobed with paler center. Angle or spur at base of costa of male valve. Spur on sacculus of male valve. Male eighth abdominal sternite distally bilobed. Sclerotized plate ventrally between female ovipositor lobes.
2. Fasciation of forewing faint, broken, thin. Angle only at base of male valve costa. Aedeagus vesica with several moderate but slender cornuti. Bursa copulatrix of female spherical, finely scobinate throughout, the scobinations thicker in patch near base. Sclerotized plate between ovipositor lobes narrow, weak.
3. Male eighth sternite distally strongly bidentate. Male eighth tergite anterior lateral projections well developed. Anterior margin of male eighth sternite with small double central projection. Spur at

base of male valve costa well developed. Bursa copulatrix longer than wide. Plate between ovipositor lobes well developed.
4. Yellow, speckled forewings with fasciae indistinct. Bursa copulatrix tapers into ductus. Bursa evenly and moderately scobinate throughout.
5. Bursa copulatrix usually very elongate, separated from ductus by asymmetric 'heel'.
6. Bursa copulatrix with oblique band of very small spines. Broad bifid plate posterior to ostium bursae.
7. Forewing reniform emphasized above fasciae, a roman rather than italic 'B'.
8. Forewing fasciae broken and reduced but more sharply defined than in other species when present. Ductus bursae with basal bulge.
9. Male eighth sternite distinctly concave on anterior margin. Male valve more weakly sclerotized than costal spur, narrower basally than apically. Male juxta setose.
10. Weakly sclerotized apical part of male valve shorter by far than sacculus, curved gently up dorsally. Saccular spur or spurs central (possibly plesiomorphic state). Ninth and tenth segments of female abdomen more or less fused. Lamella postvaginalis of female broadly sclerotized.
11. Ostium bursae of female notched, the base of the notch expanded into a part circle.

Combination of two or more of the following characters:

12. Saccular spur very long and slender; male genitalia.
13. Saccular spur displaced basally; male genitalia.
14. Saccular spur of male genitalia absent, presumed lost.
15. Costal spur of male genitalia mainly hornlike, reflexed.
16. Male uncus more or less sinuous, with long, acute, reflexed apical part.
17. Plate between ovipositor lobes with two folds angled to each other.
18. Notch of ostium bursae with acute projections laterally directed posteriorly. Yellow coloration transgresses this system, being found in *guttulosa, bracteola*, and, slightly, *flavicilia*.

Note: 13 may be referable to 11 and plesiomorphic to 14.

Reference Fig. 15. Presumed synapomorphies used in the cladogram for the genus *Ophyx*.

1. Males with a variety of secondary sexual characters but usually a scent pencil of long, straw-colored scales arising from the humeral area of the hindwing above. Uncus a slender rod on a broad tegumen. Saccus broad, rounded, voluminous. Valve with rounded apex and a subapical or apical spur on the ventral margin. A row of short setae on the interior of the ventral margin of the valve (reduced in number and enlarged in one subgroup). A mass of long, hairlike, deciduous setae on the sacculus, extending round to the exterior side of the ventral margin and along it, with larger basal blisters than for the interior margin setae. Male eighth sternite with a narrow band of sclerotization at the anterior margin, laterally produced into two basally articulated rods that link with the tergite, and centrally angled to enclose a rounded or angular area of sternite less heavily sclerotized than the rest. Male eighth tergite narrower than the sternite, constricted or waisted at one third from the anterior margin; anterior margin bilobed, the lobes slightly splayed.
2. Male patagia dark brown, concolorous with thorax in female. Male secondary sexual characters posterior to the cell on the forewing underside (though absent in *maculosus* and the New Guinea race of *inextrema*). The bilobed structure set ventrally between the female ovipositor lobes is somewhat broader than in the other group.
3. Ground color of both sexes ochreous fawn. Third segment of palps less than one third the length of the second segment. Valve apex very broad, the ventral spur set at the level of the apex rather than displaced ventrally.
4. Setae along ventral margin, inner surface, of valve enlarged (only slightly in *pratti*). Uncus twice as long as in other species (except *owgarra* which is intermediate).
5. Setae along ventral margin of valve very large, reduced to seven or eight, spaced evenly. Male coloration slightly violet gray with an oblique medial of dark brown passing through reniform, sharply defined interiorly, diffusing away exteriorly, with ground color slightly paler basad.
6. Large setae on ventral margin of valve usually eight, only diminishing slightly in size basad. Aedeagus vesica with a single distal lobe bearing a large group of long spines apically.
7. Large setae on ventral margin seven, diminishing in size basad. At least three markedly scobinate lobes in aedeagus vesica.
8. Forewing medial usually well inside reniform. Aedaegus apex with spatulate, usually rounded process at apex. Hind and sometimes mid tibiae of male usually with tufts of long, hairlike scales.

9. Hindwing reduced in size relative to forewing (only slightly and unevenly in *owgarra*). Male forewing postmedial distinctly stepped inwards over posterior half.
10. Male hindwing very much reduced to half area of that of an unmodified species. Rectangular block of paler color at costa associated with postmedial on forewing. Ventral margin of valve bulges between spur and distal end of sacculus. Prominent scobinate groove over distal third of aedeagus.
11. Forewing postmedial oblique, straight or almost so, parallel to margin. Both wings invested with dusting of mauve or pale scales, most intense at fasciae.
12. Forewing type of markings also on hindwing, especially dark brown patches in association with the postmedial.

Recent Advances in the Systematics and Biogeography of Landhoppers (Amphipoda: Talitridae) of the Indo-Pacific Region

E. L. Bousfield
National Museum of Natural Sciences
Ottawa, Canada

Abstract—Landhoppers are terrestrial amphipod crustaceans of the talitroidean family Talitridae that is subdivisible into four systematic-ecological (polyphyletic and overlapping, but pragmatically useful) units, as follows: (1) palustral talitrids–plesiomorphic, semi-aquatic (rarely terrestrial) in estuarine and some freshwater habitats of tropical and antipodean continental areas; (2) beachfleas–more advanced, semi-terrestrial and terrestrial (but non-substrate-modifying) in supralittoral and coastal rain forest habitats of tropical to boreal marine coastlines of the world; (3) sandhoppers–specialized fossorial (substrate-modifying), semi-terrestrial, supralittoral on sandy beaches of tropical and temperate marine shores; and (4) landhoppers–specialized terrestrial (but seldom substrate-modifying), in coastal continental and high-island angiosperm rain forests, mainly of tropical, Indo-Pacific, and antipodean temperate regions.

Based on more precisely defined characters of the mouthparts, peraeopods, gnathopods, and other appendages, the landhoppers are subdivided into two basic phyletic types, viz. (1) a simplidactylate (non-cuspidactylate) "ancient" group, probably derived directly from palustral talitrid ancestors, but now confined to South African, Australian, and New Guinean coastal continental regions; and (2) a cuspidactylate "modern" group, independently derived from regional beachflea progenitors of tropical warm temperate "Tethyan" coastlines of both northern and southern hemispheres. Each group encompasses both plesiomorphic (sexually dimorphic) and apomorphic (not sexually dimorphic) genera convergently derived.

Plesiomorphic landhoppers tend to be more physiologically and ecologically restricted than their apomorphic counterparts. They tend to occur nearer the coast or in areas of heavy rainfall, but retain marine dispersal capability and appear to be more successful in populating rain forests of remote volcanic high islands. Conversely, apomorphic landhoppers are generally more eurytopic and penetrate into drier and colder continental regions, but have less marine dispersal capability and are seldom native to volcanic islands.

In explaining the present disjunct world distribution of landhoppers, a plausible scenario for the origin and evolution of the Talitridae is proposed. In absence of a talitrid fossil record, the scenario is based on indirect evidence from the geochronology of continental drift and the fossil record of angiosperm forests (with which landhoppers are intimately associated), and on comparative morphology, dispersal behaviour, and ecology of component species groups within the Talitridae. In brief, talitrid amphipods evolved from primarily aquatic ancestors, allied to existing primitive genera of the intertidal talitroidean family Hyalidae, mainly during the Cretaceous Period and early Tertiary Era. Stimulus for talitrid evolution was provided by the break-up of Pangea into several smaller continents and consequent proliferation of new coastal regions of moist, stable, oceanic climates, and by contemporaneous evolution of angiosperm forests that provide continuously renewable leaf-litter food resources. The main thrust of landhopper evolution took place in tropical Tethyan and temperate antipodean "Gondwana" continental regions, little in the temperate northern hemisphere.

In order to validate the new taxa employed in these analyses, the following names are newly proposed and the taxa briefly diagnosed in the Appendix: (1) palustral talitrids—*Eorchestia*, n. g. [type species *E. rectipalma* (K. H. Barnard)], *Microrchestia*, n. g. [type species *M. macrochela* (Bousfield)], *Chiltonorchestia*, n. g. [type species *C. pusilla* (Chevreux)], *Uhlorchestia*, n. g. [type species *U. uhleri* (Shoemaker)], *Chelorchestia*, n. g. [type species *C. costaricana* (Stebbing)]; (2) beachfleas–*Tethorchestia*, n. g. (type species *T. antillensis*, n. sp.), *Floresorchestia*, n. g. [type species *F. floresiana* (Weber)]; (3) landhoppers—*Protaustrotroides*, n. g. (type species *P. victoriae*, n. sp.), *Hawaiorchestia*, n. g. [type species *H. hawaiensis* (Dana)], *Caribitroides*, n. g. (type species *C. jamaicensis*, n. sp.). New combinations for species assigned to these genera are given in Tables 5, 6, and 8 in the Appendix.

Taxonomic revisionary studies in which unnamed taxa will be formally proposed are in progress; more than 1000 species may eventually be described from mainly Indo-Pacific localities.

INTRODUCTION

I am honoured and delighted to participate in the J. Linsley Gressitt memorial symposium. Like Dr. Alison Kay, I feel perhaps somewhat of an imposter in dealing with animals that are primarily marine and aquatic and have only relatively recently solved the morphological, behavioral, and physiological problems of life on land.

Landhoppers, along with the beachfleas and sandhoppers, are classified in the Talitridae, the only family of amphipod crustaceans whose members are modified for terrestrial or semi-terrestrial life. To date, more than 200 talitrid species have been described (Bousfield, 1981b, 1982a) or are being described. Nearly half of these are landhoppers, living independent of the sea (as opposed to obligate seashore species). Perhaps more than 1000 terrestrial species remain to be described world-wide.

These carapaceless crustaceans have smooth insect-like bodies (cf. wingless Thysanura), with sessile eyes, very short first antenna, compact buccal mass, uniramous walking legs, and saltatory (jumping) form of abdomen and appendages. However, they retain primitive aquatic features such as coxal gills (modified for respiration in moist air), vestigial pleopods (abdominal swimming legs), and development of eggs in the female thoracic brood pouch.

Landhoppers are mostly nocturnal cryptozoic dwellers of the forest floor, and they range ecologically from coastal mangrove swamps to high rain forests of over 3000 m elevation (Hurley, 1968). They feed mainly on angiosperm leaf litter and organic debris and are important, often dominant, converters of organic material in energy cycles of forest floor biotic communities (Friend, 1980).

Development of knowledge of terrestrial amphipods commenced with the description of "Talitrus" *sylvaticus* Haswell (1880) from *Eucalyptus* leaf litter at Rootyhill, west of Sydney, Australia. Since that time, more than 70 terrestrial (or presumably fully terrestrial) species in a dozen genera have been recorded, mainly from coastal continental areas and volcanic islands of the Indo-Pacific, Caribbean, and antipodean regions.

Conversely, terrestrial amphipods are uncommon in the northern hemisphere and, unlike the terrestrial isopods (Oniscoidea), they are absent from North and South America and nearly totally lacking in Eurasia, and absent from continental interiors everywhere, despite apparently suitable equatorial, warm-temperate, and Mediterranean climates in some of these regions (Fig. 1).

Most landhopper species are highly regionally endemic, a phenomenon that reflects their relatively low vagility (Hurley, 1968). They are also restricted to continental coastlines and island areas of equable, moist climates, probably partly because of requirements for Na^+ and Cl^- ions in rain and soil, and/or low tolerance for hard-frost winters (Bousfield, 1968). Thus, from New Zealand, more than a dozen endemic species have been described (Hurley, 1955, 1957) and possibly 30 additional undescribed native species are at hand (Bousfield, unpublished). Sixteen species in 7 genera are known from Tasmania (no species of which occur elsewhere, even in Victoria just across the recently submerged Bass Strait) (Friend, 1980). Also, the Hawaiian high islands may contain more than 20 native species; the six described species from the archipelago include *Spelaeorchestia koloana,* the world's only known troglobitic terrestrial amphipod (Bousfield and Howarth, 1976).

However, about six landhopper species are synanthropic, carried by commerce to botanical gardens and parks in the northern hemisphere (Vader, 1972), but only two such species are widely dispersed by man. Thus, *Talitroides topitotum* has largely replaced native Hawaiian species to elevations of about 1000 m and both *T. topitotum* and *T. alluaudi* are now common along coastal regions of the Gulf States and in Florida (Biernbaum, 1980; Bousfield, unpublished). By contrast, native seashore and coastal terrestrial species and their young stages (e.g. of *Pla-*

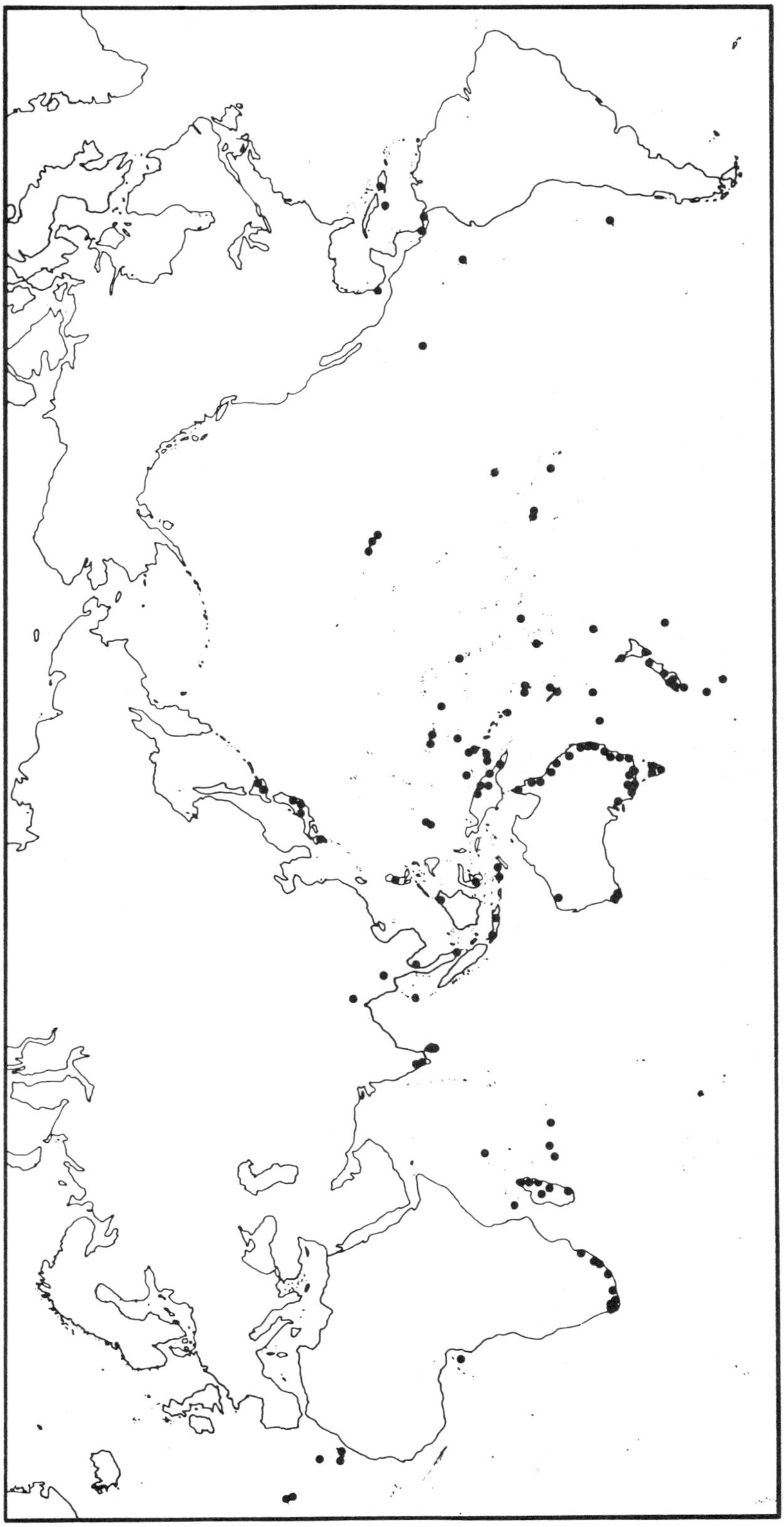

Fig. 1. Recorded distribution of terrestrial amphipods (after Friend, 1980).

torchestia platensis, Floresorchestia spp.) are almost certainly marine-dispersed by algal and log rafting, etc. (Bousfield, 1982c).

Terrestrial amphipods have long been considered taxonomically difficult (Hurley, 1975; Bousfield, 1982c). They have been frequently overlooked by students of the cryptozoic fauna of the forest floor, partly because they occur in areas relatively rarely visited by collectors (e.g. in the southern hemisphere, remote from research centers). Most collections are made by entomologists, but since landhoppers are not insects or "typical" air-breathing terrestrial arthropods, nor oniscoidean isopods, they tend to be omitted from faunal publication (e.g. the Talitridae were not included in F. M. Carpenter's revision of Brues and Melander "Classification of the Insects and Terrestrial Arthropods," 1953). Moreover, they are seldom taken by carcinologists who search for crustaceans mainly in aquatic environments.

This research hiatus is particularly surprising in view of the widespread occurrence of landhoppers in forest habitats, where they are often numerically dominant among animals of their size range in the community. Several species have proven to be lab-hardy and are useful for studies on physiological and behavioural adaptations in sea-to-land transition of living organisms (Hurley, 1968).

The classification of terrestrial amphipods has also been inadequate. Until relatively recently (e.g. Hurley, 1968; Bousfield, 1968; Barnard, 1969) land amphipods numbered about 50 species in 6 poorly defined genera. These were grouped in 2 major categories according to the morphology of the first two pairs of thoracic limbs or gnathopods, viz. (1) those with sexually dimorphic gnathopod 2 (powerfully subchelate in male), e.g., "Orchestia", *Parorchestia* spp. (names that did not distinguish terrestrial amphipods from semi-terrestrial seashore types); and (2) those with non-sexually dimorphic gnathopod 2 (weakly chelate, or "mitten-shaped" in both sexes), e.g. "Talitrus" (a name that related all terrestrial amphipods to the sandhopper *Talitrus saltator,* and to each other).

However, recent advances in character selection and sophistication of character analysis allow us more realistically to account for the evolution that has taken place in all body regions and appendages. New analytical methods (e.g. Sneath and Sokal, 1973) have permitted the species to be grouped in more "natural" genera. The many new genera reflect morphological, ecological, and biogeographical relationships that were hitherto masked by artificial and uncritical lumping of species within a very few ill-defined genera (see Friend, 1980; Bousfield, 1971, 1982c, 1983).

My purpose here is (1) to outline briefly how this methodology is applied in a world-wide revision of the Talitridae, with special reference to the landhoppers of the Indo-Pacific region; (2) to demonstrate distributional relationships within component species groups; and (3) to suggest a probable evolutionary scenario for the Talitridae, based on the geochronology of continental drift and of the evolution of angiosperm forests with which they are closely associated.

This presentation is somewhat in the nature of an interim report, since much of the analysis is still in progress. Some new taxa and new combinations are described in an appendix (p. 201) in order to form a practical basis for biogeographical discussion in the text proper. I would value your comments and suggestions for its improvement (some of which have been incorporated in this published version of the address).

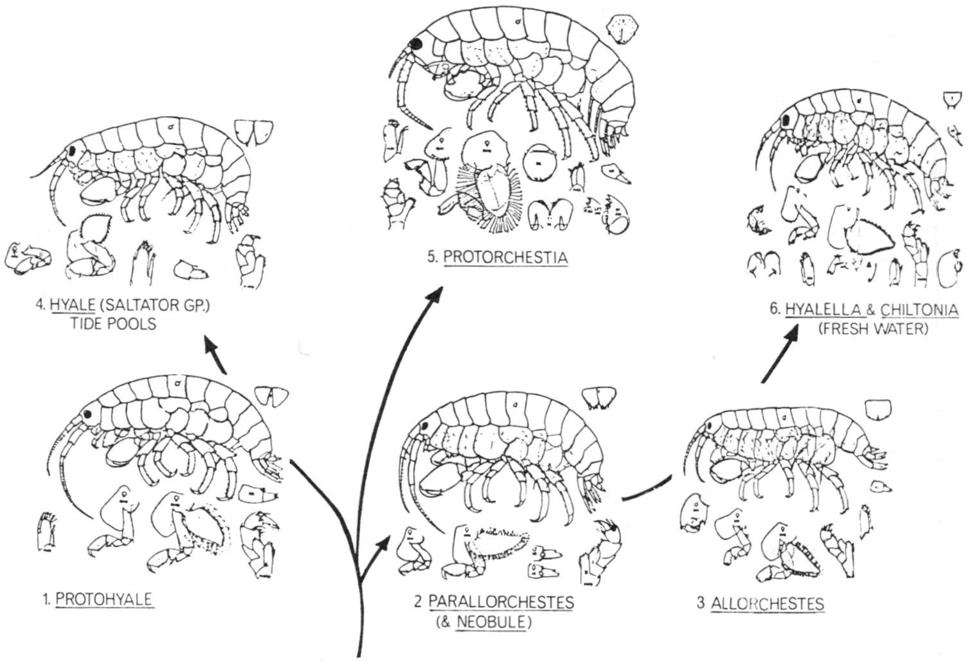

Fig. 2. Talitridae ancestry.

SYSTEMATICS

Members of the amphipod family Talitridae (and of some other families of the superfamily Talitroidea) exhibit terrestrial adaptations in morphology, physiology, behavior, and ecology (Hurley, 1968; Bousfield, 1982a,c). Conspicuous adaptive (apomorphic) morphological features include pronounced development of non-pelagic mating stages (cf. Bousfield, 1979) with strongly amplexing gnathopods (males) that may be secondarily (neotenically) lost; shortening of antenna 1 and total loss of an accessory flagellum; clustering of mouthparts in a compact buccal mass, with reduction and loss of palps (especially mandibular) and setae; modification (variously) of basic, homopodous peraeopods; reduction of pleopods (abdominal swimming legs); shortening of urosomal segments and "stiffening" of uropods; extreme reduction of uropod 3; fusion of telson lobes; reduction of large brood plates with hook-tipped setae to small, narrow plates with a few (or no) marginal setae; proliferation and lobation of coxal gills; thickening and waxing(?) of epicuticle; and development of large pores, pore spines, and other specialized surface ultra-structures (Holman, 1982; Bousfield and Halcrow, in prep.).

These features are variously shown in the six families of Talitroidea that occur commonly around the shores of the North Pacific Basin (see Bousfield, 1981b, figure 6). With respect to phyletic interrelationships, *Protorchestia,* the most primitive genus within the family Talitridae, is believed to have evolved from a hyalid-like ancestor, possibly close to the primitive extant genus *Parallorchestes* within the marine intertidal family Hyalidae (Fig. 2). *Protorchestia,* and all members of the aquatic ancestral Hyalidae, have non-cuspidactylate (simplidactylate) peraeo-

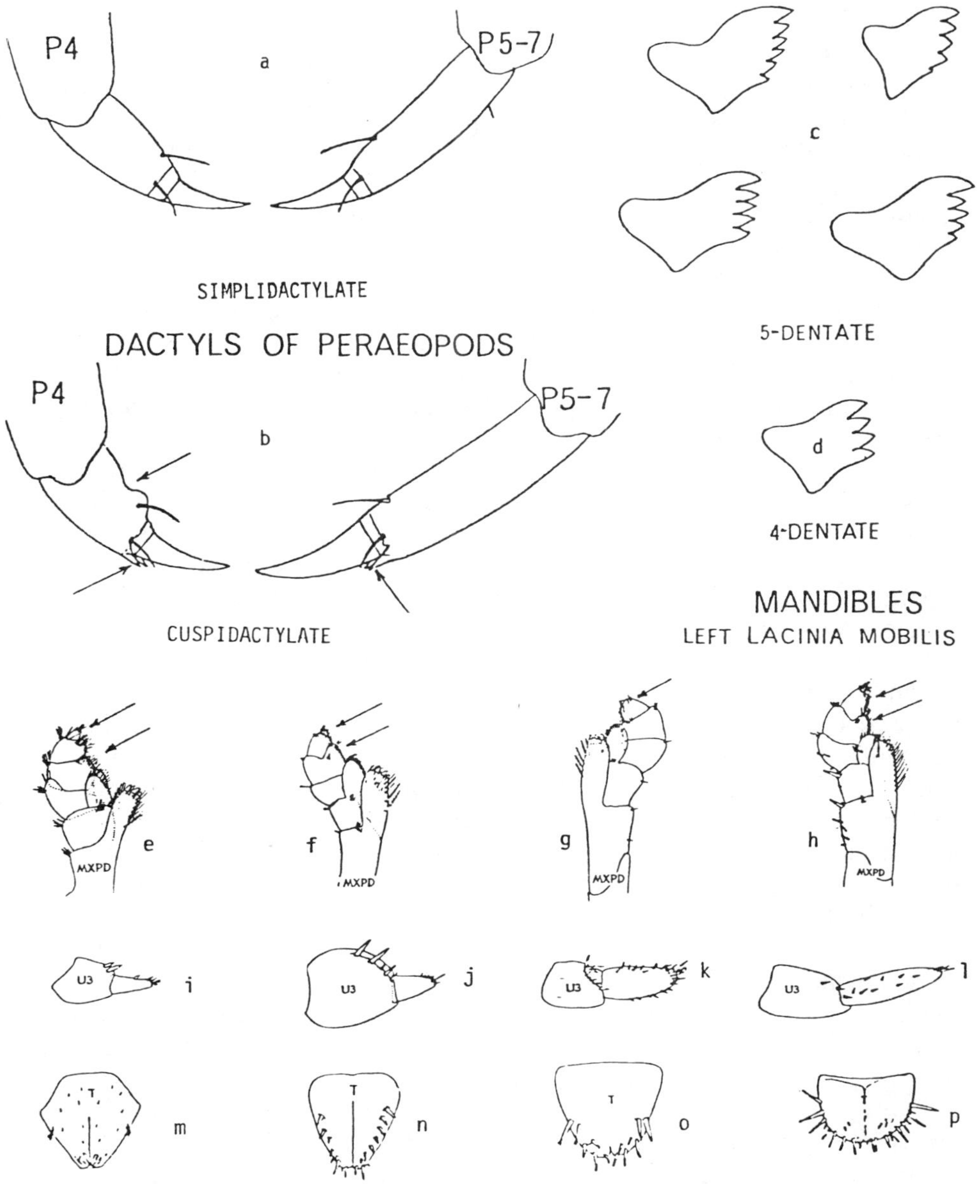

Fig. 3. Talitridae. Major taxonomic characteristics. Peraeopod dactyls (a,b); Mandible, left lacinia (c,d); Maxilliped (e–h); Uropod 3 (i–l); Telson (m–p).

pods and 5-dentate left lacinia mobilis. Talitrids are distinguished from the more plesiomorphic hyalids, however, by the short antenna 1; reduced coxa 1; non-unguiform maxilliped palp; more reduced palp of maxilla 1; mitten-shaped gnathopod 2 (female and immatures); and variously reduced and/or modified coxal gills.

The principal high-level taxonomic characters utilized in talitrid generic groupings within the family Talitridae have been described elsewhere (e.g. Bousfield,

1982c). Some of these primary characters are repeated here (Fig. 3). The phyletic range of character state or condition (plesiomorphic to the left or above and apomorphic to the right or below in the figures) is derived mainly through outgroup morphological comparison with presumed aquatic ancestral types and from obvious marine-to-terrestrial morphological trends within subgroups of the Talitridae itself. Thus, the mandibular left lacinia mobilis is primitively 5-dentate [in aquatic hyalid ancestral Talitroidea and in plesiomorphic superfamilies of Gammaridea (Bousfield, 1983)], whereas reduction to the 4-dentate (or fewer) state, or proliferation to the 6-dentate (or more) state, represents an apomorphic advanced condition as in apomorphic beachflea and sandhopper groups.

The lack of small paired cusps at the outer base of the nail of the dactyl (simplidactylate condition) is plesiomorphic, whereas their presence (cuspidactylate condition) is apomorphic (Fig. 3). A distinctly 4-segmented maxilliped palp (also lacking medio-distal lobe on segment 2) is plesiomorphic, 3-segmented (and with medio-distal lobe) is apomorphic. Plesio-apomorphic trends in uropod 3 (shape and size and armature of peduncle and ramus), and telson length:width ratios, degree of distal fusion, and dorsal armature of lobes are also utilized.

Plesio-apomorphic trends or polarization in gnathopod morphology remain important in phyletic analyses, but the characters are now more precisely defined (see also Bousfield, 1982c). Primitively, gnathopod 1 is fully subchelate in both sexes, the dactyl not exceeding the palm, and segments 4, 5, and 6 bear posterior marginal swellings, the pellucid processes; through various intermediate conditions of shortened palm, exceeded by the dactyl, and reduced pellucid processes (especially in the female), it attains the most apomorphic state, i.e. the palm is totally lacking, and pellucid processes are much reduced or lacking. Also, gnathopod 2 (male) is primitively large and powerfully subchelate, with smooth, convex palm, and closely matching dactyl; in the apomorphic state, the palm becomes more oblique, variously toothed, notched near hinge, and the dactyl is sinuous and/or toothed behind; in extreme (neotenic) apomorphy, the propod and dactyl remain "mitten"-like (similar to the female). Anterior expansion of the basis (♀) is also advanced.

Other major taxonomic characters and their trends are utilized in this analysis (cf. Bousfield, 1982c). Thus, the plesio-apomorphic status of antenna 2 ranges from simple, slender, and sublinear to swollen and/or elongate (sexually dimorphic); uropods 1 and 2 from short and stout, with postero-distal peduncular (inter-ramal) spine present and rami weakly or not marginally spinose, to elongate and slender, lacking inter-ramal spine, and rami marginally spinose; peraeopods 5-7 from primitively homopodous (alike, increasing posteriorly in length) and coxae posterolobate, to apomorphically totally unlike, with leg 6 longest, and coxae aequi- or antero-lobate; and the pleopods from primitively large and powerful, with strong peduncle and multi-segmented rami, to apomorphically short and weak with few-segmented rami, or with peduncles broadly expanded and spinose.

The taxonomic characters outlined above have been utilized in realignment of talitrid species within new and redefined generic groupings, and clustering of genera into the four main systematic-ecological family components outlined below (p. 179; Table 1; also Bousfield, 1982c, in prep). Although these groupings have some phyletic basis, the degree of overlap and convergence of characters of some

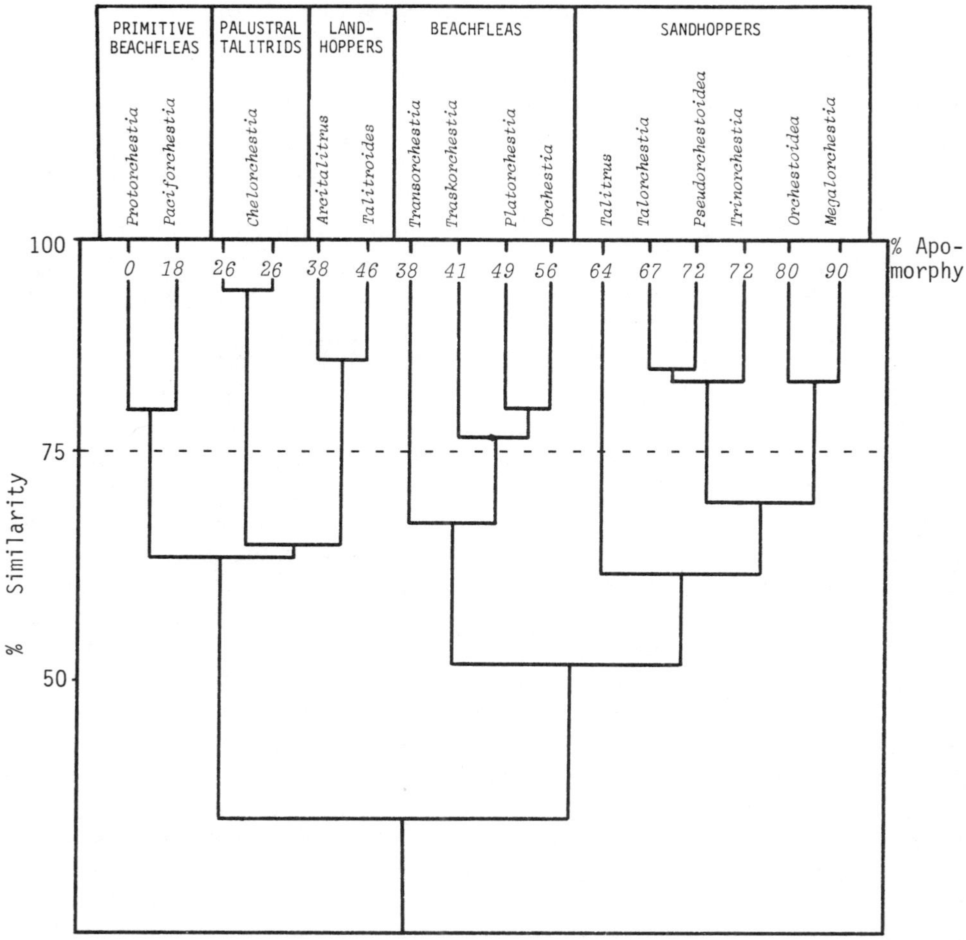

Fig. 4. Phenogram of genera of all groups of Talitridae, North Pacific Rim region, and selected related genera.

genera, and need for completion of the study (Bousfield, in prep.), renders the present recognition of these groupings pragmatic and informal, rather than formally classificatory (see Appendix). The four groupings may be diagnosed briefly as follows.

(1) Palustral talitrids, of plesiomorphic morphology (all simplidactylate, 4-dentate, with strongly sexually dimorphic gnathopods), semiaquatic (rarely terrestrial) in estuarine (mangrove, salt marsh) and some freshwater habitats of tropical and antipodean continental regions.

(2) Beachfleas, of more advanced morphology (mostly cuspidactylate, 5- and 4-dentate, but sexually dimorphic), semi-terrestrial and terrestrial [but non-substrate-modifying (of MacIntyre, 1963)], in supralittoral habitats and coastal rain forests of tropical to boreal and anti-boreal marine coastlines.

(3) Sandhoppers, of specialized fossorial (substrate-modifying) morphology (all cuspidactylate, 4- and 5(6)-dentate, nearly all with sexually dimorphic gnatho-

pods), semi-terrestrial and supralittoral on sand beaches of tropical to boreal and anti-boreal marine coastlines.

(4) Landhoppers, of specialized "terrestrial" (mostly non- substrate-modifying) morphology (simplidactylate and cuspidactylate, but all 4-dentate, and many sexually dimorphic, or trending so), fully terrestrial in coastal continental and high-island angiosperm rain forests, mainly of tropical, Indo-Pacific, and antipodean temperate regions.

The landhoppers may be subdivided, on a more rigorously phyletic basis, into two further generic groupings, viz. the simplidactylate and the cuspidactylate types (see Appendix and Tables 5–8 for component species groupings, new generic diagnoses, etc.).

Many simplidactylate landhoppers exhibit remarkably specialized "terrestrial" modifications of gills, pleopods, and other body parts (e.g. *Talitriator, Arcitalitrus, Mysticotalitrus,* and *Keratroides*). As a group, however, they are distinguished by some relatively plesiomorphic characters that relate them directly to presumed palustral talitrid ancestral types. These diagnostic characters include similar peraeopods 3 and 4 in which the respective segments 5 are subequal in length, and the dactyls simplidactylate; antenna 1 tends to be elongate, the peraeopods weakly heteropodous, and the outer ramus of uropods 1 and 2 invariably lack marginal spines. Primitive species (of Tasmanian genera "O," "N," and "T" of Friend, 1980) retain plesiomorphic, sexually dimorphic, "amplexing" gnathopods in the male, nearly homopodous peraeopods, postero- or aequi-lobate coxa 5, and relatively unmodified coxal gills; most simplidactylate landhoppers (in the other genera) are of more apomorphic (not sexually dimorphic) morphology.

Cuspidactylate landhoppers also exhibit very specialized "terrestrial" modification of appendages (e.g. *Talitroides, Spelaeorchestia*) but tend to be more eurytopic and physiologically "hardy" than simplidactylate types. Group-distinguishing characters are more apomorphic, and relate them directly to regional beachflea ancestral types. These characters include the following: dissimilar peraeopods 3 and 4, in which respective segments 5 are unequal in length, and the dactyls are cuspidactylate and unlike; the peraeopods are more strongly heteropodous, and the outer ramus of uropods 1 and 2 are usually marginally spinose. However, a higher proportion of the 13 generic groups (Table 4; Appendix, Table 8; also Bousfield, in prep.) are of the plesiomorphic, sexually dimorphic morphology.

The following biogeographic and palaeohistorical analyses treat mainly the fully terrestrial amphipods, comprising all of the landhopper group, some of the 4-dentate and a few of the 5-dentate beachflea groups, and one of the palustral talitrid group. On the basis of plesio-apomorphic states of characters used (above and Table 1), the true landhoppers are relatively highly evolved. Their plesio-apomorphic (P-A.) index of 72–80 is about equal to that of most sandhoppers; they are also somewhat more advanced than cuspidactylate beachfleas (47–68), and considerably more so than the palustral talitrids and primitive, simplidactylate beachfleas (0–47).

Relationships of representative genera of these four main groups may also be expressed in the form of a phenogram (Fig. 4). The similarity matrix for clustering by simple paired coefficients utilized a slightly broader but essentially similar set

Table 1. Summary of talitrid types and biogeographical relationships.

Taxonomic group	No. gen./spp.	World distribution	Habitat	Genera* Endem. index	Genera* P-A. index	Probable palaeohistorical origin	Anticipated new taxa
I. Palustral Talitridae	6/29+	Mainly antipodean and tropical continental	Mainly estuarine, some freshwater and terrestrial	4.3	47	Early Cretaceous	Several
II. Beachfleas							
Group 1	1/2	Antipodean; S. America and New Zealand	Stony shores, intertidal and supra-littoral	5.0	20	Early Cretaceous?	Few
Group 2	4/11	Boreal and antiboreal Pacific	Stony shores, supra-littoral	5.1	47	Cretaceous	Few
Group 3	4+/38+	Equatorial and temperate, Atlantic and Indo-Pacific	Stony shores, salt marshes, supralittoral, and terrestrial	3.4	68	Late Cretaceous; early Cenozoic?	Many
III. Sandhoppers							
Group 1	10/46+	Holarctic and equatorial	Sand beaches, supra-littoral	6.3	80	Mainly Cenozoic?	Few
Group 2	4+/21+	Austral and antipodean	Sand beaches, supra-littoral	7.0	72	Late Cretaceous?	Few
IV. Landhoppers							
Group 1	10/43+	Australia, E. New Guinea, Melanesia; S. Africa	Mesophyllic forest, leaf litter	8.8	80	Mid-Cretaceous	Few
Group 2	13+/60+	Equatorial and Indo-Pacific	Rain forest leaf litter	8.7	72	Late Cretaceous?	Very many
Totals and means	52+/250+			6.5	69		

* Endemicity index and plesio-apomorphic index defined as in Table 2.

Table 2. Summary of palustral Talitridae and biogeographical data.

Taxonomic group	Morphol. character	No. spp.	World distribution	Habitat	E.I.*	P-A. I.†
I. Palustral	4-dentate; simplidactylate					
Eorchestia, n.g.		5	New Zealand, SE Australia, South Africa	Estuarine mangrove	3	25
Microrchestia, n.g.		8	Indo-Pacific, Polynesia, India, S. Africa, W. Africa, Ecuador, New Caledonia	Mangrove swamps	2	45
Chiltonorchestia, n.g.		5	New Caledonia	Freshwater streams	10	40
Uhlorchestia, n.g.		2	N. America, Atlantic, and Mediterranean	*Spartina* marshes	7	63
Chelorchestia, n.g. (group 1)		1	Galapagos, San Cristobel	Litter and moss pads (terrestrial)	10	65
(group 2)		7	Caribbean, Florida to Brazil; also Nigeria, Galapagos, Hawaii	Mangrove swamps, salt marshes	4	69

* Endemicity index (E.I.) (see Bousfield, 1982b). Number is inversely proportional to number of continents, or island masses on which the taxon occurs. Highly regionally endemic taxa have high endemicity indices, widely distributed taxa have low E.I.'s.

† Plesio-apomorphic index (P-A. I.) (see Bousfield, 1982b,c). Number represents a percentage of total possible apomorphy of all characters used in comparing the group taxa. Taxa with low P-A. indices are primitive, those with high P-A. indices are advanced.

of primary generic characters than those of Table 1 in order to apply to all groups. Figure 4 reveals good, semiphyletic contrast between the beachfleas and sandhoppers (see Colless, 1970), but a less clear distinction between primitive beachfleas and the palustral group. Nonetheless, the phenogram validly underscores the morphological remoteness of landhoppers from sandhoppers, and the need to remove all known landhopper genera from the sandhopper genus *Talitrus.*

BIOGEOGRAPHY

Let us now examine the world distribution of terrestrial beachfleas and the true landhoppers, with specific reference to the Indo-Pacific region. The distributions were plotted following completion of the morphological (taxonomic) reorganization; the mutual independence of these analyses increases the overall significance of the following correlations.

The distribution of known palustral Talitridae is summarized in Table 2. Many

Table 3. Summary of beachfleas and biogeographical data.*

Taxonomic group	Morphol. character	No. spp.	World distribution	Habitat	E.I.	P-A. I.
II. Beachfleas (nonsubstrate modifiers)						
Group 1	5-dentate simplidactylate					
Protorchestia Bousf.		2	Southern S. America, New Zealand	stony intertidal	5	20
Group 2†	5-dentate cuspidactylate					
Paciforchestia, n.g.		4?	Amphi-North Pacific	intertidal; terrest.	5	40
Traskorchestia Bousf.		4	North Pacific Rim	HW intertidal	4	44
"Orchestia" *australis* group		1	SE Australia and Tasmania	HW intertidal	7	50
"Orchestia" (Rottnest group)		3	SW Australia, SE Australia and Tasmania	HW intertidal	6	56
Group 3	4-dentate cuspidactylate					
Transorchestia Bousf.		5+	E. Australia, New Zealand, S. South America; Juan Fernandez Is.	HW intertidal; a few terrestrial	2	48
Tethorchestia, n.g.		10+	Caribbean coasts, Florida to Brazil	HW intertidal	5	60
Orchestia Leach		13+	N. Atlantic and Mediterranean; E. S. America (syn?)	HW intertidal; 1 terrestrial	4	72
Floresorchestia, n.g.		10+	Indo-Pacific; Polynesia, Micronesia, India, S. Africa, Hawaii, Venezuela	Coastal terrestr. and HW intertidal	2	80

* Endemicity index and plesio-apomorphic index defined as in Table 2.

† *Platorchestia* (classified with the sandhoppers in Bousfield, 1982) could also be placed in Group 2 of the beachfleas.

of the component species (with distinctly 4-segmented maxilliped palp) were previously included in the "catch-all" genus *Parorchestia.* This generic name implies a terrestrial habitus, but of the palustral talitrids only *Chelorchestia vaggala* is undoubtedly terrestrial (Bowman, 1977). All other palustral genera are semi-aquatic in estuarine (mangrove and salt marsh) or freshwater habitats.

The relatively low endemicity indices of three of the genera (E.I. 2–4; Table 2) reflect the surprisingly wide distributions of these species groups over tropical, Indo-Pacific, and antiboreal continental regions; these distributions are likely to be "enlarged" as more of such relatively inaccessible estuarine environments are thoroughly sampled. This fact, and the high endemicity of the primitive *Chiltonorchestia* on the one hand (E.I. 10) and of the advanced *Uhlorchestia* on the other (E.I. 7), tends to support the hypotheses of early evolution of the group as a whole, their presumed need for isolation from more advanced talitrid competitors elsewhere, and their life-dependence on specific angiosperm habitats (in palaeohistory analysis, p. 190).

In analyzing the distribution of terrestrial beachfleas, let us first consider the more plesiomorphic (5-dentate) Groups 1 and 2 (Table 3). In the simplidactylate Group 1, *Protorchestia campbelliana,* known only from mollymawk rookeries in the subantarctic islands of New Zealand (Bousfield, 1964), is apparently mainly supralittoral and marginally terrestrial, and it will not be considered further here. Within the cuspidactylate Group 2, known terrestrial species belong to the partly supralittoral genera *Paciforchestia* (3 spp.?) and *Platorchestia* (2 spp.?), and all are restricted to forest habitats of the Japanese Archipelago in the wester Pacific. Native terrestrial species are unknown from the North American Pacific coastal region where climate is suitable but edaphic conditions are generally not (Bousfield 1982c).

The world distributions of component members of the four known 4-dentate beachflea genera are outlined in Fig. 5. Lists of component species and diagnoses of *Tethorchestia,* n. g., and *Floresorchestia,* n. g., are given in the Appendix (p. 204) and Table 6. Three of the genera (*Transorchestia, Tethorchestia,* and *Orchestia*) encompass both supralittoral and terrestrial species (at least one known species of each); the genus *Floresorchestia* contains only terrestrial (coastal) species, some of which are occasionally found in the strand.

Distributionally, three of the genera (*Tethorchestia, Orchestia,* and *Floresorchestia*) are, respectively, confined to Caribbean, Atlantic-Mediterranean, and the Indo-Pacific, along mainly tropical and warm temperate coastlines; these regions and their faunas are essentially of Tethyan origin, as discussed below (p. 195). *Floresorchestia* spp. are known to the author (unpublished) from the Caribbean region (Guadaloupe, Venezuela), and are recorded *in litt.* from Southeast Africa, as well as from several islands of the central Pacific (e.g. Samoa, Fiji, Hawaii), reflecting their marine dispersal capability in tropical oceanic regions. Significantly, the morphologically primitive *Transorchestia,* which evolved little beyond the presumed 5-dentate ancestral beachflea type (Bousfield, 1982c, in prep.), is confined to temperate antipodean continental regions (except South Africa). One species (*T. enigmatica*) may be synanthropic in the northeastern Pacific region. *Transorchestia* appears to be an antiboreal Pacific counterpart to the 5-dentate boreal Pacific beachflea genera, even to its terrestrial proliferation in a coastal

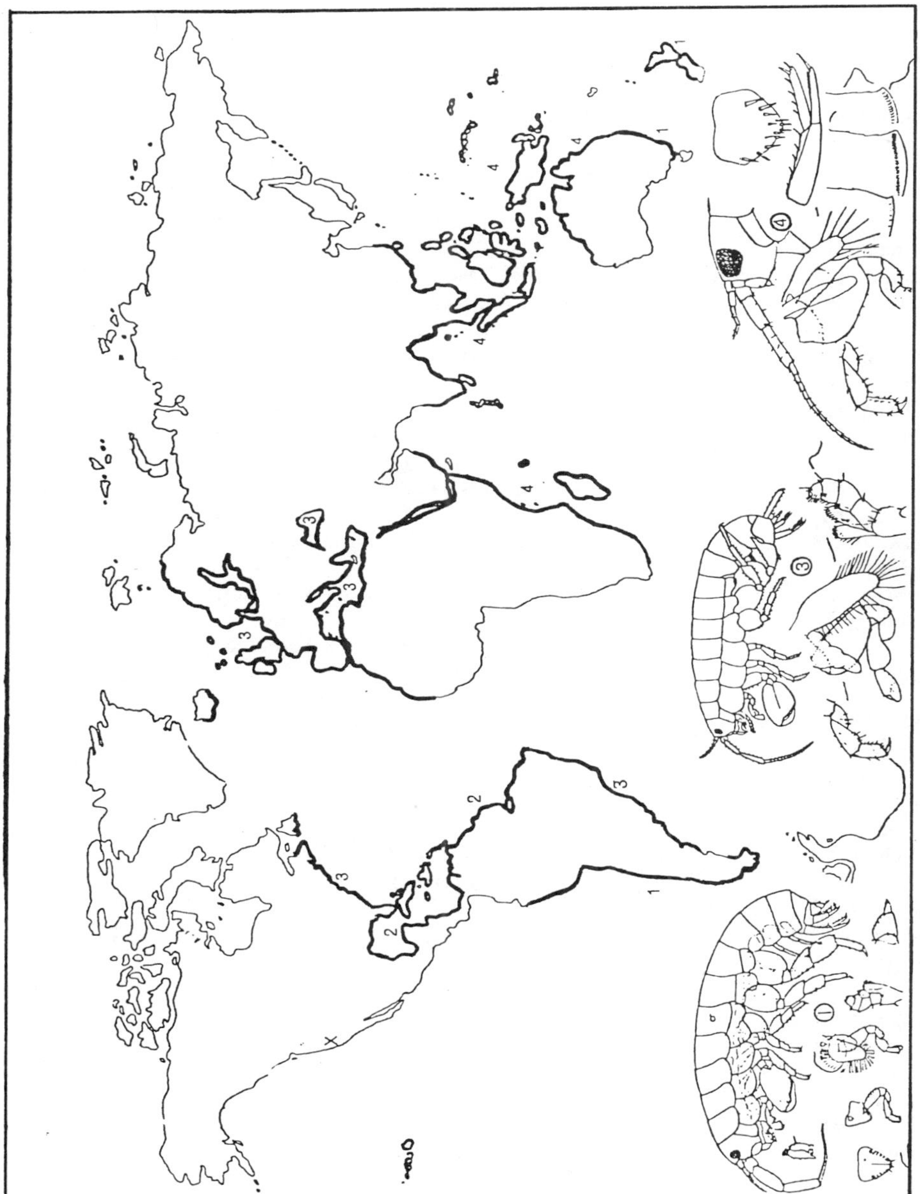

Fig. 5. World distribution of beachfleas, 4-dentate group. 1=*Transorchestia*. 2=*Tethorchestia*. 3=*Orchestia*. 4=*Floresorchestia*.

Table 4. Summary of landhoppers and biogeographic data.*

Taxonomic group	Morphol. character	No. spp.	World distribution	Habitat	E. I.	P.-A. I.
Group 1 (simplidactylate)	4-dentate					
Genus "O," Friend		2	W. Tasmania	Rain forest	10	16
Genus "N," Friend		2	W. Tasmania	Rain forest	10	71
Genus "T," Friend		3?	W. Tasmania	Rain forest	10	68
Austrotroides Friend		7	W. Tasmania, W. Australia, S. Australia	Rain forest	8	82
Protaustrotroides, n.g.		3	Victoria	Rain forest, marsh	9	77
Brevitalitrus Bousf.		4+	Queensland, Polynesia (Bismarck Archipelago)	Rain forest	7	83
Mysticotalitrus Hurley		2	SE Tasmania	Forest and marsh litter	10	84
Arcitalitrus Hurley		5+	SE Australia	Forest litter (drier)	9	80
Keratroides Hurley		10+	SE Australia and Tasmania	Wet forest litter	9	81
Talitriator Methuen		6+	S. Africa and Transvaal	Forest litter	9	78
Group 2 (cuspidactylate)	4-dentate cuspidactylate					
Parorchestia Stebb.		2	New Zealand	Rain forest	9	50
"Orchestia" groups (Hurley)		7+	New Zealand (sev. new genera)	Leaf litter	8	56
Agilestia Friend		5+	SE Australia	Leaf litter	8	70
Hawaiorchestia, n.g.		3+	Kauai, Oahu	Litter, moss	10	74
Spelaeorchestia B. & H.		1	Kauai	Lava tubes	10	75
Caribitroides, n.g.		12+	E. Mexico, Guatemala, Jamaica, Hispaniola	Litter, rain forest	9	76
"Parorchestia" (C.A.)		2	Panama, Costa Rica	Rain forest, litter	10	71
"Orchestia" (Japan)		3+	Japan, Formosa	Leaf litter	9	70
"Parorchestia" (New Guinea)		10+	(several undescribed genera)		8	72

Table 4. Continued.

Taxonomic group	Morphol. character	No. spp.	World distribution	Habitat	E. I.	P.-A. I.
"Parorchestia" (Philippines)		2+	Philippines		9	73
"Parorchestia" (Borneo)		1+	Borneo		8	74
"Parorchestia" (Indo-Pacific Is.)		10+	Polynesian, Melanesian Is.	Leaf litter or Rain forest	9	73
Talitroides Bonnier		3+	Madagascar, India, New Caledonia		7	81

* Endemicity index and plesio-apomorphic index defined as in Table 2.

island archipelago (e.g. the *T. gracilis* group of the Juan Fernandez Islands, off the coast of Chile).

As noted previously (p. 179), the sandhoppers are confined to supralittoral sand beach habitats, none are authentically terrestrial, and they are not considered further here. All previous "terrestrial" species of "Talitrus and "Talorchestia" have been removed to other genera of beachflea and landhopper groups.

With respect to the true landhoppers (Table 4), let us first consider the world distribution of the simplidactylate group (Fig. 6). These "marsupials" of the terrestrial amphipod world are native only to Australia, including Tasmania and eastern New Guinea and adjacent continental island systems, and to South Africa. They may yet be found native to New Zealand, and possibly South America and some subantarctic islands. A few species (e.g. *Brevitalitrus hortulanus, Arcitalitrus sylvaticus, A. dorrieni*) are synanthropic in greenhouses and winter-mild open gardens of temperate regions of the northern hemisphere (Vader, 1972; Bousfield, 1982c), but these tend to be relatively stenotopic and disperse slowly (Richardson, 1980).

By contrast, the distributions of component generic groups of 4-dentate cuspidactylate landhoppers are essentially Caribbean and Indo-Pacific (Fig. 7). Component species of *Caribitroides* and "Parorchestia" are known only from rain forest regions of Central America (Panama north to Chiapas and Veracruz, Mexico) and from the aluminous-lateritic high islands (Jamaica and Hispaniola) of the Greater Antilles (see Gomez-Pompa, 1973). In the Indo-Pacific region, the apomorphic genus *Talitroides* is endemic to the island of Madagascar and southern India, and occurrences elsewhere (North America, Europe, New Zealand, Australia, Hawaii, etc.) are almost certainly synanthropic (Vader, 1972; Biernbaum, 1980). The continental islands of the East Indies region, New Zealand, and temperate Australia and a few of the remote islands (Hawaii) of the west-central Pacific region encompass several genera (and probably hundreds of species) that are mostly undescribed. These include *Parorchestia* and other plesiomorphic (fully

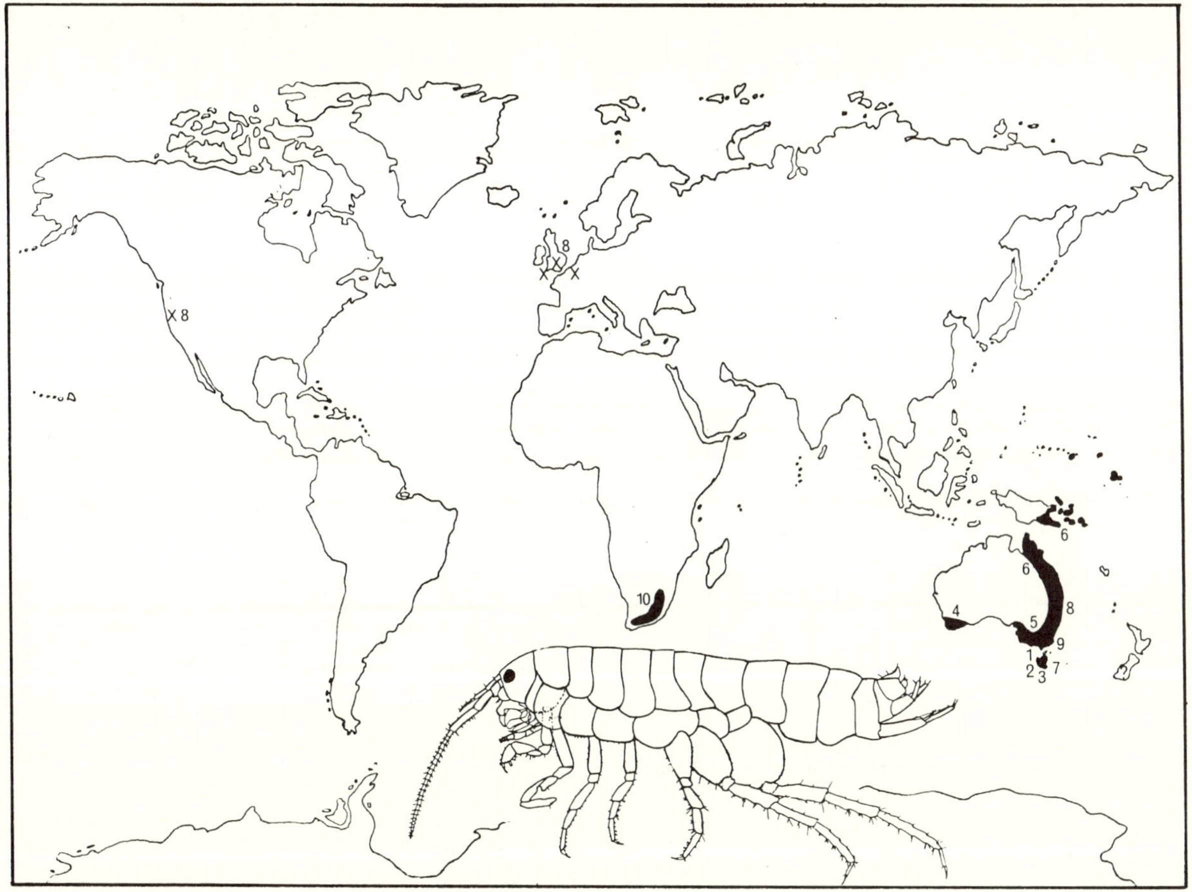

Fig. 6. World distribution of landhoppers, simplidactylate group. Genera: 1=''*O*''(Tasmania). 2=''*N*''(Tasmania). 3=''*T*''(Tasmania). 4=*Austrotroides*. 5=*Protaustrotroides*. 6=*Brevitalitrus*. 7=*Mysticotalitrus*. 8=*Arcitalitrus*. 9=*Keratroides*. 10=*Talitriator*.

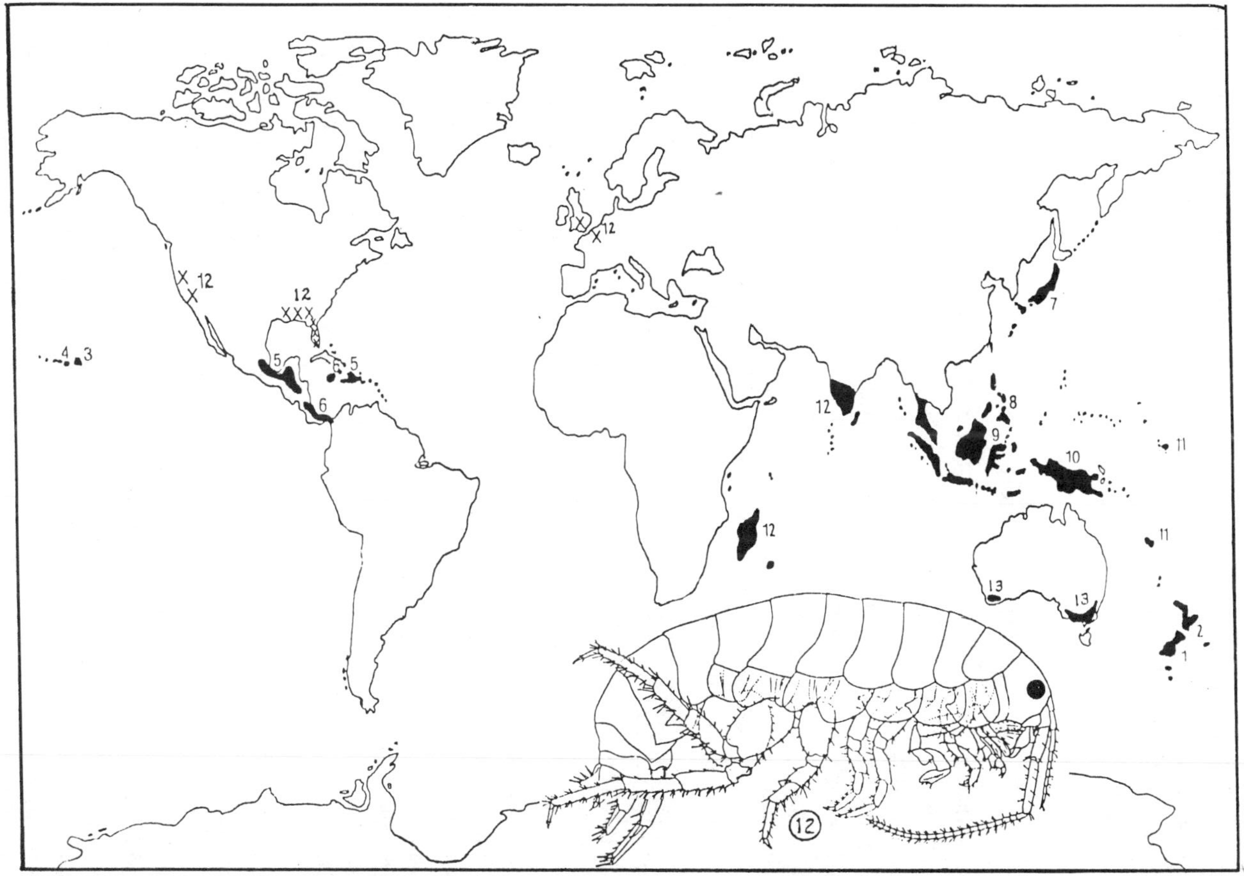

Fig. 7. World distribution of landhoppers, cuspidactylate group. Genera: 1=*Parorchestia.* 2="Orchestia" Gps (New Zealand). 3=*Hawaiorchestia.* 4=*Spelaeorchestia.* 5=*Caribitroides.* 6="Parorchestia." 7="Orchestia" Gps (Japan). 8="Parorchestia" (Philippines). 9="Parorchestia" (Borneo). 10="Parorchestia" (New Guinea). 11="Parorchestia" (Indo-Pacific high islands). 12=*Talitroides.* 13=*Agilestia.*

sexually dimorphic) genera, and some apparently intermediate types (e.g. *Spelaeorchestia*) whose phyletic relationships are yet unstudied and obscure.

Although cuspidactylate landhoppers are relatively successful synanthropically (Biernbaum, 1980), they are primarily tropical and warm temperate types that have not penetrated temperate coastal environments of the North Pacific. As noted above (p. 184, see also Fig. 9) (Group 7), the terrestrial amphipods of Japan are relatively plesiomorphic and presumably derived from local 5-dentate beach-flea types.

PALAEOHISTORY OF THE TALITRIDAE

Construction of a plausible scenario for the origin and evolution of terrestrial amphipods might assist in understanding the anomalous and disjunct world distributions of component talitrid subgroups. The lack of fossil evidence within the Talitridae and limited fossil record of amphipods in general (Bousfield, 1982b,c) renders this task essentially hypothetical. However, important clues bearing on the probable minimum age of animal groups lacking a fossil record are provided by (1) the geochronology of "drifting" continents into which species groups of low vagility and low eurytopicity are presumably "time-locked"; (2) the fossil record of organisms with which the amphipods are intimately and presumably life-dependently associated; and (3) comparative plesio-apomorphic morphology, behavior, and ecology of component amphipod generic groups.

Development of such a scenario is based on the following premises for which evidence is only partially established (e.g. in Hurley, 1968; Bousfield, 1968, 1982c; Friend, 1980). First, terrestrial talitrids developed only along coastal continental regions and on high islands, with prevailing temperate-to-tropical onshore winds; these situations presumably met with the animals' requirements for moist and equable climatic conditions and (intially at least) for significant Na^{+} and Cl^{-} ions in the rain and substrata. Second, landhoppers have never penetrated continental interiors where such combinations of favorable climatic and edaphic conditions are (and were) lacking. Third, angiosperm forest environments were able to provide suitable renewable substrata and food resources for these cryptozoic detritivores.

Palaeohistorically, the most recent development of extensive continental islands and oceanic archipelagos coincided with the evolution and diversification of angiosperm forests, mainly during the late Mesozoic and early Cenozoic Eras. In the early Mesozoic, continental coastlines were presumably contiguous and characterized by relatively few embayments or offshore archipelagos (Howarth, 1981). Coastal forests were then mainly cycads and other gymnosperms, ferns, and other primitive vascular plants (Dilcher and Taylor, 1980). Food resources from these trees were presumably exploited by insects, myriapods, isopods, and other cryptozoic arthropods. Fossil records indicate that these invertebrate groups were widely distributed over the continents in both hemispheres, variously since the Carboniferous (Hessler, 1969; Kukalova-Peck, 1973).

With respect to the pertinent record of continental drift and plate tectonics, we might first consider the distribution of continental masses during the Palaeozoic/Mesozoic transition, i.e. during Permo-Jurassic times, about 250-180 m.y.b.p. At

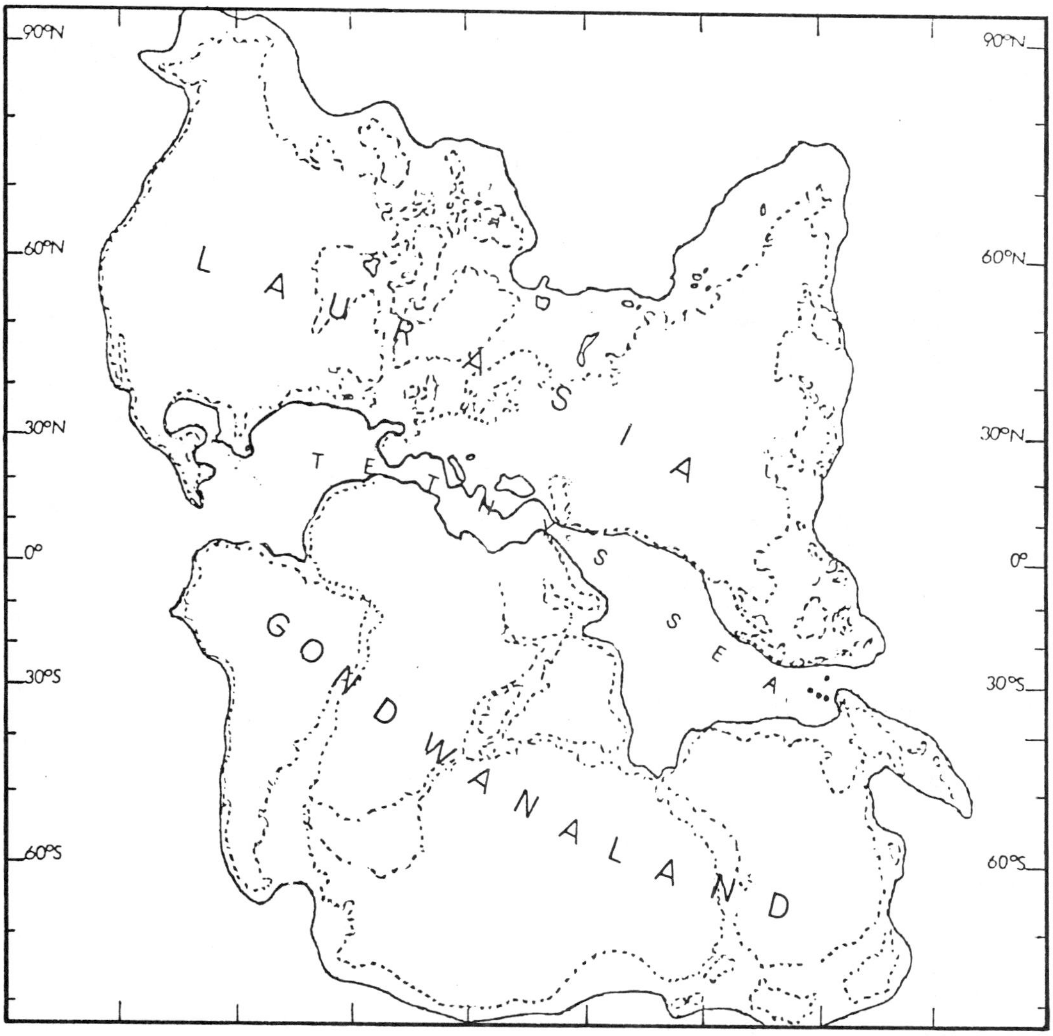

Fig. 8. Distribution of continents, middle Cretaceous, 110 (±10) M.Y.B.P., Laurasia, and Gondwanaland.

that time, land masses of the world were merged about the equator in one large continental area, Pangea, but deeply dissected in the east-central region by a large marine embayment, the Tethys Sea (Howarth, 1981). By late Jurassic times, Pangea had pulled apart more or less equatorially into two smaller but very large supercontinents. These consisted of a northern Laurasia that comprised the loosely connected North American and Eurasian plates, and a southern Gondwanaland that combined the present South American, African, Madagascar, Indian, New Guinea, Australian, Antarctic, New Zealand, Falkland, and other lesser plates (Fig. 8). The two supercontinents were then separated in the west by a relatively narrow, tropical and warm temperate, and presumably somewhat brackish, epicontinental extension of the Tethys Sea (Howarth, 1981).

Direct landward penetration by talitrids would presumably have been facilitated along coastlines of high rainfall, blowing in from the sea, or indirectly via surface streams also containing significant content of Na^+ and Cl^- ions (premises above). The former route is postulated for beachflea progenitors, and the latter appears more probable for estuarine palustral ancestors. Upriver penetration would seem more readily achieved in the absence of continent-endemic freshwater hyalellid competitors (e.g. in New Caledonia, northern New Zealand, and on volcanic islands such as the Hawaiian Archipelago, but not in South America where hyalellids are the dominant freshwater amphipod group).

During formation of the Indian Ocean (Cretaceous Period), diversification of flowering plants continued rapidly on the adjacent continents. Both monocotyledonous and dicotyledonous trees proliferated well into continental interiors, and rhizophoroid mangrove swamps developed throughout the coastal tropics (Retallack and Dilcher, 1981). The earliest true terrestrial amphipods apparently evolved from simplidactylate palustral ancestors in high-rainfall coastal regions of Gondwanaland. Even if originally contiguous, these populations would soon have become isolated from one another by continuing continental separations, first between the Africa-India-Madagascar and Australia-Antarctica-South America plates, and shortly thereafter between Africa and India-Madagascar, and between South America, Australia-Antarctica, and New Zealand-New Caledonia (Fig. 9).

By Palaeocene times of the early Tertiary (ca. 60 m.y.b.p.), most of the present continental masses, now richly vegetated with "modern" angiosperm groups, had become separated by broad seas. Despite sporadically intervening volcanic islands, these seas formed significant barriers to dispersal of most terrestrial and freshwater biotas (Fig. 9), but less so to estuarine and intertidal species that retained some capability for marine dispersal. Although Australia and Antarctica remained fused, New Zealand and New Caledonia had become isolated, and India and Madagascar were separating. As India drifted northwards, on a collision course with Eurasia, it spun off in its wake a few small continental islands (e.g. Seychelles), each supporting a microsample of the original continental biota. The cuspidactylate apomorphic landhopper genus *Talitroides* may be representative of this fauna. Widening of the Atlantic Ocean had by then broadly separated North America from Eurasia, but land connections in the Beringian region were established soon after, during the late Eocene and Oligocene periods (Adams, 1981).

The early Tertiary was marked by increasing "modernization" of dicotyledonous and monocotyledonous taxa (including mangroves) that were evolving in synchrony and in specialized endemicity on the various continental masses. The extent of regional floral diversity was roughly proportional to the size of contiguous continental areas (Raven and Axelrod, 1981) and to nearness of other sources of readily dispersed and recently evolved vicariants. Ecological replacement and/or restriction of conifers, ginkgos, and other primitive vascular plant groups continued, especially in continental interiors. Correspondingly, the tropical and temperate coastal areas presumably became increasingly ecologically and nutritionally suitable for penetration by terrestrially adapted shore-dwelling and estuarine talitroids.

However, the proportion of primitive angiosperm taxa remained high on rel-

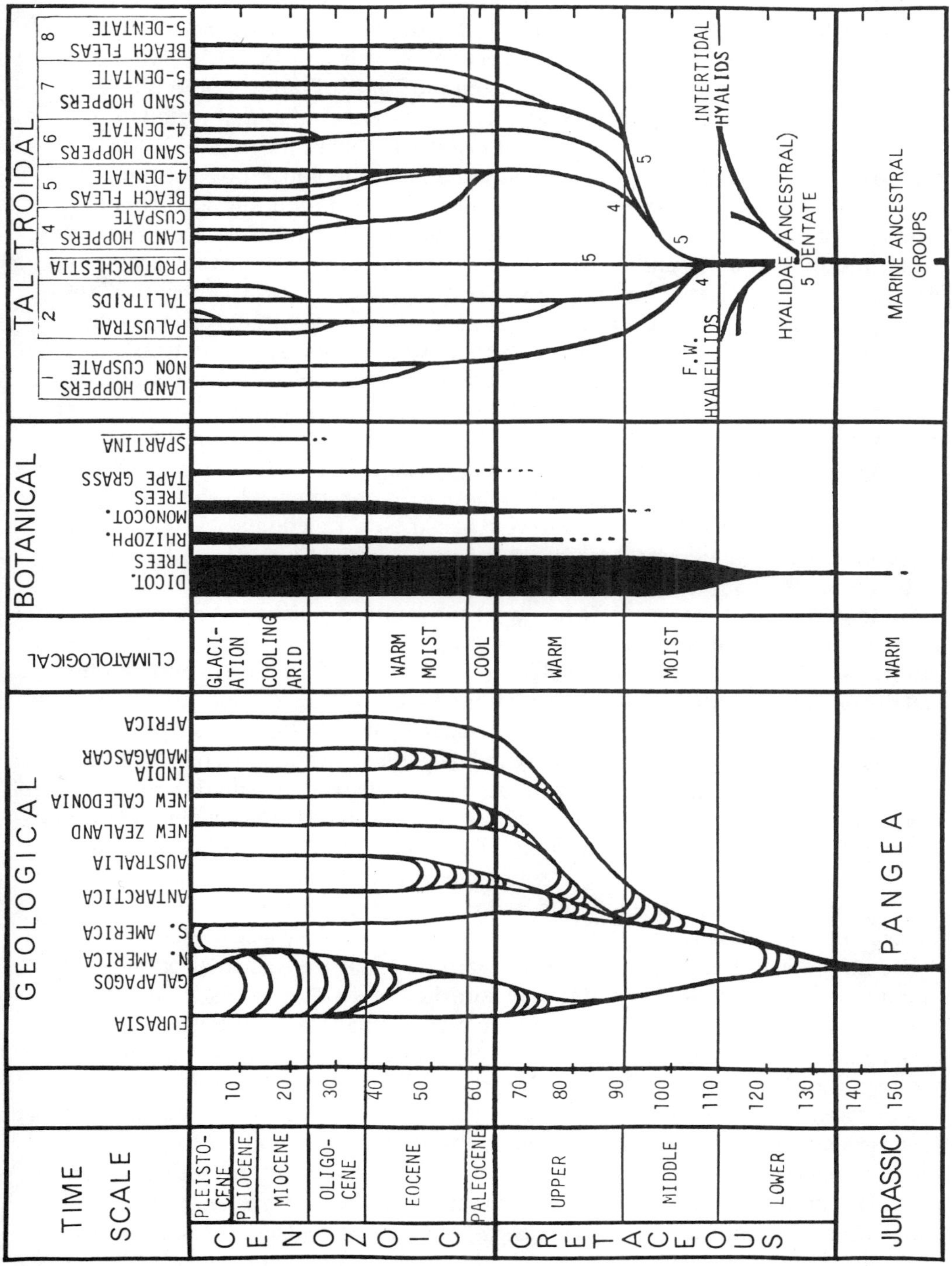

Fig. 9. Principal continental, phytological, and hypothetical talitroideal events of the Mesozoic and Cenozoic Eras.

atively small continental islands (e.g. New Caledonia, New Zealand) that had been isolated early. These primitive floras (and corresponding faunas) persisted relatively unchanged throughout the Tertiary, largely because of the stable, highly equable oceanic climates of the islands (Raven and Axelrod, 1981). Also, widening marine dispersal barriers shielded them from the influx of more apomorphic and eurytopic biotas that evolved subsequently in adjacent continental regions.

Thus, also by the early Tertiary, the four principal ecological groups of Talitridae had diversified (Fig. 9). Primitive palustral talitrids had apparently become dispersed throughout Caribbean Tethyan, eastern Tethyan, and antipodean continental estuaries and mangrove swamps, from South America to the New Guinea shelf, and in freshwater and leaf-litter environments of New Caledonia. Simplidactylate, fully terrestrial landhoppers had evolved from them earlier (by late Cretaceous), but apparently only in Gondwanaland regions. Almost certainly they once occurred in Antarctica, and possibly more widely in the subantarctic islands, but they were subsequently eliminated there by deteriorating climatic conditions and glaciation in the late Tertiary and Holocene. Simplidactylate landhoppers proliferated throughout the New Guinea shelf [e.g. *Brevitalitrus* spp. (Bousfield, 1971)] although records from some outlying volcanic islands (Samoa, Carolines, etc.) may also represent artifacts of Polynesian colonization. However, simplidactylate landhoppers seem not to have reached New Caledonia, New Zealand, or South America, evidence that suggests a late Cretaceous (70-80 m.y.b.p.) time of origin for them.

By mid-Jurassic time (ca. 150 m.y.b.p.) the benthic marine ancestors of terrestrial amphipods, possibly similar to extant genera within the nearly worldwide family Hyalidae, had probably evolved (Bousfield, 1982b). These benthic and saltatory hyalids had occupied intertidal habitats, utilizing as food mainly particulate organic material (and *aufwuchs*) of macroalgae. The algal divisions Phaeophycophyta (browns), Chlorophycophyta (greens), and Rhodophycophyta (reds) extend back in the fossil record variously to the Devonian and Precambrian (Taylor, 1981) and are assumed to have been well developed along cool temperate coasts of the mid-Mesozoic Era (Bousfield, 1981b, 1982c). The most plesiomorphic known hyalid genera (e.g. *Parallorchestes* and *Neobule*) today occur mostly in open coastal macroalgal habitats, and these genera are restricted to the North Pacific and southeastern Australian coasts respectively (Bousfield, 1982c).

We may now attempt a chronological correlation of the principal palaeogeographic, phytological, and hypothetical talitroidal events of the later Mesozoic and the Cenozoic Eras (Fig. 9). During the early Cretaceous, the ferns, treeferns, and gymnosperms were being replaced by early woody angiosperms, especially in the region of West Gondwanaland (Raven and Axelrod, 1981). By middle Cretaceous and coincident with widening of the Tethys Sea between Laurasia and Gondwanaland, primitive dicotyledonous forests were evolving in tropical and warm temperate coastal lowland regions, especially in the eastern Tethys region (Retallack and Dilcher, 1981). Coastal fern mangroves (e.g. *Weichselia*) and freshwater coastal swamp ferns (e.g. *Tempskya*) were being replaced by angiosperm types that competed more effectively for the same coastal depositional environments. The leaf litter of such deciduous trees, serving as an abundant and frequently renewed primary talitrid food resource (Friend, 1980), could then have stimulated evolution of primitive terrestrial detritivores from palustral and beachflea progenitors (cf. *Protorchestia* and *Eorchestia*) along rocky and estuarine shores of those regions.

Undoubtedly pertinent to this palaeohistorical scenario is the fact that the more primitive (5-dentate and simplidactylate) and presumably more ancient aquatic Hyalellidae still occur in temperate fresh and brackish waters of most southern continental areas, but they are apparently absent from the more tropical New Caledonia and India-Madagascar plates (Bousfield, 1980, 1982a). Also, hyalellids appear to have reached the Nearctic region only recently, following establishment of late-Tertiary land connections between South America and North America (Adams, 1981), but they remain totally absent from Eurasia (Bousfield, 1981a).

By the mid-Tertiary, all the southern continents had become isolated. The separation of Australia from Antarctica, including the Tasman shelf, was completed by about 45 m.y.b.p. However, the northward drift of Africa had closed the middle Tethyan Sea and, by early Miocene time, the continent was reunited with Eurasia via Spain and Arabia. The closure cut off relatively small and landlocked seas (the Mediterranean and Paratethyan) in the center, but opened the large Caribbean and central Atlantic basins to the west, and the much expanded Indian Ocean to the east and south (Adams, 1981). Both of these large oceanic regions became centers of landhopper evolution.

Terrestrial talitrids that proliferated throughout the Tethyan and post-Tethyan coastal rain forests are morphological derivatives of the 4-dentate cuspidactylate beachflea group (*Orchestia* spp.) that evolved and diversified around the shores of the newly formed Atlantic Ocean. The relatively primitive shore-dwelling *Tethorchestia* subgroup apparently gave rise to the more advanced, non-sexually-dimorphic species of the *Caribitroides* landhopper group. These became fully terrestrial in high rain forests of continental islands of the Mexican-Antillean Miogeosyncline (see Dengo, 1973), possibly as early as the late Cretaceous but more probably during the early and middle Tertiary. Absence of landhoppers from some of the Greater Antilles (e.g. Puerto Rico and Cuba), despite apparent ecological and nutritive suitability of habitat (Howard, 1973), is enigmatic.

Relatively primitive terrestrial derivatives of Tethyan beachflea progenitors, probably evolving during late Cretaceous, include marine-dispersed species of *Floresorchestia,* the more apomorphic cuspidactylate landhopper complexes such as *Talitroides* (Madagascar and southern India) and the unnamed generic groups of Borneo, Sumatra, the Philippines, and continental high islands of the East Indies and western New Guinea regions (Table 1; Appendix, p. 204).

Proliferation of more advanced palustral talitrids (*Uhlorchestia* and *Chelorchestia*) apparently coincided with the evolution of *Spartina* salt marshes in the Caribbean and western Atlantic regions during the middle to late Tertiary (Jarzen, pers. comm.) (Fig. 1). Although these groups are endemic to the American Atlantic and Caribbean regions, at least one species of *Chelorchestia* had reached the African coast and two species (one terrestrial) had colonized the Galapagos (Table 2). In southern Gondwana regions, weakly cuspidactylate palustral progenitors also gave rise to virtually the entire landhopper fauna of New Zealand (apparently subsequent to its separation from Australia and New Caledonia) and to faunas of smaller continental islands (e.g. Kermadecs) (Table 5; Appendix p. 201).

Terrestrial amphipods colonized ecologically suitable volcanic islands of the Indo-Pacific region probably continuously throughout the Tertiary. All species appear to be direct descendents of apomorphic 4-dentate beachflea progenitors that retained capability for long-distance marine dispersal. Thus, *Floresorchestia* and various "Parorchestia" types colonized the Hawaiian Chain, and *Transorchestia* proliferated on Juan Fernandez. Perhaps significantly in this respect the nearly cosmopolitan-tropical and warm-temperate 5-dentate beachflea species *Platorchestia platensis* (and variants) apparently penetrated continental rain forests only in the Northwest Pacific (archipelago of Japan), a region apparently devoid of more apomorphic 4-dentate beachflea types.

In contrast to antiboreal coastal continental areas and islands of the southern hemisphere, boreal counterparts of the northern hemisphere apparently played only a minor role in landhopper evolution. Reconstruction of terrestrial amphipod faunas of changing coastlines and island areas of the North Pacific, from late Jurassic to early Tertiary (see Howarth, 1981), is highly problematical. Even if such faunas were formerly extensive, only vestiges remain (e.g. in Japan, but none in western Alaska). However, these very plesiomorphic 5-dentate generic types (e.g. *Paciforchestia* group) may be of considerable antiquity, perhaps dating from the late Cretaceous when parts of the Japanese Archipelago first appeared (Howarth, 1981). Their survival on these northern continental islands may be attributed to much the same factors that ensure the long-term stability of primitive biotas of New Zealand and New Caledonia in the southern hemisphere (above).

Elsewhere in the boreal northern hemisphere, the marginally terrestrial species *Orchestia cavimana* of western Europe and the primitive sexually dimorphic landhoppers of the volcanic high islands of the Azores, Madeiras, and Canaries may represent the sole (remaining) impact of Tethyan 4-dentate beachflea progenitors on landhopper evolution in the temperate North Atlantic region.

The 4-dentate terrestrial amphipod groups of cool temperate (antiboreal) regions of the southern hemisphere tend to be very ecologically restricted and exhibit low vagility (Friend, 1980) and presumably low marine dispersal capability. They have apparently not been able to cross the dispersal barrier to the northern hemisphere imposed by high surface temperatures of the tropical oceans. Tropical cuspidactylate apomorphic (not sexually dimorphic) landhoppers are relatively eurytopic and "continent-hardy," a conclusion warranted by their successful synanthropic introduction into suitable northern continental regions such as the U. S. Gulf Coast (Richardson, 1980; Biernbaum, 1980). Likewise, however, their low natural marine dispersal capability (e.g. absence from nearby volcanic high islands) may have prevented their spread into temperate northern oceanic regions. The nearly total replacement of plesiomorphic native Hawaiian landhoppers (to altitudes of about 1000 m) by the introduced *Talitroides topitotum* (Bousfield and Howarth, 1976) mirrors the probable fate of primitive endemic Japanese landhoppers had those islands been invaded, in the past, by more specialized 4-dentate Tethyan counterparts.

SUMMARY AND CONCLUSIONS

The foregoing considerations of comparative morphology, biogeography, and palaeohistory lead us to the following conclusions concerning the probable origin and evolution of landhoppers of the Indo-Pacific region.

(1) The talitroidean family Talitridae, comprising the terrestrial and semi-terrestrial amphipods of the world, encompasses four systematic-ecological (polyphyletic) units, viz. (a) the morphologically primitive palustral talitrids, mainly semi-aquatic in marine, estuarine, and freshwater habitats of tropical and antipodean continental regions; (b) the morphologically more advanced beachfleas, mostly semi-terrestrial and non-substrate-modifying on rocky or sedimentary, tropical to temperate, marine shores, but often terrestrial in coastal rain forests; (c) the highly apomorphic sandhoppers, fossorial (substrate-modifying) and semi-terrestrial on sand beaches of tropical and temperate marine shores; and (d) the apomorphic landhoppers, fully terrestrial but rarely substrate-modifying, in coastal continental and high-island rain forests, mainly in tropical, Indo-Pacific, and antipodean temperate regions of the world.

(2) Landhoppers may be subdivided into several natural phyletic groups, based on the plesio-apomorphic status of the mandibular left lacinia mobilis, the dactyls of the peraeopods (walking legs), and other newly discovered or more precisely redefined taxonomic characters (e.g. Bousfield, 1982c, in prep.).

(3) In attempting to explain the present anomalous and disjunct world distribution of landhoppers, a plausible scenario for the origin and evolution of the Talitridae is proposed. Without a talitrid fossil record, the ideas presented are based on indirect evidence provided by the geochronology of drifting continents on which endemic talitrid populations are "time-locked"; the fossil record and palaeohistory of angiosperm forests to which the landhoppers are presumably nutritively and ecologically bound; and the comparative morphology, dispersal behavior, and ecology of component species groups within the Talitridae.

(4) Talitrid amphipods evolved from primarily aquatic ancestors, allied to primitive genera of the marine-intertidal Hyalidae. Stimulus was provided by the drifting apart of the early Mesozoic supercontinent Pangea into several smaller continents that greatly increased coastal regions of moist, stable, oceanic climates and by contemporaneous evolution of angiosperm forests that provided continuously renewable leaf-litter food resources, mainly during the Cretaceous and early Cenozoic Periods.

(5) Terrestrial amphipods are of two basic phyletic types: (a) simplidactylate landhoppers derived from estuarine palustral talitrid ancestors in southern (Gondwana) continental regions only, and (b) more apomorphic cuspidactylate landhoppers derived mainly from 4-dentate beachflea progenitors, along tropical and warm temperate Tethyan and post-Tethyan shorelines of both hemispheres.

(6) The two phyletic landhopper groups (above) encompass plesiomorphic (sexually dimorphic) and apomorphic (not sexually dimorphic) genera; the latter were apparently derived independently and convergently within each phyletic group as a result of neotenic loss of ancestral "amplexing" male gnathopod 2.

(7) In general, the plesiomorphic landhoppers tend to be more ecologically restricted on land, occurring near the coast and/or in areas of very heavy precipitation, but retain greater marine dispersal capability and thus readily populate rain forests of geologically ephemeral and isolated volcanic islands. The apomorphic (not sexually dimorphic) landhoppers tend to be more eurytopic and penetrate into significantly drier and colder coastal continental regions, but have virtually lost marine dispersal capability and rarely occur on volcanic islands.

(8) The main evolutionary thrust and diversification of terrestrial amphipods occurred in tropical Tethyan and temperate antipodean "Gondwana" continental

regions. Native landhoppers of the temperate northern hemisphere (e.g. *Pacifor-chestia* of Japan) are today relict and isolated, and morphologically advanced little beyond their primitive, presumed beachflea progenitors of nearby seashores.

ACKNOWLEDGMENTS

The writer is grateful to scientific colleagues for stimulating exchanges of ideas, loans of specimens, and supplying pertinent references that contributed significantly over the years to the development of concepts herein. Especially helpful have been T. E. Bowman and J. L. Barnard, National Museum of Natural History, Washington; R. J. Lincoln, British Museum (Natural History); Dr. E. Hurley, DSIR, Wellington, New Zealand; F. G. Howarth, Bishop Museum, Honolulu; A. J. Friend and A. M .M. Richardson, University of Tasmania, Hobart; Peter Raven, Missouri Botanical Gardens; F. R. Schram, San Diego Natural History Museum; and National Museum of Natural Sciences (Ottawa) colleagues D. M. Jarzen, and J. J. Dickinson. Illustrations have been prepared with the assistance of National Museum of Natural Sciences (Ottawa) zoological illustrators C. Douglas and John Crosby; Susan Bourque, Ottawa; and Floy E. Zittin, Vancouver.

LITERATURE CITED

Adams, C. G. 1981. An outline of Tertiary palaeogeography. p. 221–36, *In:* Greenwood, P. H., ed., *The evolving earth.* Part V. Br. Mus. Nat. Hist., Cambridge Univ. Press.

Barnard, K. H. 1940. Contributions to the crustacean fauna of South Africa. XII. *Ann. S. Afr. Mus.* 32(5), No. 18: 381–543.

Barnard, J. L. 1955. Gammaridean Amphipoda (Crustacea) in the collections of the Bishop Museum. *Bull. Bernice P. Bishop Mus.* 215: 1–46.

———. 1969. The families and genera of marine gammaridean Amphipoda. *U.S. Natl. Mus. Bull.* 271: 1–535.

Biernbaum, C. K. 1980. Occurrence of the "tramp" terrestrial amphipods *Talitroides alluaudi* (Chevreux) and *T. topitotum* (Burt) (Amphipoda:Talitridae) in South Carolina. *Brimleyana* 3: 107–11.

Bousfield, E. L. 1964. Talitrid amphipod crustaceans. p. 45–57 *In:* J. L. Gressitt, ed., Insects of Campbell Island. *Pac. Insects Monogr.* 7.

———. 1968. Transition to land in amphipod crustaceans. Discussion. Symposium on Terrestrial Adaptations in Crustacea, AAAS, New York. *Am. Zool.* 8: 393–98.

———. 1971. Amphipoda of the Bismarck Archipelago and adjacent Indo-Pacific islands (Crustacea). *Steenstrupia* 1: 255–93.

———. 1979. A revised classification and phylogeny of amphipod crustaceans. *Trans. R. Soc. Can.*, 1978, Ser. IV, 16: 343–90.

———. 1980. Studies on freshwater amphipods of New Zealand and Tasmania. *International Conference on Biology and Evolution of Crustacea.* Sydney, Australia, May, 1980. (Abstract.)

———. 1981a. Biogeographic indicators within the higher Crustacea of North America. Symposium on biogeography of North America. *Annu. Meet. Can. Soc. Zool.,* Waterloo, Ont., May, 1981. (Abstract.)

———. 1981b. Evolution in North Pacific coastal marine amphipod crustaceans. p. 69–89 *In:* Scudder, G. G. E. & J. L. Reveal, eds., *Evolution today.* Proc. Int. Congr. Syst. Evol. Biol.

———. 1982a. Amphipoda: Gammaridea. p. 254–85 *In:* Parker, S. P., ed., *Synopsis and classification of living organisms.* Vol. 2. McGraw-Hill, New York.

———. 1982b. A palaeohistory of the Amphipoda. p. 96–100 *In: Yearbook of science and technology for 1982-1983.* Vol. 5. McGraw-Hill, New York.

———. 1982c. The amphipod superfamily Talitroidea in the northeastern Pacific Region.

I. Family Talitridae: systematics and distributional ecology. *Natl. Mus. Nat. Sci., Ottawa, Publ. Biol. Oceanogr.* No. 11. 75 p.

———. 1983. An updated classification and palaeohistory of the Amphipoda. p. 257–78 *In:* Schram, F. R., ed., *Crustacean issues, 1: Crustacean phylogeny.* Balkema, Rotterdam.

Bousfield, E. L. & E. Hendrycks. 1985. The amphipod superfamily Talitroidea in the Northeastern Pacific Region. II. Family Hyalidae: systematics and distributional ecology. *Natl. Mus. Nat. Sci.*, Ottawa, Publ. Nat. Sci. No. 7. (In press.)

Bousfield, E. L. & F. G. Howarth. 1976. The cavernicolous fauna of Hawaiian lava tubes. 8. Terrestrial Amphipoda (Talitridae), including a new genus and species with notes on its biology. *Pac. Insects* 17(1): 144–54.

Bowman, T. E. 1977. *Orchestia vaggala,* a new land-hopper from the Galapagos Islands (Crustacea: Amphipoda: Talitridae). *Proc. Biol. Soc. Wash.* 90(3): 658–68.

Burt, D. R. R. 1934. On the amphipod genus *Talitrus,* with a description of a new species from Ceylon, *Talitrus (Talitropsis) topitotum* subgen. et sp. nov. *Ceylon J. Sci., Sect. B. Zool.,* 18(2): 181–91.

Chevreux, E. 1915. Amphipodes de la Nouvelle-Calédonie et des îles Loyalty. *In:* Sarasin, F. & J. Roux, eds., *Nova Caledonia, Zoologie II.* L. 1, No. 1. 14 p., 3 figs.

Chilton, C. 1920. On a small collection of Amphipoda from Juan Fernandez and Easter Island. p. 82–92 *In:*, Skottsberg, C., ed., Stockholm, Sweden. 3.

Colless, D. H. 1970. The phenogram as an estimate of phylogeny. *Syst. Zool.* 19(4): 352–62.

Dana, J. D. 1853. On the geographical distribution of Crustacea. p. 689–1018 *In: United States exploring expedition during the years 1838-1842, under the command of Charles Wilkes, USN.* Vol. 14, No. 2.

Dengo, G. 1973. *Estructura geologica, historia tectonica y morfologia de America Central.* 2nd ed. Centro Regional de Ayuda Tecnica, A.I.D. Mexico. 52 p.

Dilcher, D. L. & T. N. Taylor. 1980. *Biostratigraphy of fossil plants.* Academic Press, New York. xii, 260 p., illus.

Friend, A. J. 1980. The taxonomy, zoogeography, and aspects of the ecology of the terrestrial amphipods (Amphipoda: Talitridae) of Tasmania. PhD Thesis, Dept. Zool., Univ. Tasmania, Hobart. ca. 350 p., illus.

———. 1981. New terrestrial amphipods (Amphipoda: Talitridae) from Australian forests. *Aust. J. Zool.,* 1982, 30: 461–81.

Gomez-Pompa, A. 1973. Ecology of the vegetation of Veracruz. p. 73–148 *In:* Graham, A., ed., *Vegetation and vegetational history of northern Latin America.* Elsevier Sci. Publ. Co., Amsterdam.

Haswell, W. A. 1880. On the Australian amphipods. *Proc. Linn. Soc. N.S.W.* 4: 246–49, pl. VIII, fig. 1.

Hessler, R. R. 1969. Peracarida. *In:* Moore, R. C., ed., *Treatise on invertebrate paleontology, Part R, Arthropoda 4,* Vol. 1. Geol. Soc. Am. Inc., and Univ. Kansas. 398 p.

Holman, H. 1982. Cuticular microstructures in the amphipod superfamily Talitroidea. *Contrib. Pap., Crustacea.* I. Dallas, Texas, 28 December 1981. (Abstract.)

Howard, R. A. 1973. The vegetation of the Antilles. p. 1–38 *In:* Graham, A., ed., *Vegetation and vegetational history of northern Latin America.* Elsevier Sci. Publ. Co., Amsterdam.

Howarth, M. K. 1981. Palaeogeography of the Mesozoic. p. 197–220 *In:* Greenwood, P. H., ed., *The evolving earth.* Part V. Br. Mus. Nat. Hist., Cambridge Univ. Press.

Hurley, D. E. 1955. Studies on the New Zealand amphipodan fauna. No. 8. Terrestrial amphipods of the genus *Talitrus* Latr. *Pac. Sci.* 9: 144–57.

———. 1957. Terrestrial and littoral amphipods of the genus *Orchestia* family Talitridae. *Trans. R. Soc. N. Z.* 85(1): 149–99.

———. 1968. Transition from water to land in amphipod crustaceans. *Am. Zoologist,* 8: 327–53.

———. 1975. A possible subdivision of the terrestrial genus *Talitrus* (Crustacea Amphipoda: family Talitridae). *N.Z. Oceanogr. Trust.* Vol. 2 (14): 157–70.

Karaman, G. 1970. XXIX. Beitrag Kenntnis de Amphipoden–Genus *Orchestia* (Talitridae) in Adriatischem Meer. *Glas. Repub. Zavoda Prir.–Prir. Mus. Titogradu* 3(970): 5–36.

Kukalova-Peck, J. 1973. A phylogenetic tree of the animal kingdom (including orders and higher categories). *Natl. Mus. Nat. Sci., Ottawa, Publ. Zool.* No. 8. 78 p.
MacIntyre, R. J. 1963. The supra-littoral fringe of New Zealand sand beaches. *Trans. R. Soc. N.Z.* 88(4): 89–103.
McCoy, E. D. & K. L. Heck, Jr. 1976. Biogeography of corals, seagrasses, and mangroves: an alternative to the center of origin concept. *Syst. Zool.* 25(3): 201–10.
Methuen, P. A. 1913. Description of an amphipod belonging to the family Talitridae, from the Woodbush, Transvaal. *Proc. Zool. Soc. Lond.* 1913: 109–12, pls. 10, 11.
Raven, P. H. & D. I. Axelrod. 1981. Angiosperm biogeography and past continental movements. *Ann. Missouri Bot. Gard.* 61: 539–673.
Retallack, G. & D. L. Dilcher. 1981. A coastal hypothesis for the dispersal and rise to dominance of flowering plants. p. 27–77 *In:* Niklas, K. J., ed., *Palaeobot., Palaeoecol., Evol.* Vol. 2. Praeger Publ., New York.
Richardson, A. M. M. 1980. Notes on the occurrence of *Talitrus dorrieni* Hunt (Crustacea: Amphipoda: Talitridae) in south-west England. *J. Nat. Hist.* 14: 751–57.
Ruffo, S. & G. Vesentini Paiotta. 1972. Etudes hydrobiologiques en Nouvelle Calédonie (Mission 19765 du Premier Institut de Zoologie de l'université de Vienne). *Cah. O.R.S.T.O.M., sér. Hydrobiol.* 6(3-4): 247–60.
Sandell, R. 1978. Some comments on Victorian landhoppers. *Victorian Nat.* 94: 72–73.
Shoemaker, C. R. 1930. Descriptions of two new amphipod crustaceans (Talitridae) from the United States. *J. Wash. Acad. Sci.* 20(6): 107–14.
Sneath, P. H. & R. R. Sokal. 1973. *Numerical taxonomy. The principles and practices of numerical classification.* W. H. Freeman and Co., San Francisco. 573 p.
Stebbing, T. R. R. 1899. Amphipoda from the Copenhagen Museum and other sources. Part II. *Trans. Linn. Soc. Lond. Zool.*, ser. 2, 7(8): 395–432. illus.
———. 1906a. Amphipoda I. Gammaridea. *Das Tierreich,* Lief 21, Berlin. 806 p.
———. 1906b. A new Costa Rican amphipod. *Proc. U.S. Natl. Mus.* 31(1490): 501–04, pl. 9.
Taylor, T. N. 1981. *Palaeobotany. An introduction to fossil plant biology.* McGraw-Hill, New York. xvi, 590 p., illus.
Vader, V. 1972. Terrestrial amphipods collected in greenhouses in the Netherlands. *Zool. Bidr.* 13: 32–36.
Weber, M. 1892. Die Susswasser-Crustacean des Indischen Archipels. *Zool. Ergebn. Reise Niederl. Ostindien* 2: 528–71, illus.

APPENDIX

In order to validate taxonomic names within the main text that are new to science or newly reorganized, the pertinent taxonomic units are herewith diagnosed and/or redefined.

Group I. Palustral talitrids.

Diagnosis: Small to medium, estuarine intertidal and freshwater (rarely terrestrial) talitrids characterized by *Hyale*-like bodies. Eyes often large, nearly dorsally contiguous. Antenna 2 not sexually dimorphic in form. Mandibular left lacinia 4-dentate; maxilliped palp usually distinctly 4-segmented, segments 1–3 broad, segment 2 lacking distinct mediodistal process.

Coxae medium, 2–4 variably cuspate behind. Gnathopods strongly sexually dimorphic; gnathopod 1 distinctly subchelate (less strongly in female), segments 5 and 6 tumescent behind (less in female); gnathopod 2 strongly subchelate or chelate (male), micro-chelate (mitten-like) in female, with basis sublinear. Peraeopods 3–7 basically homopodous, occasionally sexually dimorphic, simplidactylate, dactyls short.

Pleopods slender, rami well developed. Uropods short; uropod 1 peduncle with strong disto-lateral (inter-ramal) spine; outer ramus of uropods 1 and 2 usually lacking marginal spines; uropod 3, ramus usually shorter than peduncle. Telson lobes tapering and separated apically, apical (rarely lateral) spines only.

Coxal gills similar, sac-like, medium large, simple or lobate. Brood plates large, margins with numerous hook-tipped or simple setae. Generic Composition: Five known genera, newly described herewith (see also Table 2). These are divisible into two groups, as follows.

Group A. *Eorchestia*, n. g., *Microrchestia*, n. g., and *Chiltonorchestia*, n. g. Primarily aquatic, in mangrove estuaries and fresh waters. Characterized by antenna 1 elongate (equals peduncle A2); maxilliped palp distinctly 4-segmented; coxal plates medium deep; gnathopod 1 deeply subchelate, posteriorly tumescent (both sexes); peraeopods and uropods short; coxal gills variously lobate; brood plates broad, margin setae long, numerous.

Group B. *Uhlorchestia*, n. g., and *Chelorchestia*, n. g. Primarily semiterrestrial, intertidal, about HW level of *Spartina* salt-marshes and mangrove swamps. Characterized by body elongate, shallow; antenna 1 short; maxilliped palp obscurely 4-segmented; coxae shallow; gnathopod 1 shallowly subchelate (male), weakly subchelate and non-tumescent (female); peraeopods (especially P6 and P7) and uropods elongate; coxal gills simple, sac-like; brood plates medium, margins few-setose.

Remarks: Palustral talitrids exhibit mainly plesiomorphic characters; these reflect their semi-aquatic life style, and their close relationship to presumed aquatic ancestral types. Through the primitive genus *Eorchestia,* the group is closely related to the 5-dentate simplidactylate beachflea genus *Protorchestia* Bousfield, 1982. Superficially, *Eorchestia* appears similar to *Parorchestia* (*tenuis* type) of New Zealand, but the latter is a cuspidactylate beachflea derivative, of advanced gill and brood plate morphology. In most generic characters, *Eorchestia* appears phyletically close to the new Tasmanian genera *"O," "T," and "N"* (of Friend, 1980), the most primitive of the simplidactylate subgroup of landhoppers. However, the landhopper genera differ in having small, sub-linear, simple, apically setose brood plates; reduced and modified pleopods; and terrestrial habitus.

KEY TO GENERA OF PALUSTRAL TALITRIDAE

1. Maxilliped palp distinctly 4-segmented; gnathopod 1 (both sexes) fully subchelate, segments 4, 5 and 6 posteriorly tumescent; telson lobes with apical spines only; coxae 2–4 about as wide as deep; antenna 1 elongate, about equal to peduncle of antenna 2 2

\- Maxilliped palp appearing 3-segmented (obscurely 4-segmented); gnathopod 1 slender subchelate or weakly subchelate, segments 4, 5, and 6 posteriorly tumescent in male only; telson lobes with apical and lateral spines; coxae wider than deep; antenna 2 short, rarely longer than peduncle 4 of antenna 2 4

2. Coxal gills simple, plate-like or sac-like; gnathopod 1 strongly more subchelate in male; brood plates very broad; marginal setae numerous, hook-tipped; .... *Eorchestia*, n. g.

\- Coxal gills lobate, convoluted or processiferous; gnathopod 1 subsimilar in male and female; brood plates subovate; marginal setae not hook-tipped 3

3. Coxal gills singly (occasionally doubly) lobate; gnathopod 2 (male), palm strongly toothed and incised; peraeopods 6 and 7 sexually dimorphic (segments 4 and 5 broadened in male); uropod 2 not sexually dimorphic; uropod 3, ramus very much shorter than peduncle and weakly armed .... *Microrchestia*, n. g.

\- Coxal gills conspicuously trilobate; gnathopod 2 (male) palm evenly convex; peraeopods 6 and 7 not sexually dimorphic; uropod 2 usually sexually dimorphic; uropod 3 ramus about equal to peduncle .... *Chiltonorchestia*, n. g.

4. Gnathopod 1 distinctly subchelate in both sexes; gnathopod 2 subchelate in male; uropod 3 ramus nearly equal in length to peduncle .... *Uhlorchestia*, n. g.

\- Gnathopod 1 fully subchelate in male only (weakly subchelate, dactyl greatly exceeding small palm in female); gnathopod 2 chelate or parachelate in mature males; uropod 3, ramus distinctly shorter than peduncle .... *Chelorchestia*, n. g.

Note: Some primitive simplidactylate landhoppers (e.g., *"O," "T," and "N"* of Friend, 1980) might also be included in the pragmatic "palustral" category but these are separable on their sublinear, apically setose brood plates, variably reduced and modified pleopods, and truly terrestrial habitus.

Eorchestia, **new genus**

Diagnosis: Estuarine, salt-marsh talitrids. Characterized by antenna 2 longer in male; maxilliped palp distinctly 4-segmented, segments 2 and 3 broad; gnathopod 1 deeply subchelate, strongly tumescent behind; gnathopod 2 (male), palm smoothly oblique, posterior angle with prominence and spine; gnathopod 2 (female), basis sublinear; peraeopods 5–7 homopodous, P7 longest; coxa 5 aequilobate; pleopod peduncles with 2 retinacula; uropods 1 and 2, outer ramus marginally bare; uropod 3, ramus with apical spines only; telson lobes with apical spines only; gills large, simple, sac-like; brood plate margins with long, hook-tipped setae.

Type species: *Orchestia rectipalma* K. H. Barnard, 1940

Additional Species: Several undescribed estuarine species from New Zealand and south-eastern Australia (including Tasmania) currently under study (Table 5).

Etymology: Combines the Greek *eos* ("dawn" or "early") and *Orchestia,* together symbolizing the primitive morphology of this talitrid species group.

Microrchestia, **new genus**

Parorchestia: Bousfield 1971, p. 281 (in part)

Diagnosis: Small, semi-aquatic and intertidal talitrids of Indo-Pacific mangrove swamps. Characterized by body relatively short and deep; antenna 1 long (equal to or greater than peduncle A2); antenna 2 slender; maxilliped palp distinctly 4-segmented, segments 2 and 3 medium-broad, weakly spinose; gnathopod 1 (both sexes) fully subchelate, segments 4, 5, and 6 short, tumescent posteriorly; gnathopod 2 (male) large and powerfully subchelate, palm oblique, irregular, with strong hinge tooth, dactyl smoothly curved; gnathopod 2 (female) basis narrowly linear, segment 3 short, 4, 5, and 6 tumescent posteriorly; coxa 5 aequilobate, coxa 6 shallowly posterolobate; peraeopods 3 and 4 closely subequal, simplidactylate; peraeopods 5–7 weakly heteropodous, segments 4 and 5 variously expanded; peraeopods 6 and 7 sexually dimorphic; pleopods strong, peduncles with 3–8 retinacula; uropods 1 and 2 short, outer ramus usually lacking marginal spines; uropod 3, ramus short, weakly armed; telson short, lobes nearly fused and setose (not spinose) apically; coxal gills large, lobate, sac-like; brood plates medium broad, marginal setae simple, medium-long.

Type species: *Parorchestia macrochela* Bousfield, 1971.

Additional species: 8 species, in material at hand and in litt., all from inter-tidal habitats of mangrove swamps in continental regions, from tropical South America (Ecuador) and Africa (Nigeria), Madagascar, through south-eastern India (Porto Novo region) to the Bismarck Archipelago (New Britain, Hermit Is.).

Etymology: Combines the Greek *mikros* and *Orchestia* to denote the relatively small size of component members of this distinctive plesiomorphic talitrid genus.

Chiltonorchestia, **new genus**

Parorchestia: Chevreux 1915, p. 11 (in part)

Orchestia: Ruffo and Paiotta, 1972, p. 253 (in part)

Diagnosis: Small to medium-sized, primarily aquatic talitrids, endemic to freshwaters of New Caledonia. Characterized by antenna 1 long (nearly equal to peduncle of antenna 2); antenna 2, flagellum long; gland cone minute; maxilliped palp strong, distinctly 4-segmented, 4th with slender spine-like unguis; segments 1–3 broad, spinose medially; gnathopod 1 fully subchelate (both sexes), segments 4, 5, and 6 posteriorly tumescent (weakly in female), segment 5 elongate; gnathopod 2 powerfully subchelate (male), of "Chiltonia" (Hyalellidae) form, palm smoothly and obliquely convex, spinose; gnathopod 2 (female), basis sublinear, segments 4 and 5 tumescent behind, distal lobe of propod short, with dactyl appearing parachelate; coxae 2–4 deep, weakly cuspate behind; coxa 5 deep, aequilobate; peraeopods 5–7 short, basically homopodous, dactyls short, cusps lacking (or trace only); pleopods well developed, peduncles each with 2 retinacula; urosome segments 2 and 3 telescoping dorsally into urosome 1; urosome 3 with lateral flanges masking telson; uropods 1 and 2 short; uropod 1, peduncular disto-lateral spine short, outer ramus marginally bare; uropod 2, rami marginally spinose, outer ramus sexually dimorphic; uropod 3, ramus long, apically and marginally spinose; telson lobes, separated slightly, apically spinose; coxal gills lobate, and/or attenuated, with (three) accessory lobes; sternal gills

Table 5. Component genera and species of palustral Talitridae.*

Eorchestia	*Microrchestia*	*Chiltonorchestia*	*Uhlorchestia*	*Chelorchestia*
rectipalma (K. H. Bnd.)	*macrochela* (Bousf.)	*pusilla* (Chevreux)	*uhleri* (Shoemaker)	*costaricana* (Stebbing)
sp. A. [New Zealand]	*similis* (Bousf.)	*sarasini* (Chevreux)	sp. "S" (Bousf. and Heard)	*darwinii* (Müller)
sp. B. [New Zealand]	*notabilis* (K. H. Barnard	*starmühlneri* (Ruffo)		*vaggala* (Bowman)
sp. C. [New Zealand]	sp. A. [Asari]	sp. A. [New Caledonia]		sp. A. [Florida]
sp. D. [Victoria]	sp. B. [India]	sp. B. [New Caledonia]		sp. B. [Florida, east]
sp. E. [Tasmania]	sp. C. [Nigeria]	sp. C. [New Caledonia]		sp. C. [Galapagos]
	sp. D. [Ecuador]			sp. D. [Nigeria]
	sp. E. [Madagascar]			

* All named species are **new combinations.**

lacking; brood plates large, broad, marginal setae numerous, long, simple, or split-tipped.

Type species: *Parorchestia pusilla* Chevreux, 1915.

Additional species: Six species, including *C. pusilla* (Chevreux), *C. sarasini* (Chevreux), *C. starmuhlneri* (Ruffo and Paiotta), and 3 undescribed species in material at hand collected by W. L. Beck, S. B. Peck, and others, all endemic to freshwaters of New Caledonia (Table 5).

Etymology: A combining name that refers to the *Chiltonia*-like appearance of the body and some appendages (maxilliped palp, gnathopod 2 in both sexes, and sexually dimorphic uropod 2) of this distinctive, mainly aquatic, talitrid genus.

Uhlorchestia, new genus

Uhlorchestia: Bousfield and Heard, in prep.

Diagnosis: Slender-bodied, shallow-plated, semiterrestrial talitrids, endemic to salt marshes of the North Atlantic region. Characterized by antenna 1 short (little beyond peduncle 4 of A2); antenna 2 slender, geniculate at flagellum; maxilliped palp short, obscurely 4-segmented, segments 1–3 broadened, medially spinose; gnathopod 1 (male) slender-subchelate, segments 4, 5, and 6 posteriorly tumescent; gnathopod 1 (female) weakly subchelate, dactyl exceeding palm, distal segments not tumescent; gnathopod 2 (male) strongly subchelate, propod typically large, posterior margin short, palm hinge-toothed and oblique, dactyl sinuous; gnathopod 2 (female) minutely chelate, basis slightly expanded antero-proximally, segments 4 and 5 tumescent behind, 6 elongate; coxae 2–4 medium deep and cuspate behind; coxa 5 deep, aequilobate; coxa 6 shallowly posterolobate; peraeopods 3 and 4 closely subequal, simplidactylate; peraeopods 5–7 basically homopodous, longest posteriorly; pleopods slender, peduncles each with 2 retinacula; uropods 1 and 2 short, rami marginally spinose, disto-lateral spine short; uropod 3, ramus medium with apical and marginal spines; telson lobes narrowing distally, armed with apical and dorsolateral spines; coxal gills large, sac-like, not lobate; brood plates medium-broad, marginal setae long, simple.

Type species: *Orchestia uhleri* Shoemaker, 1930.

Additional species: *Uhlorchestia* sp. S., Bousfield and Heard, in prep. (Table 5).

Chelorchestia, new genus

Diagnosis: Small, slender-bodied, semiterrestrial, rarely terrestrial, talitrid amphipods, endemic to salt marshes and mangrove swamps (occasionally coastal terrestrial habitats) of tropical and warm-temperate Atlantic and eastern Pacific regions. Characterized by antenna 1 very short (not exceeding peduncle 4 of A2); antenna 2 slender, geniculate at flagellum; maxilliped palp short, appearing 3-segmented, segments 1–3 medium-broad, weakly medially spinose; gnathopod 1 (male) slender subchelate, segments 4, 5, and 6

Table 6. Component genera and species of 4-dentate beachfleas.

Transorchestia	*Orchestia*	*Tethorchestia*	*Floresorchestia**
chiliensis (M.-E.)	*gammarellus* Pallas	*antillensis*, n. sp.	*floresiana* (Weber)
enigmatica (B. & C.)	*mediterranea* Costa	sp. B. [Florida]	*anomala* (Chevreux)
serrulata (Dana)	*montagui* (Audouin)	sp. C. [Caribbean]	sp. A. [Venezuela]
gracilis (Chilton)	*cavimana* Heller	sp. D. [Caribbean]	*vitilevana* (J. Bnd.)
sp. A. [Australia]	*gambierensis* Chevr.	sp. E. [Oceania]	*samoana* (Bousf.)
miranda (Chilton)	*magnifica* Vecchi	sp. F. [Jamaica]	*anoquesana* (Bousf.)
?*bollonsi* (Chilton)	*remyi* Schell	sp. G. [Jamaica]	*monospina* (Steph.)
Several species (*gracilis* group) [Juan Fernandez Is.]	*r. roffensis* Wildish		*pectenispina* (Bousf.)
	grillus (Bosc)		*ancheidos* (K. Bnd.)
	stephenseni Cecchini		*pickeringi* (Dana)
	microphtalma A. & S.		sp. [Bonia I.]
	ghigii Vecchi		
	kosswigi Ruffo		

* All named species of *Floresorchestia* are **new combinations.**

tumescent posteriorly; gnathopod 1 (female) nearly simple, dactyl greatly exceeding short palm, distal segments not tumescent behind; gnathopod 2 (mature male) strongly chelate or parachelate, propod produced distad at posterior angle into a finger-like process against which closes the sinuous dactyl, segment 5 usually fused with segment 6; gnathopod 2 (female), basis short, sublinear, segment 3 elongate, 4 and 5 posteriorly tumescent; coxae 2–4 shallow, elongate, cuspate behind; coxa 5 anterolobate; coxa 6 shallowly posterolobate; peraeopods 3 and 4 closely similar, simplidactylate; peraeopods 5–7 slender, elongate, weakly heteropodous, dactyls short, with trace of outer marginal cusps; pleopods slender, slightly reduced, peduncles with 2 retinacula; uropods 1 and 2 of medium length, outer ramus of U1 usually without, of U2 usually with, marginal spines; inter-ramal spine prominent; uropod 3, ramus very short or partly fused with peduncle, weakly armed; telson lobes narrowing distally, with apical and dorso-lateral spines; coxal gills medium-large, sac-like, weakly lobate; brood plates subovate, marginal setae simple, medium length, distal.

Type species: *Orchestia costaricana* Stebbing, 1906b

Additional Species: *C. vaggala* (Bowman); and several undescribed species in material at hand from mangrove swamps and *Spartina* marshes, of Florida and the Caribbean region, Brazil, Argentina, Galapagos, Hawaii (possibly synanthropic), and Nigeria (Table 5).

Group II. Beachfleas.

The group has been briefly diagnosed in the text (p.179) and in greater detail elsewhere (Bousfield, 1982c, in prep). Three sub-groupings have been made (Table 1), viz: (a) plesiomorphic, semi-aquatic, 5-dentate, simplidactylate, intertidal species of antipodean continents; (b) more apomorphic, mainly semiterrestrial, 5-dentate, cuspidactylate, intertidal and terrestrial species, of temperate north and south Pacific coastal regions; (c) most apomorphic, semiterrestrial and terrestrial, 4-dentate, cuspidactylate, intertidal and terrestrial species, of Indo-Pacific and Tethyan regions.

Component genera and species of 4-dentate beachfleas are given in Table 6. Two of the genera (*Transorchestia* Bousfield and *Orchestia* Leach) have been diagnosed or redescribed elsewhere (Bousfield, 1982c). *Tethorchestia* and *Floresorchestia* are newly described below. Some members have become terrestrial in coastal rain forests (e.g. *Floresorchestia* spp.) but have retained their plesiomorphic amplexing gnathopod morphology.

Tethorchestia, **new genus**

Diagnosis: Small to medium-sized, mainly strand and lagoon beachflea species. Closely allied to *Orchestia* Leach (as diagnosed by Bousfield, 1982c) but differing in the following characters: eyes large, tending to dorsal contiguity; antenna 2 slender, finely spined, geniculate at flagellum, weakly or not sexually dimorphic; maxilliped palp 4-segmented (4th minute, masked by spines of 3); gnathopod 1 sexually dimorphic (weakly subchelate in female), but segment 4 often tumescent behind (male); gnathopod 2 (male), propod subovate, palm smoothly convex (not indented, sinuous, or toothed), dactyl smoothly curved, attenuated at nail; gnathopod 2 (female), basis expanded antero-proximally (not medially), segments 4, 5, and 6 posteriorly tumescent; peraeopods 3–7 slender, dactyls slender; peraeopods 3 and 4, segments 5 not markedly unequal; coxa 5 strongly anterolobate; peraeopods 5–7 slender, moderately heteropodous; P6 and P7 (male) weakly (or not) incrassate; in P7, distal setation of segment 6 tends to special development [somewhat as in *Traskorchestia ochotensis* (Brandt)]; pleopods (mainly Pl3) tending to reduction; uropod 1, disto-lateral (inter-ramal) peduncular spine usually well developed, outer ramus lacking marginal spines; uropod 3, peduncle deep. Telson short, with apical and dorso-lateral spine groups.

Type species: *Tethorchestia antillensis*, new species.

Additional species: Six other species have been identified in material at hand from coastal localities in Florida, the West Indies, and Caribbean regions (Table 6). Full descriptions of these species are planned elsewhere (Bousfield, in prep., etc.).

Remarks: Character states (within antennae, gnathopod, uropods) of some member species overlap with those of some members of the genus *Orchestia.* Thus, *Tethorchestia* may be linked to *Orchestia* via the N. Atlantic salt marsh and estuarine species *O. grillus* (Bosc) and *O. stephenseni* Cecchini; in balance of character states, however, the latter are clearly assignable to the genus *Orchestia.* Some species of *Tethorchestia* bear a superficial resemblance to the plesiomorphic 5-dentate species *Paciforchestia klawei* Bousf. In the latter genus, however, gnathopod 2 (female) is fully subchelate and tumescent behind; gnathopod 2 basis (female) is sublinear; peraeopods 5–7 are homopodous; pleopod peduncles are multi-retinaculate; and the telson is not dorsally spinose, among other differences.

Tethorchestia antillensis, **new species**

Diagnosis: With the characters of the genus. Antenna 2, flagellum about equal to peduncle (male); gnathopod 2 (male), segment 4 not tumescent behind; peraeopod 7 (male) segments 4 and 5 weakly incrassate, segment 6 armed antero-distally and distally with 4–5 clusters of stiff setae, 3–5 setae (and 1–2 spines) per cluster; pleopods subequal, rami 10-segmented, slightly shorter than peduncles; uropod 1, interramal spine medium, inner ramus lacking inner marginal spines.

Material examined: Ascension Bay, Quintana Roo (Yucatan), Mexico; sand beach inside reef, April 7, 1960, J. G. Clarke and E. L. Bousfield coll. Male 12.0 mm. HOLOTYPE; female (ov.), 11.0 mm. ALLOTYPE; several male, female, and immature PARATYPES; specimens and slide mounts, NMNS collections, Ottawa.

Remarks: The species is closely similar to the common (yet undescribed) "checker back" beachflea of semi-protected shores of Florida and the Gulf Coast of the United States, but differs mainly in the armature of the gnathopods, peraeopod 7, and uropod 1 (Bousfield and Thomas, in prep.).

Floresorchestia, **new genus**

Orchestia floresiana group: Bousfield, 1971

Diagnosis: Small to medium, slender-bodied, mainly terrestrial species of rain forests of tropical and warm-temperate Indo-Pacific coastal regions. Characterized by eyes large, tending to dorsal contiguity; antenna 1 very short; antenna 2 slender, delicately spinose, geniculate, not sexually dimorphic; maxilliped palp appearing 3-segmented (4th fused with 3rd), outer plate not arcuate, subapically spinose; gnathopod 1 sexually dimorphic, fully subchelate and distal segments tumescent behind (male), extremely weakly subchelate (or simple) and segments not tumescent (female); gnathopod 2 (male) powerfully subchelate, propod subovate, palm often sinuous and/or with median notch; dactyl often sinuous and toothed behind, nail attenuated; gnathopod 2 (female), basis sublinear, slightly expanded antero-proximally, distal segments tumescent behind; peraeopods 3 and 4 unequal, cuspidactylate; coxa 5 variably anterolobate; peraeopods 5–7 heteropodous, slender, dactyls slender; P7 (male) not incrassate, posterior margin of basis smoothly convex, lined with small, evenly spaced, beaded knobs which are apparently stroked across the lateral submarginal row of "pits" of adjacent abdominal side plates 2 (and 3, occasionally also 1),

Table 7. Component genera and species of simplidactylate (non-cuspidactylate) landhoppers.

Genus "O" (Tasmania)	Genus "N" (Tasmania)	Genus "T" (Tasmania)	*Protaustrotroides*	*Austrotroides*	*Arcitalitrus*	*Keratroides*
sp. N. (Friend, in prep.) sp. Q. (Friend, in prep.)	sp. P. (Friend, in prep.)	sp. A. (Friend, in prep.)	*victoriae,* n. sp. sp. A. [Grampian Mts., Victoria]	sp. L. (Friend, in prep.) sp. M. (Friend, in prep.) sp. N. (Friend, in prep.) *pectinalis* Friend *occidentalis* Friend *crenatus* Friend	*sylvaticus* (Hasw) *dorrieni* (Hunt) sp. A. [Victoria] sp. B. [Victoria] sp. C. [Mt. Skene, Victoria]	*kershawi* Sayce *angulosus* Friend *vulgaris* Friend sp. A. (Friend, in prep.) sp. B. (Friend, in prep.) sp. C. (Friend, in prep.) sp. D. (Friend, in prep.) sp. E. (Friend, in prep.) sp. F. (Friend, in prep.) sp. G. (Friend, in prep.)
Mysticotalitrus	*Brevitalitrus*	*Talitriator*				
tasmaniae (Ruffo) *crypticus* Friend	*hortulanus* (Bousf.) *stephenseni* Bousf. *wolffi* Bousf. *dyaulanus* Bousf. *toli* (J. L. Barnard) *nesius* (J. L. Barnard) sp. A [Queensland]	eastwoodae Methuen *cylindripes* K. H. Barnard *setosa* K. H. Barnard *calva* K. H. Barnard *macronyx* K. H. Barnard				

Note: New Tasmanian taxa of Friend, in prep., are named in thesis copies (1980) but are not yet formally published.

Table 8. Component genera and species of cuspidactylate landhoppers.

Parorchestia	"Orchestia" [New Zealand]	*Agilestia*	*Hawaiorchestia* [Hawaii]	*Spelaeorchestia*	*Caribitroides* [West Indies]	"Parorchestia" [Central America]
tenuis (Dana) *sylvicola* (T. & C.)	*lesliensis* Hurley *sinbadensis* Hurley *simularis* Hurley *rubroannulata* Hurley *improvisa* Chilton *insularis* Steph. *parva* Chilton	*hyperocha* Friend *hylaea* Friend	*hawaiensis* (Dana)* [numerous undescribed species from Kauai, Oahu]	*koloana* (B. & H.)	*jamaicensis*, n. sp. sp. B [Jamaica] sp. C [Jamaica] sp. D [Haiti] sp. E [E. Mexico] sp. F [Central America]	sp. A [Panama] sp. B [Costa Rica]
"Parorchestia" [misc. regions]		**"Paraorchestia" [New Guinea]**	**"Parorchestia" [Philippines]**	***Talitroides***		
marquesana Steph. *kermadecensis* Chil. *kinabuluensis* Shoe *solifuga* Iwasa *gowerensis* Boust. sp. A [Mauritius] c. 20 additional undescribed spp.	*maynei* Chilton *patersoni* Steph. *stewarti* Stephensen sp. A sp. B	sp. A sp. B sp. C sp. D sp. E sp. F sp. G	*laguanae* Baker *luzonensis* Baker sp. A	*alluaudi* Chevr. *topitotum* Burt	sp. G [Central America] sp. H [Central America] sp. I [Central America] sp. J [Central America] sp. K [Central America]	

*** New combination.**

together forming a primitive stridulating apparatus; pleopods slender, rami normal, multi-segmented; uropod 1, peduncular interramal spine usually developed, outer ramus lacking marginal spines (occasionally armed with "amplexing" spine in male); uropod 3 small, ramus short; telson lobes finely marginally and apically spinose; coxal gills enlarged, weakly convoluted and/or lobate; brood plates sublinear, margins distally simple-setose.

Type species: *Orchestia floresiana* Weber, 1892

Additional Species: Eleven species to date, mostly described, widely occurring throughout the tropical Indo-Pacific and in the Caribbean (Table 6).

Remarks: The genus is highly distinctive, demarcated by a curious mixture of apomorphic character states (of antennae, gnathopods, peraeopods, abdominal side plates, coxal gills and brood plates) and plesiomorphic features (of pleopods, uropods, and telson). These, in combination, presumably enable component species to invade coastal rain forests, yet retain high marine dispersal capability.

Group III. Sandhoppers.

No new taxa are proposed herein.

Group IV. Landhoppers.

Small to medium-large, truly terrestrial (rarely supralittoral) talitrids characterized by "grasshopper"-like bodies and hyaline integument; usually with modified (not sexually dimorphic) life cycle; coxal gills convoluted, lobate, attenuated; peraeopods heteropodous, anterolobate; and pleopods variously reduced or modified. Typically also the eyes are small to medium-large; antenna 2 slender, geniculate at flagellum; buccal mass slightly prognathous; mandibular left lacinia 4-dentate; maxilliped palp appearing 3-segmented, 4th minute or fused to 3rd, segment 2 lacking medio-distal lobe; gnathopod 1 weakly (or not) subchelate, distal segments usually not posteriorly tumescent; gnathopod 2 (female), basis linear, segment 3 variously elongate; uropod 1, prepeduncle moderately developed, peduncle with inter-ramal spine; telson short, lobes fused apically (or nearly so).

Landhoppers are divisible in two groups.

(A) Simplidactylate landhoppers: characterized by similar peraeopods 3 and 4 with similar non-cuspate dactyls (see text, p.180). Generic inclusions provided in Table 7.

Protaustrotroides, **new genus**

Protaustralitroides: Bousfield, ms., 1964

Diagnosis: Body medium-large, deep-plated. Antenna 1 elongate (about equal to peduncle A2), peduncular segment 3 longest. Maxilliped, palp 3-segmented, outer plate not arcuate, sub-apically broadly spinose; gnathopod 1 weakly subchelate, dactyl exceeding palm (female), segments 4, 5, 6, tumescent behind (weakly in female); gnathopod 2 (male) strongly subchelate (or partly reduced neotenically), in female, distal segments slender, tumescent behind; peraeopods 3–7 elongate, 5–7 strongly heteropodous, coxa 5 strongly anterolobate; pleopods vestigial, rami minute or lacking, without setae; uropods 1 and 2, rami subequal, all marginally spinose; uropod 3 ramus distinct, with lateral spines; telson lobes nearly fused, with a few apical and lateral spines; coxal gills variously convoluted, those of P2 and 6 attenuated; brood plates medium broad, marginal setae short.

Type species: *Protaustrotroides victoriae*, n. sp.

Additional species: at least one other species is known from Victoria (Grampians) (see Sandell, 1978, and Table 7).

Remarks: The genus is very close to *Austrotroides* Friend (1981), from Tasmania and western Australia, but differs primarily in the plesiomorphic retention of sexually dimorphic gnathopod 2, and the more spinose rami of uropod 1.

Protaustrotroides victoriae, **n. sp.**

Diagnosis: With the characters of the genus. Uropod 1 (male), outer ramus with spines on inner and outer margins; uropod 2, inner ramus with 1 distal marginal spine; uropod 3, ramus with 1 marginal spine; pleopod 1, rami very short, slender, bare; pleopods 2 and 3 shorter, lacking rami; coxal gill of P6, attenuated apex with 2 ventral notches.

Material Examined: Apollo Bay, Victoria, under leaves, Dec. 25, 1904, G. Coghill coll. Male HOLOTYPE (14.0 mm); Female ov. ALLOTYPE (14.0 mm); 1 male, 6 female

PARATYPES; collections of the National Museum of Victoria, Melbourne; slide mounts, collections of the National Museum of Natural Sciences, Ottawa; additional material from Otway Ranges, Victoria, October 26, 1978, R. Sandell and E. L. Bousfield coll.

Talitriator Methuen, 1913

Remarks: The formae of *T. eastwoodae* Methuen, described and figured by K. H. Barnard, 1940, are herewith elevated to full species status (Table 7). The move accords with the views of Friend (1980) and is based on reexamination of material from East Cape Province, kindly supplied by the late J. Omer-Cooper, South Africa. Full redescription and refiguring of these species is in preparation.

(B) Cuspidactylate landhoppers: characterized by dissimilar peraeopods 3 and 4, with dissimilar, cuspate dactyls (see text, p.180). Generic and generic-group inclusions are provided in Table 8.

Parorchestia Dana, 1853

Remarks: Rediagnosis of this genus, based on the full redescription of the *tenuis*-type species by Hurley (1957), has been discussed by Friend (1980). The full taxonomic range of associated genera, within the New Zealand subcontinent (Table 8), awaits analysis of material at hand, which should match the thorough and perceptive analysis of the terrestrial amphipods of Tasmania by Friend (1980).

Hawaiorchestia, **new genus**

Parorchestia: Stebbing, 1906a; Barnard, 1955 (in part)

Diagnosis: Large, slender-bodied, deep-plated, long-legged landhoppers, endemic to the Hawaiian high islands. Characterized by elongate antenna 1, peduncle 3 longest; maxilliped palp 4-segmented, 2 and 3 not expanded, outer plate arcuate, apex sub-acute; gnathopod 1 subchelate (male), distal segments tumescent behind, weakly subchelate (female), dactyl exceeding palm, distal segments not tumescent; gnathopod 2 sexually dimorphic, powerfully subchelate (male), palm smooth; mitten-like (female), basis slightly expanded proximally; coxa 5 anterolobate; peraeopods 3–7 elongate, cuspidactylate, 6 and 7 slightly sexually dimorphic; pleopods vestigial, lacking rami; uropods 1 and 2, outer ramus marginally smooth; uropod 3 not shortened; telson lobes fused apically, marginally spinose; coxal gills modified, P6 attentuated; brood plates sublinear, marginal setae simple.

Type species: *Orchestia hawaiensis* Dana, 1853.

Additional species: At least two other species are present in material from Oahu and Kauai, collected by F. G. Howarth, Bishop Museum, and colleagues, and the author, during January, 1973, 1974.

Remarks: In most characters, *Hawaiorchestia* bears a remarkably close resemblance to the simplidactylate landhopper genus *Protaustrotroides* from south-eastern Australia. However, the latter has a very different form of the maxilliped, the uropod rami are spinose, and the dactyls are not cuspate. *Spelaeorchestia koloana* B. and H. is probably a neotenic troglobitic derivative of *Hawaiorchestia* (Bousfield and Howarth, 1976).

Caribitroides, **new genus**

Diagnosis: Medium-sized, slender-bodied, long-legged landhoppers of Antillean and Central American high rain forests. Characterized by antenna 1 short, peduncular segments short; maxilliped palp 3-segmented (4th fused), segments 2 and 3 not broad, outer plate variously arcuate; gnathopods not sexually dimorphic; gnathopod 1 simple, lacking palm, segment 6 not shortened, distal segments not tumescent behind; gnathopod 2, basis slightly expanded proximally, segment 3 elongate, distal segments tumescent behind; peraeopods 3 and 4 unequal, dactyls unequal, cuspate; coxa 5 anterolobate; peraeopods 5–7 heteropodous, dactyl not elongate; pleopods slightly reduced, functional, 3rd shortest; peduncles slender, each with 2–5 retinacula; uropod 1, outer ramus lacking marginal spines; uropod 2, rami subequal, both marginally spinose; uropod 3 not shortened, ramus with marginal spine; telson lobes fused, with apical and dorso-lateral spines; coxal gills modified but not attenuated; brood plates sublinear, marginal setae short, simple.

Type species: *Caribitroides jamaicensis*, new species.

Additional species: several undescribed species have been noted in material at hand from high rain forests of Jamaica, Hispaniola, southern Mexico, and Honduras; formal descriptions of these are in preparation.

Remarks: For some time, landhoppers of Jamaica have been known to the author and to colleague E. L. Mills; initial communication (in litt.) was made via Hurley (1968) and Friend (1980). *Caribitroides* is distinct from *Talitroides* and *Spelaeorchestia,* the only other genera of non-sexually-dimorphic, cuspidactylate landhoppers described to date. It differs from them mainly in its short antenna 1, less reduced pleopods, more spinose uropods, and less modified coxal gills. Its resemblance to simplidactylate genera, especially to the South African genus *Talitriator* (e.g. Hurley, 1968), is apparently superficial and convergent.

Caribitroides jamaicensis, **new species**

Diagnosis: With the characters of the genus. Maxilliped, outer plate weakly arcuate, with subapical spines; pleopods little reduced, peduncles each with smooth margins, and bearing 2–4 retinacula; pleopods 1 and 2, rami 7–9 segmented, inner ramus shorter; pleopod 3 about one-third shorter, rami subequal, 5–6 segmented; telson lobes each with 2–3 apical spines and 3 dorsolateral spines.

Material Examined: Hardwar Gap, near Newcastle, Jamaica, 1200 m, leaf litter, May, 1957, E. L. Mills coll. Female ov. HOLOTYPE 12.0 mm; several female and immature PARATYPES. Additional material from this general locality collected by T. H. Farr and E. L. Bousfield, March 24, 1960. NMNS collections, Ottawa.

Talitroides Bonnier, 1898

Friend (1980) maintained this genus distinct from the sandhopper genus *Talitrus,* and known terrestrial genera, including his simplidactylate genera from the Australian Region. An updated generic diagnosis of *Talitroides* should include the following combination of characters (see phenogram, Fig. 5): body small to medium in size; antenna 1 long; maxilla 1, palp 1-segmented; maxilliped outer plate not arcuate, broadly subapically spinose, palp 3-segmented; gnathopod 1, segment 6 short; gnathopod 2, lobe of segment 6 extended well beyond dactyl; peraeopods 3–7 short; pleopods 1 and 2 reduced but functional, inner ramus the shorter, peduncles marginally plumose; pleopod 3 vestigial; uropod 2, peduncle distally spinose, rami subequal, outer marginally bare; uropod 3 very short; telson lobes fused, margins stoutly spinose; coxal gill of P6 reverse L-shaped. Component species: *T. alluaudi* Chevreux (type) and *T. topitotum* Burt (Table 8).

INDEX

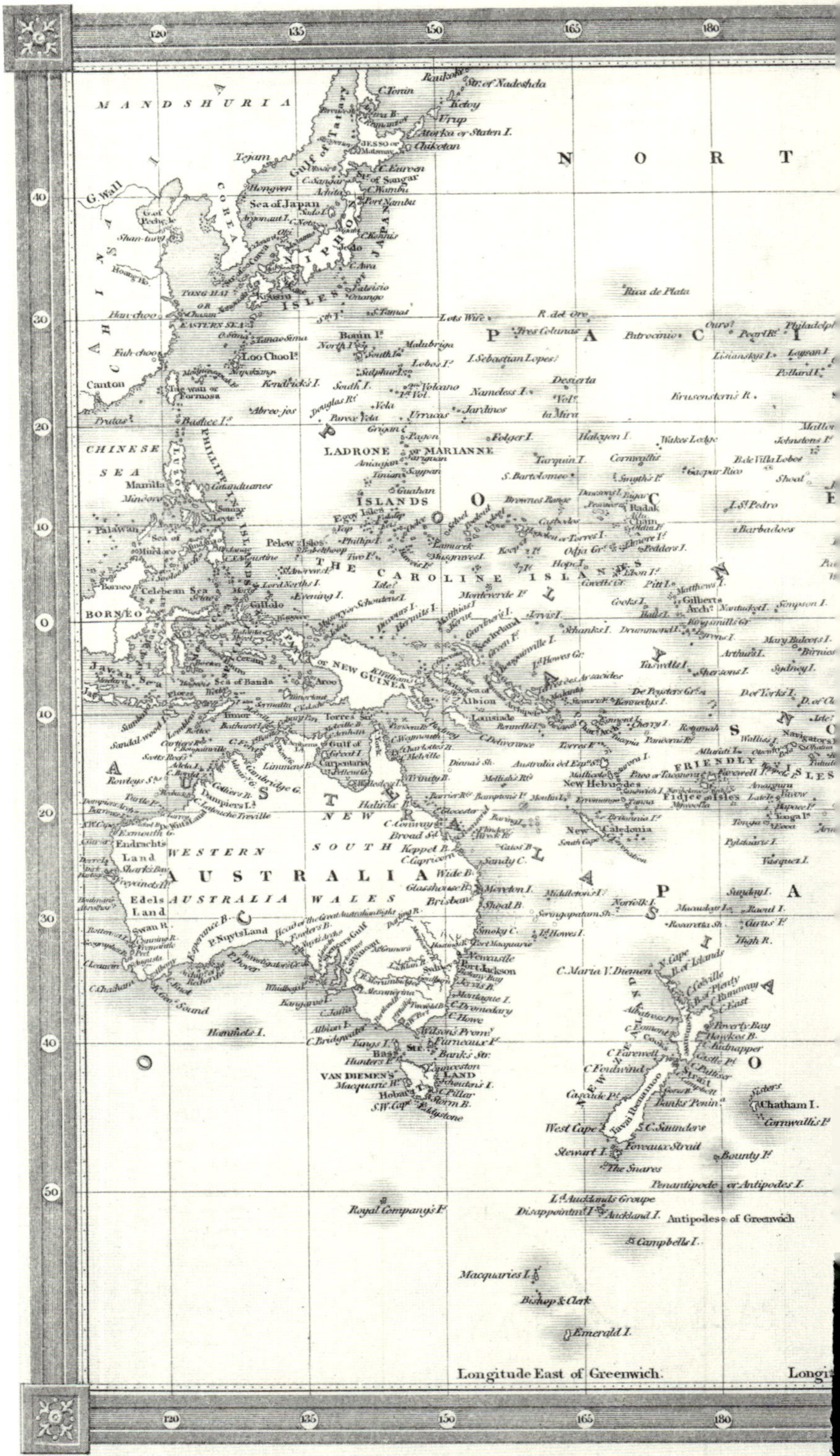

London: Publishea